Achieving sustainable cultivation of oil palm

Volume 1: Introduction, breeding and cultivation techniques

It is widely recognised that agriculture is a significant contributor to global warming and climate change. Agriculture needs to reduce its environmental impact and adapt to current climate change whilst still feeding a growing population, i.e. become more 'climate-smart'. Burleigh Dodds Science Publishing is playing its part in achieving this by bringing together key research on making the production of the world's most important crops and livestock products more sustainable.

Based on extensive research, our publications specifically target the challenge of climate-smart agriculture. In this way we are using 'smart publishing' to help achieve climate-smart agriculture.

Burleigh Dodds Science Publishing is an independent and innovative publisher delivering high quality customer-focused agricultural science content in both print and online formats for the academic and research communities. Our aim is to build a foundation of knowledge on which researchers can build to meet the challenge of climate-smart agriculture.

For more information about Burleigh Dodds Science Publishing simply call us on +44 (0) 1223 839365, email info@bdspublishing.com or alternatively please visit our website at www.bdspublishing.com.

Related titles:

Achieving sustainable cultivation of oil palm Volume 2: Diseases, pests, quality and sustainability
Print (ISBN 978-1-78676-108-8); Online (ISBN 978-1-78676-111-8, 978-1-78676-110-1)

Achieving sustainable cultivation of soybeans Volume 1: Breeding and cultivation techniques
Print (ISBN 978-1-78676-112-5); Online (ISBN 978-1-78676-114-9, 978-1-78676-115-6)

Achieving sustainable cultivation of soybeans Volume 2: Diseases, pests, food and other uses
Print (ISBN 978-1-78676-116-3); Online (ISBN 978-1-78676-119-4, 978-1-78676-118-7)

Managing soil health for sustainable agriculture Volume 1: Fundamentals
Print (ISBN 978-1-78676-188-0); Online (ISBN 978-1-78676-191-0, 978-1-78676-190-3)

Managing soil health for sustainable agriculture Volume 2: Monitoring and management
Print (ISBN 978-1-78676-192-7); Online (ISBN 978-1-78676-194-1, 978-1-78676-195-8)

Precision agriculture for sustainability
Print (ISBN 978-1-78676-204-7); Online (ISBN 978-1-78676-206-1, 978-1-78676-207-8)

Chapters are available individually from our online bookshop: https://shop.bdspublishing.com

climate•SMART•publishing
IN AGRICULTURAL SCIENCE

BURLEIGH DODDS SERIES IN AGRICULTURAL SCIENCE

NUMBER 27

Achieving sustainable cultivation of oil palm

Volume 1: Introduction, breeding and cultivation techniques

Edited by Professor Alain Rival, Center for International Cooperation in Agricultural Research for Development (CIRAD), France

burleigh dodds
SCIENCE PUBLISHING

Published by Burleigh Dodds Science Publishing Limited
82 High Street, Sawston, Cambridge CB22 3HJ, UK
www.bdspublishing.com

Burleigh Dodds Science Publishing, 1518 Walnut Street, Suite 900, Philadelphia, PA 19102-3406, USA

First published 2018 by Burleigh Dodds Science Publishing Limited

Library of Congress Control Number: 2017956485

British Library Cataloguing in Publication Data
A catalogue record for this book is available from the British Library

ISBN 978-1-78676-104-0 (print)
ISBN 978-1-78676-106-4 (online)
ISBN 978-1-78676-107-1 (online)
ISSN 2059-6936 (print)
ISSN 2059-6944 (online)

Typeset by Deanta Global Publishing Services, Chennai, India
Printed by Lightning Source

Contents

Series list

Title	Series number
Achieving sustainable cultivation of maize - Vol 1 From improved varieties to local applications Edited by: Dr Dave Watson, CGIAR Maize Research Program Manager, CIMMYT, Mexico	001
Achieving sustainable cultivation of maize - Vol 2 Cultivation techniques, pest and disease control Edited by: Dr Dave Watson, CGIAR Maize Research Program Manager, CIMMYT, Mexico	002
Achieving sustainable cultivation of rice - Vol 1 Breeding for higher yield and quality Edited by: Prof. Takuji Sasaki, Tokyo University of Agriculture, Japan	003
Achieving sustainable cultivation of rice - Vol 2 Cultivation, pest and disease management Edited by: Prof. Takuji Sasaki, Tokyo University of Agriculture, Japan	004
Achieving sustainable cultivation of wheat - Vol 1 Breeding, quality traits, pests and diseases Edited by: Prof. Peter Langridge, The University of Adelaide, Australia	005
Achieving sustainable cultivation of wheat - Vol 2 Cultivation techniques Edited by: Prof. Peter Langridge, The University of Adelaide, Australia	006
Achieving sustainable cultivation of tomatoes Edited by: Dr Autar Mattoo, USDA-ARS, USA & Prof. Avtar Handa, Purdue University, USA	007
Achieving sustainable production of milk - Vol 1 Milk composition, genetics and breeding Edited by: Dr Nico van Belzen, International Dairy Federation (IDF), Belgium	008
Achieving sustainable production of milk - Vol 2 Safety, quality and sustainability Edited by: Dr Nico van Belzen, International Dairy Federation (IDF), Belgium	009
Achieving sustainable production of milk - Vol 3 Dairy herd management and welfare Edited by: Prof. John Webster, University of Bristol, UK	010
Ensuring safety and quality in the production of beef - Vol 1 Safety Edited by: Prof. Gary Acuff, Texas A&M University, USA & Prof.James Dickson, Iowa State University, USA	011
Ensuring safety and quality in the production of beef - Vol 2 Quality Edited by: Prof. Michael Dikeman, Kansas State University, USA	012
Achieving sustainable production of poultry meat - Vol 1 Safety, quality and sustainability Edited by: Prof. Steven C. Ricke, University of Arkansas, USA	013
Achieving sustainable production of poultry meat - Vol 2 Breeding and nutrition Edited by: Prof. Todd Applegate, University of Georgia, USA	014
Achieving sustainable production of poultry meat - Vol 3 Health and welfare Edited by: Prof. Todd Applegate, University of Georgia, USA	015

Introduction

Oil palm is widely cultivated in tropical countries for use in food and feed, personal care products and other applications such as biodiesel. Oil palm cultivation faces a range of challenges, such as its environmental impact (deforestation and biodiversity loss) as well threats from pests and diseases. There is an urgent need to make oil palm cultivation more efficient and environmentally sustainable.

These challenges are addressed in the two volumes of *Achieving Sustainable Cultivation of Oil Palm*:

- Volume 1: Introduction, breeding and cultivation techniques
- Volume 2: Diseases, pests, quality and sustainability

The chapters in Volume 1 review the latest developments in conventional and marker-assisted breeding, as well as transgenic approaches, as well as assessing ways of assessing and optimising yields through integrated approaches relying on precision agriculture aimed at fine tuning nutrient and soil management.

Part 1 Setting the scene

The first part of the volume begins by reviewing trends in production and key challenges facing the sector. Chapter 1 addresses the growth of the palm oil market , key actors and trends. For many years palm oil has proved to the most productive, highest yielding oil crop, with a yield 4 to 10 times higher than any other oil crop. This has driven palm oil to become the most consumed vegetable oil in the world. The chapter provides an overview of oil palm cultivation and palm oil production throughout the world, with a particular focus on Indonesia and Malaysia, and describes the growth of the palm oil industry and the nature of the global market. The chapter also addresses the diverse ecological and social impacts of palm oil and suggests future directions for sustainable palm oil production.

Chapter 2 moves from the commercial market to consider research trends in oil palm cultivation. Since its establishment as a commercial crop in Malaya in 1917, palm oil has become the most important vegetable oil traded in the world. Its standing in the world market is attributed to the oil's versatile applications, stable supply and affordability. With the estimated global world population reaching nine billion by 2050, the supply of palm oil must grow to help meet the demand for vegetable oils. The chapter summarises the trends in research which have been driven by key challenges faced by the industry. It covers the establishment of oil palm as an economic crop, the improvement of yield by selective breeding, the use of genomics to expedite research and a holistic approach to remodeling cultivation systems for eco-efficiency, with the goal of achieving cultivation of oil palm with minimal environmental impact.

Following on from the aspirations for oil palm cultivation outlined in Chapters 1 and 2, Chapter 3 addresses the issues of achieving sustainable oil palm cultivation on the ground via detailed case studies drawn from Indonesia, Colombia and Cameroon, which are compared and contrasted. The chapter discusses the conceptualization of sustainable pathways in oil palm production and the background to oil palm production in three contrasted producing countries which involve different groups of stakeholders with different production goals. The chapter outlines the challenges of sustainable production pathways and sustainability initiatives. Finally, it considers the actors, regulations and practices in the dynamics of sustainability that may make sustainable production possible in Asia, South America and Africa.

Building on Chapter 3, Chapter 4 deals with the role of governance and land tenure in regulating the oil palm boom. Oil palm expansion has delivered economic development in host countries, including indirect benefits for local infrastructure development and rural poverty reduction, as well as multiplier effects for the national economies. However, its development has often come at the cost of basic rights and to the detriment of biodiverse, carbon-rich tropical forests, with local communities sometimes evicted from their lands and precious ecosystems destroyed.

Chapter 4 identifies several structural constraints to better governance, and argues that these must be addressed in order to build more sustainable and inclusive oil palm supply chains and landscapes. The chapter lays out the conceptual framework of palm oil governance, identifying the key action arenas and actors. The chapter explores the institutional architecture and critical problems affecting the palm oil governance complex, while highlighting major gaps in the governance of the commodity chain.

Part 2 Plant physiology and breeding

The focus of the second part of the volume is on developments in understanding oil palm physiology, genetics and genetic diversity, as well as their application to improved breeding techniques. Chapter 5 examines advances in understanding oil palm reproductive development. As for many other crops, yield components of the oil palm rely partly on the optimal implementation of its reproductive development. As a result, studies focusing on the mechanisms underlying sex ratio determination, inflorescence development or fruit maturation have rapidly multiplied, significantly improving our understanding of these critical processes. The chapter describes the impact of sex ratio, inflorescence and flower development on oil palm yields and describes the processes of fruit development and shedding and their importance, as well as oil acidification. The chapter looks ahead to the likely future impact of climate change and developments in this area that may increase the sustainability of oil palm cultivation.

Chapter 6 develops Chapter 5's focus on oil palm breeding by examining the importance of genetic resources in the implementation of efficient strategies of genetic improvement. Oil palm planting materials have been generated from an extremely narrow genetic base, and it has been generally recognized that the narrowness of the genepool is a major obstacle towards increasing yields in many crops. The chapter discusses the work undertaken by the Malaysian Palm Oil Board to broaden this genetic base. It also examines ways of assessing genetic diversity in oil palm, through the analysis of fruit forms and types, morphological traits and genetic markers. The chapter also reviews methods for the conservation of oil palm collections and ways of utilizing germplasm in order to develop improved varieties which constitute the basis of sustainable plantations.

Chapter 7 follows on from Chapter 6's focus on oil palm genetics to consider advances in conventional oil palm breeding techniques. At the beginning of the twentieth century, oil palm was a semi-wild crop which had not been subjected to modern breeding techniques. The chapter discusses oil palm breeding objectives, breeding methods and data collection methods currently in use; it then reviews the impacts and progress of oil palm breeding programs. Finally, the chapter considers seed production and future trends in oil palm breeding.

Returning to the theme of genetics, the focus of Chapter 8 is on advances in marker-assisted breeding of oil palm. Oil palm, with its long breeding cycle and large land requirement for breeding trials, can be considered a suitable candidate crop for marker assisted selection (MAS). The chapter explores the applications of MAS in oil palm breeding, including pedigree testing and the prediction of monogenic and quantitative traits. The chapter shows that MAS enables accurate tagging of markers related to polygenic traits, a prerequisite for their integration into in oil palm breeding strategies, and suggests future lines or research.

Complementing the themes of the preceding chapter, Chapter 9 addresses advances in genetic modification of oil palm. Transgenic crop are generated through recombinant DNA methods to alter gene expression in order to create new varieties for breeders that may be either difficult or impossible to produce using conventional approaches. Over the past few decades, transgenic methods have been successfully applied to develop genetically modified (GM) varieties of the major oilseed crops. The chapter reviews research efforts for production of GM oil palm plants over the past 20 years. Although these efforts have yet to result in stable lines of commercially useful GM varieties of oil palm, there are good prospects that the greatly improved knowledge of genomics coupled with advanced technologies such as genome editing will be successful in the future, provided it meets public approval.

Part 3 Cultivation techniques

The final part of the book discusses developments in oil palm cultivation practices. Chapter 10 examines the modelling of crop growth in order to achieve higher oil yields. Since the development of the first semi-mechanistic oil palm model, OPSIM, the development of new models has increased in frequency, aiming to take into account aspects of oil palm physiology and the physical processes and causal relationships between the environment and the crop. The chapter describes the development of a new oil palm growth and yield model called PySawit. The chapter discusses the evaluation of PySawit's accuracy when its predictions were compared with several measured parameters of growth and yield in oil palm. The chapter includes a full explanation of the methodology of the model and discussion of its results, and explores the modeling of meteorology, photosynthesis, energy balance, soil water and crop growth

Continuing the theme of improved agricultural practices, Chapter 11 addresses efforts to improve soil and nutrient management in oil palm plantations. Soil health is increasingly regarded as a key factor in oil palm nutrition and productivity. The chapter presents the current state of knowledge about soil and nutrient management through several examples of efficient and productive oil palm cultivation. The chapter highlights the challenges which still need to be addressed in order to make the crop more sustainable in the long term while considering some new concepts in plant nutrition and soil fertility management. The chapter also considers the potential of new technologies such as sensors and drones. The chapter considers key issues relating to soil fertility, nutrient management, mineral nutrition and planting materials, and focuses particularly on issues affecting smallholders.

Complementing Chapter 11's focus on soil management, Chapter 12 focusses on maintaining soil health in oil palm cultivation. Many existing oil palm plantations were established after clearing tropical rainforests. In tropical soils under forest, the topsoil is where the fertility lies due to its physico-chemical properties, developed through an accumulation of organic matter and intense biological activity. The properties of this topsoil change rapidly in the first four years after felling, but it has been reported that the topsoil then evolves towards a new, stable chemical state. A central challenge is therefore to ensure that this new stable state possesses the properties that enable high yields. The chapter considers the key issues and challenges in maintaining soil health. It then reviews the effects of management practices in oil palm plantation on soil biological activity, and considers the potential benefits of biomass recycling.

The volume's final chapter, Chapter 13, considers the increasingly important issue of palm oil's use as biomass for biofuel. The use of fossil fuels depletes the world's limited supply of coal, oil and gas, and releases stored CO_2 into the atmosphere. Biofuels, derived from biomass, are renewable and carbon neutral, because consumption merely releases CO_2 taken from the atmosphere by the growing biomass. However, increasing the use of biofuels could also increase the pressure on ecosystems which must support the production of additional biomass. The chapter assesses the sustainability of biofuels, discusses their production and economics, and considers levels of government support for production and related geopolitical issues.

Dedication

The two volumes of *Achieving sustainable cultivation of oil palm* are dedicated to the memory of our very esteemed colleague Hubert de Franqueville. Hubert was a renowned and brilliant scientist with an elegant and precise mind who was highly appreciated by the whole palm community. I would have loved sharing the editing of these two volumes with Hubert. We all miss his sharp understanding and exquisite humor.

Alain Rival
Editor

Part 1

Setting the scene

Chapter 1

The palm oil market: growth and trends

Stefano Savi, Roundtable on Sustainable Palm Oil, Malaysia

1 Introduction

For many years now, palm oil has proved to be the most productive, highest yielding oil crop. Its yield is 4–10 times higher compared to that of any other vegetable oil, a factor that has helped palm oil in becoming what is today the most consumed vegetable oil in the world (Alias et al., 2014, see Fig. 1). Palm oil is derived from the fruit of a specific species of palm tree, known as *Elaeis guineensis*. There are two types of oil that can be produced from the fruit of oil palm. The first type of oil is extracted from the flesh of the fruit and is used in a variety of applications, mostly in food, where its properties make it especially suited to be used as cooking oil, or in margarines, as substitutes of cocoa butter, and also as a replacement for milk fat. The second type of oil is derived from the palm fruit kernel. Derivatives of this palm kernel oil can be utilized in cosmetics and other speciality applications. Of the goods that are present on the shelves of the stores, over half of them contain palm oil as an ingredient.

The vegetable oil market is divided into three major segments: food, industrial non-food and biodiesel feed. A major share of the consumption market is taken by the applications of food; however, the biofuels segment is considered the fastest growing one due to its use as biodiesel feedstock. Recent data have shown that, in Europe, use of palm oil in the biofuel market has overtaken both food and cosmetics, and now represents the majority of the palm oil use in the region (RSPO, 2016). It is forecasted that biodiesel will reach the highest growth of approximately 8.8% from 2015 till 2022 (Grand View Research, 2016). With governments trying to limit reliance on conventional fossil fuels, it is expected

http://dx.doi.org/10.19103/AS.2017.0018.01

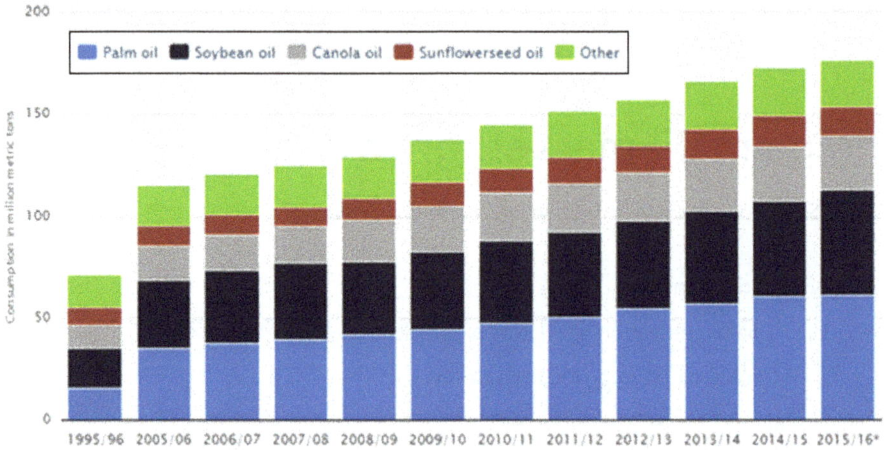

Figure 1 Consumption of vegetable oils worldwide from 1995/1996 to 2015/2016, by oil type (in million metric tons). Source: Statista.com.

that biodiesel fuel will witness a rapid growth in the forthcoming years, at least initially contributing to an increase in demand of first-generation biofuels, such as those derived from palm oil.

2 Oil palm cultivation and palm oil production

The expansion of palm oil plantations globally started in the early twentieth century, driven by the Industrial Revolution in Europe and expansion of overseas trade. Although oil palm is originally from West Africa, the first commercial-scale oil palm plantations were established by the British in South East Asia (initially in Malaysia) and saw great expansion in the 1960s when land settlement schemes for planting oil palm were introduced as a means to eradicate poverty for landless farmers and smallholders in the region. Today, Indonesia is the largest palm oil producer, followed by Malaysia and Thailand (USDA, 2016, see Table 1).

Table 1 Global palm oil production by country

(Values in metric tons)	
Indonesia	35,000,000
Malaysia	21,000,000
Others	4,945,000
Thailand	2,300,000
Colombia	1,280,000

Source: (Globalpalmoilproduction.com, 2016).

3 Regional oil palm production trends

3.1 Oil palm in Indonesia

There has been a vigorous growth in the palm oil industry of Indonesia, where very few industries have experienced the same growth as that of palm oil. The growth is clearly visible, not only in the exports and production statistics but also in the estate areas converted to oil palm. The industry originated in Sumatra during the Dutch colonial period and around 70% of the plantations are situated there. The rest of the 30% can be located in the island of Kalimantan (Investments, 2016). Over the years, palm oil has proven to generate high yields and, due to increasing demand, both large corporations and medium and small farmers have shown interest in the industry and have switched from other crops (Castiblanco et al., 2013). This switch has lead Indonesia to become the largest producer and exporter of palm oil worldwide. India, Europe (via the Netherlands), Malaysia, China and Singapore are among the most important destinations where Indonesian palm oil is exported (Investments, 2016). The export and production statistics of palm oil in Indonesia are given in Table 2.

Today, the processing industry and plantation of oil palm in Indonesia are the most important industries of the country in terms of contribution to the national economy. Moreover, this industry provides numerous employment opportunities to millions of individuals in Indonesia.

The data extracted from the Indonesian Ministry of Agriculture show that the overall area covered by palm constitutes about eight million hectares (Investments, 2016). It is expected that by the end of 2020, this amount will increase to 13 million hectares (Investments, 2016).

Private organisations produce approximately half of the overall production of Indonesian palm oil, whereas the government-owned plantations play a relatively modest role. The remaining 35% of the production comes from smallholder farmers (Investments, 2016).

3.2 Oil palm in Malaysia

In 1870, the oil palm was introduced for the first time as an ornamental plant in Malaysia. The area of land planted with oil palm experienced a very rapid growth since the 1960s. In 1985, the area of land planted with oil palm reached about 1.5 million hectares, which later increased to 4.3 million hectares by 2007. By 2011, the amount of planted area reached 4.917 million hectares, and today, Malaysian oil palm is considered as one of the most useful and important commodity crops for this country.

Table 2 Export and production statistics of palm oil in Indonesia

	2008	2009	2010	2011	2012	2013	2014	2015	2016
Production (million tons)	19.2	19.4	21.8	23.5	26.5	30.0	31.5	32.5	32.0[1]
Export (million tons)	15.1	17.1	17.1	17.6	18.2	22.4	21.7	26.4	27.0[1]
Export (in USD billion)	15.6	10.0	16.4	20.2	21.6	20.6	21.1	18.6	18.6[1]

[1]Indicates forecast.
Sources: Indonesian Ministry of Agriculture & Indonesian Palm Oil Producers Association (Gapki).

Table 3 Export and production statistics for palm oil in Malaysia

	2008	2009	2010	2011	2012	2013	2014	2015	2016
Production (million tons)	17.3	17.8	18.2	18.2	19.3	20.2	19.9	18.8	21.0[1]
Export (million tons)	16.0	16.6	17.2	17.6	18.5	17.3	17.4	17.5	18.0[1]

[1]Indicates forecast.
Source: Malaysia Palm Oil Board.

Until 2012, Malaysia was the leading producer of oil palm, but more recently, due to the unavailability of new cultivable land in Malaysia, Indonesia has overtaken Malaysia to become the top producer of palm oil (Table 3). While in 1999 the production of palm oil in Malaysia contributed to about 51% of global palm oil output, as of year 2011 this percentage has decreased to 38% (Palmoilworld.org, 2016).

Currently, a replanting programme has been launched by Malaysia, where old trees are replaced so that the excess of existing stock can be cleared out, in an effort to increase currently lowering prices for crude palm oil. Moreover, there are efforts made by the government to open up new markets for palm oil in countries such as Turkey, Turkmenistan, Iran and Kazakhstan (www.thesundaily.my/news/1545769).

3.3 Oil palm in the rest of the world

Following the crop's success and the increasingly limited availability of land in South East Asia, the oil palm industry has in more recent years expanded to other regions, particularly in Latin America and Africa. Particularly in West Africa, governments are seeing oil palm development as a potential source of tax and export revenue. A growing number of investors, including some of the world's largest plantation companies, are experiencing purchase of new concessions in Africa. This said, developments for oil palm in this region are still at the very early stage, and often, it is slowed down by high operational costs linked to the lack of appropriate infrastructure.

Yields are found to be much lower in Africa than those in South East Asia for various reasons, including climate and infrastructural limitations and a predominantly smallholder approach to production. In fact, in Africa, where oil palm is traditionally cultivated as a subsistence crop for food, smallholders account for between 70% and 90% of growers, depending on the country (sustainablepalmoil.org, 2016).

In Latin America, both production and consumption of palm oil are increasing. In that region, palm oil is grown in 12 countries, thus contributing nearly 6% of global production per annum. Colombia is the region's largest palm oil producer and is among the top five producers worldwide (sustainablepalmoil.org, 2016).

4 Palm oil industry and market

4.1 Growth in the palm oil industry

It is forecasted that the market for palm oil on a global level will exceed 80 million metric tons by 2020, a figure driven by a growing demand for both non-edible and edible applications (Grand View Research, 2016). With a growth rate of 3.2% per annum, it is considered as one of the fastest growing markets. Among the reasons for the growth in global market size of palm oil (and vegetable oils) is the ever-growing global demand for food, driven by fast development, especially in the Asian subcontinent, and the demand for biodiesel feedstock.

The recent changes in biofuel policies globally have in fact increased the supply requirements of the biodiesel industry and have diverted large volumes of raw materials, traditionally meant to enter the food supply chain, from the food sectors towards the biofuel industry. This has at times resulted in a shortage of food supply, with food prices hiking up resulting in a constraint on the vegetable oil market (Flammini, 2008).

4.2 Palm oil applications

The physical and chemical properties of palm oil allow for it to be utilised in a variety of applications, which have led to the expansion of its market on a global level. Palm oil has traditionally been used in Asia and Africa as cooking oil. Additionally, it is used as an ingredient in the manufacture of margarine, non-dairy creams and ice creams. It is used in products such as lubricants, grease and candles and in biodiesel production, replacing mineral oils both in the transport industry and for power generation. Its derivatives are used for the manufacturing of soaps and detergents, cosmetics, pharmaceuticals, water-treatment products and production of bactericides.

In 2015, global demand was dominated by crude palm oil followed by palm kernel oil and palm kernel meal (USDA, 2016). A rapid increase in edible oil, biodiesel, lubricants, surfactants and cosmetics use is expected to further increase demand in the future. There has also been a growth in the demand of animal feed (where palm kernel cakes are mostly utilised) in areas such as Asia Pacific and North America. The overall market of palm oil derivatives is also expected to continue growing in volumes as the demand for cosmetics increases globally (Grand View Research, 2016).

In recent years, the global market has been dominated by edible oil and followed by lubricants, surfactants and biodiesel, with the latter gaining enormous traction in Europe (RSPO, 2016). A global tendency to move away from *trans*-fats in food has benefited the palm oil market, resulting in palm oil taking over soybean as the most consumed vegetable oil in the years after 2004/2005 (USDA, 2016). With the increase in energy needs, consumer preference and regulations have shifted energy production towards the use of biofuels. As bio-based lubricants and surfactants are encouraged to be used by national governments and industry environmental regulations, expectations are that such trends will help in the growth and development of the Industry (Grand View Research, 2016).

4.3 Global demand

The oil crops sector has been in the past years the fastest growing agricultural sector, with growth rates exceeding those of livestock products. One of the main reasons for this growth has been the increasing demand for vegetable oils in developing countries, where development and the population's higher access to disposable income have caused an increase in food consumption and a shift towards diets richer in oils and fats (FAO, 2003).

Consumption in each country tends to favour locally produced oils and fats. Although in temperate regions annual seed crops are the main sources of oil, in tropical countries coconut oil and palm oil, together with groundnut oil, are the most produced and consumed oils. The high yield and relatively lower costs of production of palm oil have resulted in the price of crude palm oil being considerably lower compared to those of other vegetable oils, making it the favoured choice of consumers particularly in regions such as South East Asia and West Africa. From 2015 to 2022, it is expected that crude palm oil will observe the highest growth in the sector at approximately 7.5% compound annual growth rate. Availability of key raw material due to increased planting, especially in South East Asia, together with growing disposable income levels in China, India and Indonesia are expected to drive the market. Asia Pacific remains in fact the largest palm oil consumer region, covering 65% of the overall market volume in 2014, with India remaining the largest importer. Besides, recent studies forecast an 8% growth in the South and Central American markets from 2015 to 2022 (Grand View Research, 2016).

4.4 Comparison with other vegetable oil markets

In the last decade, the production growth rates of major fats and oils have varied significantly. Among these vegetable oils, the fastest growth was in the palm kernel oil and palm oil with production rates reaching about 70% and 82%, respectively, while rapeseed increased by 70%, soybean oil increased by 43% and sunflower oil increased by 64%. Although in 2005 palm oil and palm kernel oil represented 25% of global oils and fats production (including animal fats), by 2015, their combined share grew to 33% (Oil World, 2016).

As a result of a move towards the replacement of *trans*-fatty acids with palmitic acid due to health concerns, palm oil took over soybean, and it became the most used vegetable oil in the mid-2000s. In 2015, production of soybean oil was 47 million tonnes, whereas production of palm oil was at around 61 million. The two vegetable oils together accounted for a total of approximately 50% of 202 million global productions of oils and fats in the year (USDA, 2016). This said, the oil palm (together with rapeseed and sunflower) is considered among those vegetable oil crops that are grown mainly for their oil content, and its production is based directly on the demand of oils and fats. Soybean oil, on the other end, is a by-product in the production of soybean meal, which remains the main driver of the crop's production growth. In 2014–15, around 78 million tonnes of fats and oils were exported on a global level. Among these, palm kernel oil and palm oil made up the biggest single export at approximately 47 million tonnes with a significant lead over that of soybean, which ranked second as the most exported oil, at 10 million tonnes. The production level of palm kernel and palm oil was also significantly higher than any of the other fats and oils at 71% (USDA, 2016).

5 Ecological impact of palm oil production

5.1 Economic and climatic factors

The economic conditions, the changing and different eating habits of consumers in developing countries and the rise in living standards and the growth in the demand of vegetable oils are a few of the factors that hint that palm oil is here to stay. However, there are several factors that inhibit the growth and expansion of the palm oil market such as changing climatic conditions, fluctuations in the price and adverse effects on the people and environment due to unsustainable oil palm cultivation, such as exploitation of local communities and labour, deforestation and loss of habitat of different endangered species (Obidzinski et al., 2012).

There have been significant movements in the price of palm oil, which played an important role in leading fats and oils imports and exports on the global platform. In 1998, the world market suffered a decrease in supply of palm oil due to the harmful effect on the crop caused by the 1997 drought due to El Niño, which was coupled with the Indonesian Government's imposed ban on the export of palm oil (which was later converted to an export tax). After the drought due to El Niño in 1997 and 1998, a downfall in the world's stock related to fats and oils was observed, where stocks reached a low of 12 million tons in September 1998. Changes in climatic conditions, such as the ones associated with the El Niño events, can have tremendous effects in the production of the oil palm. A recurrence of the El Niño phenomena in 2015 has led some to speculate that palm oil prices will rise as much as 17% in 2017 (Bloomberg, 2016).

In the beginning of 1999, a significant downfall in the prices of palm oil was observed because of the increased production of palm oil in Indonesia and Malaysia as the market recovered from the effects of the drought and uncontrolled bushfires, which blanketed the whole region with haze. Moreover, as the Indonesian Government reduced export taxes, the Indonesians producers returned to the export markets (Rea.co.uk., 2016). There is a need for the skilful prediction of such climatic changes so that precautions can be taken before their harmful effects come into observation, while the development of adaptation strategies may also help in building resilience against climate changes in the future (Slingo et al., 2005).

5.2 The issue of deforestation

One of the advantages of the oil palm plant is that it does not require a lot of effort to grow and produce oil. Oil palm is also a perennial crop, which about three years after planting starts yielding fruits, with a continual productive lifespan of 25–30 years. Moreover, the land required by oil palm can be relatively little, and it can grow in small places as it is land efficient and generates profits even at a small scale. Farmers who cultivate oil palm can have beneficial levels of income that are comparatively more than those of the farmers who cultivate other crops such as rice, rubber and cassava, even if on a small scale. Even though the production of palm oil can be a benefit for all stakeholders, including governments and local populations (Obidzinski et al., 2012), these benefits come at a cost. Consequences of the unsustainable production of palm oil have in fact been vastly documented in the past years, including deforestation (Mukherjee and Sovacool, 2014). It has been shown that tropical deforestation, as associated with unsustainable oil palm cultivation, results in warmer, drier conditions at the local scale, and that large-scale deforestation under the tropics could result in global warming equivalent to that caused by burning of fossil fuels since 1850, with more warming and considerably dryer weather in the tropics (Nature Climate Change, 2015). This is not only an issue for the environment but also an important risk factor for the security of supply and market of palm oil itself, considering the effects of droughts and El Niño in the past years.

5.3 Loss of habitat of endangered species

The oil palm is a tropical plant and as such can only be grown successfully in an area spanning from 10 degrees South to 10 degrees North of the equator. This is not only one of the areas with the highest levels of poverty in the world (World bank, 2009), but also where most of the world's remaining forests and biodiversity hot spots can be found. As tropical forests are being converted into oil palm plantations on a large scale, the habitat of animals and plants are being affected adversely. Due to increased deforestation, the animals are isolated in specific areas while majority of the land is being used for plantation purposes. Most of the habitats that are being converted for the production of oil palm are home to animals and species that are endangered (Castiblanco et al., 2015). Moreover, mainly due to a lack of proper governance in some of the countries where palm oil is cultivated, there have been cases of illegal expansion of palm oil plantations, both by large- and middle-sized companies and smallholders. There have been cases in which national parks have been negatively affected, an example being the Tesso Nilo National Park in the island of Sumatra, Indonesia, where 43% of its area has been cleared off and overrun with illegal plantation of oil palms (World Wildlife Fund, 2016).

5.4 Air pollution

In traditional land clearing practices, such as those used for the plantation of oil palm especially in South East Asia, burning is commonly used to clear out the vegetation and prepare the soil for planting. Due to the burning of the vegetation and, in some cases, of the peatlands where plantations are being developed, smoke is released resulting in huge amounts of emissions of carbon dioxide into the air, thus polluting the environment and as a consequence has contributed to climate change (World Wildlife Fund, 2016).

5.5 Water and soil pollution

Waste generated from palm oil mills has been highlighted as one of the major single sources of environmental pollution. It has been ascertained that for every metric ton of palm oil produced, 2.5 metric tons of palm oil mill effluent (POME) are generated in the process. In recent years, the industry has put a lot of effort in coming up with solutions in order to minimise and control both solid and liquid wastes generated by palm oil mills' operations. Direct release of this effluent causes freshwater pollution, which can negatively affect both biodiversity and local population activities such as fishing and farming. Direct release of chemicals used in plantations is also a major factor of concern. Although palm oil productions require fewer fertilisers than other crops, the uncontrolled discharge of these materials in the environment can affect underground water and pollute the surface (World Wildlife Fund, 2016).

5.6 Soil erosion

The impacts of oil palm plantation on the land not only involve soil quantity but also soil quality. When land is prepared and vegetation is removed in preparation of planting, soil erosion can start occurring, adversely affecting the soil quality due to the reduction of nutrients and organic matter levels contained in the soil and also by affecting soil properties such as infiltration rates (World Agriculture & Environment, 2004). Erosion occurs mainly during forest clearing and plantation establishment when the soil is left uncovered. Effects are most visible on steep slopes and are accentuated by planting trees in rows up and down hillsides instead of contours. Later in the oil palm cycle, soil erosion can have the additional burden of transporting fertilisers and pesticides, which adhere to the suspended solids, which can further contaminate waterways (NBPOL, 2011).

5.7 Climate change

Tropical forests such as those covering large parts of Indonesia and Malaysia are considered to be the last remaining global 'carbon sinks'. Their conversion into oil palm plantations, especially when the biomass is cleared by burning, release high amounts of carbon dioxide in the environment, thus contributing to climate change. The practice of draining and converting tropical peatland forests is particularly damaging as these 'carbon sinks' store more carbon per unit area than any other ecosystem in the world. For example, in 1997, peatland fires in Indonesia may have been one of the main sources of global carbon emissions in the country. An estimated 0.81–2.57 gigatons of carbon were released into the atmosphere by the fires representing values in a range from 13% to 40% of the mean annual global carbon emissions from fossil fuels that year. The impact has such a scale that Indonesia is considered to be the third largest emitter of greenhouse gases on a global level (World Wildlife Fund, 2016).

6 Social impacts

Oil palm developments can also create social conflicts if local communities and workers are not appropriately taken into consideration when new developments are initiated and during ongoing operations. Issues can occur not only between the companies and the workers, or the companies and the local community, but also within neighbouring communities, especially in the case of customary land disputes, such as in those incurred by the Dayak communities in Borneo (Colchester, 2012). For example, in Indonesia, indigenous lands cover between 40 and 70 million hectares; however, only 1 million hectares are legally recognised by the government. The non-legal recognition of indigenous peoples' customary land rights can lead to issues in the assignment of land concessions if villagers are pushed out of their customary lands.

Agriculture is one of the three most dangerous sectors in terms of work-related fatalities, non-fatal accidents and occupational diseases. About 59% of all children aged 5–17 years involved in hazardous work are in agriculture (ILO, 2016). Oil palm cultivation is not immune to these issues, and the use of forced, child and trafficked labour has been reported in the industry (US Department of Labour & UNICEF 2012).

7 Conclusion and future trends

Negative impact of unsustainable oil palm cultivation is and will play a fundamental role in the growth and development of the industry itself. Palm oil, being the world's most consumed vegetable oil, has received a lot of attention from global environmental and social NGOs due to the magnitude of effect, frequency of occurrence and duration of its impacts. A number of cases have been highlighted of companies not operating according to civil society's expectations. This has not only harmed these companies' social license to operate but also have proven to define the perceived level of sustainability of the industry, based on its lowest common denominator. The issue of unsustainable production is in fact one that has already demonstrated to have been linked to the industry as a whole, despite the efforts from virtuous players, leading to movements by both the market and civil society to advocate towards a shift away from the use of palm oil. This is an issue that the industry must continue to tackle, not only by showing its sustainable credentials, but also by becoming more transparent in acknowledging these issues and allow for improved transparency and public scrutiny of its progress. Although some industry players are making strides in this direction, a lot more needs to be done by the industry in acknowledging the issues (both in terms of environmental and social negative impacts) and by all players including governments in catalysing a transformation leading to greater transparency (e.g. in resolving the issues surrounding the publication of concession boundaries, especially in Indonesia and Malaysia), shared responsibility and trust.

In view of the above, there are two major factors that will shape the trajectory of growth of the palm oil market in the near future, namely its substitutability (driven by the availability, affordability and acceptability of its substitutes) and its sustainability when compared to other oils.

The growth of palm oil has happened for good reasons. Its low land intensity, combined with the lowest input for energy, fertiliser and pesticides per tonne of production, has made it the most economically viable vegetable oil in the market. This said, recent favourable harvests of US soybean and Eastern European sunflower oil seeds have made such oils more competitive when compared to palm oil, while lacking the notorious negative perception (if not the environmental and social issues) of the first, and the production of which would not be constrained to tropical forest regions. Furthermore, recent efforts by several biotech companies have led to the development of new vegetable oils produced by algae or yeasts that have similar chemical and physical properties as those of palm oil (and palm kernel oil) (*The Guardian*, 2014). Several consumer goods companies are investing in research and development of these substitutes which, while failing to currently demonstrate viability at scale, certainly pose a risk for the outlook of the palm oil market in the future, and the connected value chain.

Although substitutes could become possible, negative impacts (such as those linked to unsustainable agriculture and oil palm cultivation specifically) are common to all human activities and are effected to a greater degree by operation practices, rather than by the nature of the activity itself. In view of this, a more common position has arisen in the past years among NGOs working on palm oil that focuses on the production of palm oil that is more sustainable, rather than promoting substitution. For this purpose, several of the industry players, together with social and environmental NGOs, and with the involvement of banks and investors, have worked towards developing and promoting methods of production that limit the negative impacts of oil palm cultivation. Of the several initiatives (both industry and government led) that have arisen in the last years, the Roundtable on Sustainable Palm Oil (RSPO), a global multi-stakeholder initiative on sustainable palm oil founded in the year 2004, is the longest standing and the most widely accepted by the market. The RSPO has now almost three thousand members, which include plantation companies, processors and traders, consumer goods manufacturers and retailers of palm oil products, financial institutions, environmental NGOs and social NGOs, from many countries that produce or use palm oil. Following the principles and criteria that are set by the RSPO, the expectation is that the organisation's voluntary members need to work in the process of producing and using palm oil that is sustainable, according to the RSPO standards. In the years from its creation, sales of sustainable palm oil have continued increasing at a pace higher than the global production of palm oil. As such, RSPO has grown to certifying about 20% of the global palm oil production in 2016. The RSPO also reported increase of sales of approximately 162% for January 2016 (RSPO, 2016).

Sustainability standards are not only guiding the decision of producers on how to carry out developments and manage operations but also are more and more importantly guiding investors in deciding which sectors and companies present the lower investment and reputational risk, thus influencing the market. Financial institutions, both through their involvement in the RSPO and in other initiatives pushing for sustainable investment, such as the United Nations Principles for Responsible Investment (UN PRI), have made moves to stop investing in companies that are accused of environmental or social issues. Responsible finance principles have also been applied at government level. In 2013, the Norwegian Government Pension Fund announced that they would be pulling investments from 23 Asian palm oil companies due to the concerns related to deforestation and climate change. The United Kingdom (Defra, 2012) also decided to use only certified sustainable palm oil (CSPO) in 'Central Government Food and Catering Services' starting from October 2012. In the following years, industry groups in several European countries have joined forces to ensure a fully sustainable palm oil supply chain in Europe by 2020 by forming national initiatives for sustainable palm oil in the Netherlands, Germany, the United Kingdom, France, Belgium, Norway, Denmark, Sweden and Italy (RSPO, 2016). In response, the governments of six key European Union countries have recently declared their support towards the project by signing the Amsterdam Palm Oil Declaration. Although major progress in the uptake of CSPO has been made in Europe, India and China – the first and third largest importers of palm oil – are also making strides in the uptake of CSPO. In China, the three companies who are the major importers of palm oil now have plans towards 100% CSPO uptake (RSPO ACOP data, 2016). Other companies operating in China, such as Mars, have announced that they have already covered all their palm oil supplies with certified materials.

The influences of large corporations and governments who have committed to sustainable palm oil sourcing, together with the interrelation of the global markets for palm oil, are factors that continue to influence the move towards a future where sustainable palm oil will be the norm.

8 Where to look for further information

For further information on the author and related work, please visit www.rspo.org

9 References

Alias, E. F., Radam, A., Fen, Y. P., Yacob, M. R. and Alam, M. F., 2014. Growth in Malaysia's export food market: a shift-share analysis. *Asian Soc. Sci.*, 10(3), 26–43.

Bloomberg, 2016. *'Enormous' Impact of El Nino on Palm Seen Fueling 17% Rally.* [online] Available at: http://www.bloomberg.com/news/articles/2016-03-09/-enormous-impact-of-el-nino-on-palm-oil-seen-fueling-20-rally.

Bsigroup.com, 2016. *RSPO and MSPO for Sustainable Palm Oil | BSI Group.* [online] Available at: http://www.bsigroup.com/en-MY/RSPO-MSPO-Certification/ [Accessed 1 August 2016].

Carrington, D., 2014. Ecover adopting algae-based laundry liquid to cut palm oil use. The Guardian. Available at https://www.theguardian.com/environment/2014/apr/02/ecover-algae-laundry-liquid-palm-oil.

Castiblanco, C., Etter, A. and Aide, T. M., 2013. Oil palm plantations in Colombia: a model of future expansion. *Environ. Sci. Policy*, 27, 172–83.

Castiblanco, C., Etter, A. and Ramirez, A., 2015. Impacts of oil palm expansion in Colombia: what do socioeconomic indicators show? *Land Use Policy*, 44, 31–43.

Choong, C. G. and McKay, A., 2014. Sustainability in the Malaysian palm oil industry. *J. Clean. Prod.*, 85, 258–64.

Colchester, M., Jalong, T. and Meng Chuo, W., 2012. *Sarawak: IOI-Pelita and the community of Long Teran Kanan*. Available at http://www.forestpeoples.org/sites/fpp/files/publication/2012/10/sarawak-ioi-pelita-and-long-teran-kanan-september-2012-2.pdf.

Consumption of vegetable oils worldwide from 1995/1996 to 2015/2016, o, 2016. *Vegetable Oils: Global Consumption, 2016 | Statistic*. [online] Statista. Available at: http://www.statista.com/statistics/263937/vegetable-oils-global-consumption/[Accessed 1 August 2016].

Earthisland.org, 2016. *What's Fueling the Demand for the Palm Oil Destroying the Rainforests of Indonesia?* [Online] Available at: http://www.earthisland.org/journal/index.php/elist/eListRead/whats_fueling_the_demand_for_the_palm_oil_destroying_the_rainforests_of_ind/ [Accessed 7 June 2016].

FAO, 2003. *World Agriculture: Towards 2015/2030 an FAO Perspective*. [online] Available at: http://www.fao.org/3/a-y4252e.pdf.

Flammini, 2008. *Biofuels and the Underlying Causes of High Food Prices*. [online] Available at: http://www.globalbioenergy.org/fileadmin/user_upload/gbep/docs/BIOENERGY_INFO/0810_Flammini_-_Biofuels_and_the_underlying_causes_of_high_food_prices_GBEP-FAO.pdf.

Futuremarketinsights.com, 2016. *Sustainable Palm Oil Market – Global Industry Analysis, Size and Forecast, 2016 to 2026*. [online] Available at: http://www.futuremarketinsights.com/reports/sustainable-palm-oil-market [Accessed 5 June 2016].

Global Legal Insights, 2015. *Mergers and Acquisition*. [online] Global Legal Group, Kuala Lumpur, 94–7. Available at: http://www.taypartners.com.my/tp/2015/download/articles/GLI-MA2_Malaysia.pdf [Accessed 5 June 2016].

Globalpalmoilproduction.com, 2016. *Palm Oil | Global Palm Oil Production 2016/2017*. [online] Available at: http://www.globalpalmoilproduction.com/[Accessed 1 August 2016].

Gprholdings.com, 2016. *Prospects | Global Palm Resources*. [online] Available at: http://www.gprholdings.com/prospects.html [Accessed 7 June 2016].

Grand View Research, 2016. *Palm Oil Market Analysis By Derivative (Crude Palm Oil, Palm Kernel Oil, Palm Kernel Cake), By Application (Edible Oil, Cosmetics, Bio-diesel, Lubricants, Surfactants) And Segment Forecasts To 2022*. [online] California. Available at: http://www.grandviewresearch.com/industry-analysis/palm-oil-market [Accessed 7 June 2016].

Investments, I., 2016. *Palm Oil in Indonesia - Indonesian CPO Production & Export | Indonesia Investments*. [online] Indonesia-investments.com. Available at: http://www.indonesia-investments.com/business/commodities/palm-oil/item166 [Accessed 5 June 2016].

Kaye, L., 2014. *Sustainable Palm Oil Sales Surge as Demand Meets Supply*. [online] Triple Pundit: People, Planet, Profit. Available at: http://www.triplepundit.com/2014/08/sustainable-palm-oil-demand-meets-supply/# [Accessed 5 June 2016].

Kilian, L. and Hicks, B., 2013. Did unexpectedly strong economic growth cause the oil price shock of 2003–2008?. *J. Forecasting*, 32(5), 385–94.

Kircher, M., 2015. Sustainability of biofuels and renewable chemicals production from biomass. *Curr. Opin. Chem. Biol.*, 29, 26–31.

Levin, J., Ng, G. and Fortes, D., 2012. *Profitability and Sustainability in Palm Oil Production: Analysis of Incremental Financial Costs and Benefits of RSPO Compliance*. WWF, Washington, DC.

Lord, S. and Clay, J., 2011. Environmental Impacts of Oil Palm – Practical Considerations in Defining Sustainability for Impacts on the Air, Land and Water. [online] Available at: http://www.nbpol.com.pg/wp-content/uploads/downloads/2011/02/EnvironmentalImpactOfOilPalm.pdf

Mukherjee, I. and Sovacool, B. K., 2014. Palm oil-based biofuels and sustainability in southeast Asia: A review of Indonesia, Malaysia, and Thailand. *Renew. Sustainable Energy Rev.*, 37, pp.1–12.

Nature Climate Change, 2015. *Effects of tropical deforestation on climate and agriculture*. [online] Available at: http://www.nature.com/nclimate/journal/v5/n1/full/nclimate2430.html

Naylor, R. L., 2016. Oil crops, aquaculture, and the rising role of demand: A fresh perspective on food security. *Glob. Food Sec.*, 11, 17–25.

Obidzinski, K., Andriani, R., Komarudin, H. and Andrianto, A., 2012. Environmental and social impacts of oil palm plantations and their implications for biofuel production in Indonesia. *Ecol. Soc.*, 17(1), 25.

Pacheco, P., 2012. *Soybean and Oil Palm Expansion in South America: A Review of Main Trends and Implications* (No. CIFOR Working Paper no. 90), Center for International Forestry Research (CIFOR), Bogor, Indonesia.

Palmoilworld.org., 2016. *Welcome to the Malaysian Palm Oil Board//Malaysian Palm Oil Industry// Washington, DC//1-202-572-9768*. [online] Available at: http://www.palmoilworld.org/about_malaysian-industry.html [Accessed 1 August 2016].

Rea.co.uk., 2016. *World Palm Oil Prices | R.E.A. Holdings plc*. [online] Available at: http://www.rea.co.uk/rea/en/markets/oilsandfats/worldprices [Accessed 5 June 2016].

Rspo.org., 2016. *Sustainable Palm Oil*. [online] Available at: http://www.rspo.org/about/sustainable-palm-oil [Accessed 1 August 2016].

RSPO-Roundtable on Sustainable Palm Oil, 2015. *Principles and Criteria for The Production of Sustainable Palm Oil 2013*, RSPO, Kuala Lumpur, 8–56.

Schuster Institute Investigative Journalism, 2016. *Schuster Institute Investigative Journalism*. [online] Available at: http://www.schusterinstituteinvestigations.org/#!palm-oil-industry-response/cgm5 [Accessed 1 August 2016].

Slingo, J. M., Challinor, A. J., Hoskins, B. J. and Wheeler, T. R., 2005. Introduction: food crops in a changing climate. *Philos. Trans. R. Soc. Lond. B Biol. Sci.*, 360(1463), 1983–9.

Strategyr.com, 2016. *Palm Oil Market Trends*. [online] Available at: http://www.strategyr.com/MarketResearch/Palm_Oil_Market_Trends.asp [Accessed 5 June 2016].

Sustainablepalmoil.org, 2016. *Biofuels*. [online] Available at: http://www.sustainablepalmoil.org/transport-biofuels/[Accessed 7 June 2016].

Umar, M. S., Jennings, P. and Urmee, T., 2013. Strengthening the palm oil biomass renewable energy industry in Malaysia. *Renew. Energy*, 60, 107–15.

Umar, M. S., Jennings, P. and Urmee, T., 2014. Generating renewable energy from oil palm biomass in Malaysia: the feed-in tariff policy framework. *Biomass Bioenergy*, 62, 37–46.

USDA, 2016. *Oilseeds: World markets and Trade*. [online] Available at: http://apps.fas.usda.gov/psdonline/circulars/oilseeds.pdf

World Bank, 2006. *World poverty map*. [online] Available at: http://www.ruralpovertyportal.org/region

World Wildlife Fund, 2016. *Palm Oil | Industries | WWF*. [online] Available at: http://www.worldwildlife.org/industries/palm-oil [Accessed 7 June 2016].

Wu, T. Y., Mohammad, A. W., Jahim, J. M. and Anuar, N., 2009. GA holistic approach to managing palm oil mill effluent (POME): biotechnological advances in the sustainable reuse of POME. [online] Available at: http://www.ncbi.nlm.nih.gov/pubmed/18804158

Yeong, E. 2015. Malaysia expanding palm oil into new markets. *The Sun Daily*. Available at: http://www.thesundaily.my/news/1545769

Chapter 2

Research trends in oil palm cultivation

Yuen May Choo, formerly The International Society for Fat Research (ISF), USA; and YewAi Tan, formerly Malaysian Palm Oil Board, Malaysia

1 Introduction

The role of agriculture in ensuring continued economic development is crucial. Although the agricultural sector was neglected for a short period when Malaysia, for example, focused on industrialisation in the 1980s and 1990s, it was promoted as the third engine of growth (Wong, 2007) in the Ninth Malaysia Plan which covered the 2006–2010 period. In this respect, the oil palm industry played a key role because oil palm is currently the most important commodity crop in Malaysia. Malaysia also has world-renowned expertise in the management and research and development (R&D) of tree crop agriculture, first in the rubber industry and then the oil palm industry and plantation forestry as well.

Besides its role as an important source of revenue to Malaysia, the oil palm industry provides a source of income for smallholders and plantation workers, and in so doing helps rural development and improves livelihoods for current and future generations of rural Malaysians. Despite its vital role in the economy, oil palm has been described as a forest risk commodity because of allegations that significant areas of tropical forest have been destroyed to clear land for commercial planting of oil palm.

Trends in oil palm cultivation evolved in response to changing landscapes associated with development (Malaysia is a developing country hoping to reach the status of a developed nation in 2020), urbanisation, sensitivity to the issues of environmental protection, social equity and economic progress; and increased agricultural activity resulting from increasing demand for food to feed the growing world population. Thus, at the initial phase of establishing oil palm as a commodity crop, the focus was on selection of suitable oil palm cultivars, followed by the improvement of oil palm yield using selective breeding of the Malaysian Palm Oil Board's

http://dx.doi.org/10.19103/AS.2017.0018.02

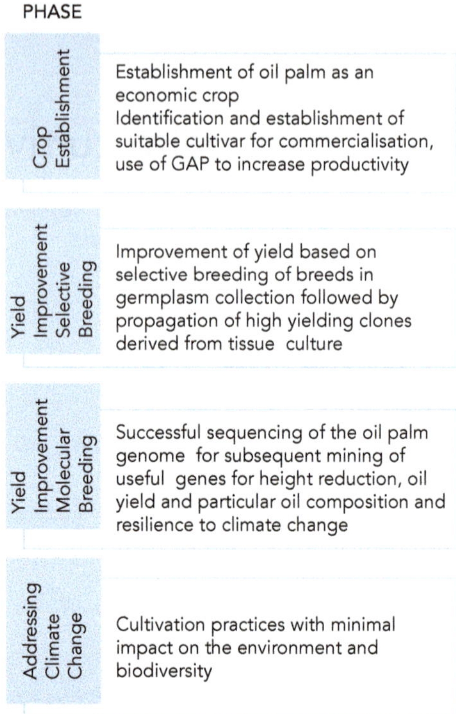

Figure 1 Phases denoting trends in R&D for oil palm cultivation.

(MPOB) large germplasm collection from Africa. At the same time, experience gained from cultivation of oil palm was further refined using best management systems based on good agricultural practices (GAP), where soil fertility and enhancement of soil nutrients to maximise the land yield and therefore oil palm yield were main concerns. Control of pests and diseases was also important in setting the trend for oil palm cultivation.

The next phase of new cultivation patterns was led by advances in biotechnology where the unravelling of the oil palm genome presented opportunities for the mining for genes associated with yield, resistance to environmental stress, height, and chemical and physical properties of the palm oil.

At the same time, with the increasing impact of climate change on global weather patterns, growing attention on the sustainability of the oil palm industry resulted in issues on conservation and protection of the environment becoming key drivers which impact not only the palm oil trade but also the way oil palm is cultivated. This is because all agricultural activities involve the consumption of natural resources and land use change, both of which influence greenhouse gas (GHG) emissions that trigger climate change. In short, the trend in research on oil palm cultivation has been propelled firstly by the need to establish oil palm, native to Africa, as an agricultural crop in Malaysia; secondly, the requirement to ensure its success as a commercial crop by implementing GAP and nurturing breeds with good yields; thirdly, by the successful discovery of the oil palm genome when genes can now be mined for designer oil palms; and last but not least, by pressure on the oil palm industry to demonstrate commitment to sustainability while addressing issues pertaining to conservation and protection of the environment and biodiversity.

In essence, the trend is really reflective of the demand to address key issues and challenges which oil palm faced at particular phases of the industry since its establishment as a commercial crop in 1917.

The chronological and main phases of R&D in oil palm cultivation trend are depicted in Fig. 1.

It was observed that the trend in R&D on cultivation of oil palm was defined by challenges and opportunities presented at the time.

2 Establishment of oil palm as an economic crop

Oil palm is from the genus *Elaeis* with two taxonomically well-defined species, the African oil palm *Elaeis guineensis* and the American oil palm *E. oleifera*. As the yield of the American palm is significantly lower compared to the African oil palm, *E. guineensis* is the species which is commercially exploited. This native palm from Africa has three common varieties, that is, *pisifera*, *dura*, and *tenera*. The latter is a result of hybridisation between the *pisifera* and the *dura* and commercial oil palm plantations are dominated by *tenera* palms because of their higher content of reddish-orange palm oil in the mesocarp oil.

As both male and female flowers are produced by the oil palm, it is therefore a monoecious crop. In nature, cross-pollination of the palm is performed by insects or wind. In order to increase yield of fruit bunches in commercial plantations, assisted or manual pollination was carried out until the 1970s. This practice changed dramatically in 1981 when pollinating weevils, *Elaeidobius kemerunicus*, were released to replace laborious hand pollination. The introduction of pollinating weevils not only reduced the workforce needed for manual pollination but also increased the fruit set and thereby oil yields per surface unit. This remarkable change in oil yield is attributed to Leslie Davison (http://www.merdekaaward.my/Recipients/By-Category/Outstanding-Contribution-to-the-people-of-Malaysia/Datuk-Leslie-Davidson.aspx), who noticed the extremely efficient pollination rate of palms in Africa and hypothesised through keen observation that an insect and not wind was the pollinating agent for the oil palm. He lobbied for investigation to confirm his hypothesis and subsequently research by scientists from the Commonwealth Institute of Biological Control led by Dr. Rahman Syed confirmed that oil palm is indeed insect pollinated and the most effective pollinator is the African weevil *Elaeidobius kamerunicus*.

In addition to the amazing success of the pollinating weevil in increasing yield, replacement of manual pollination greatly benefited smallholders who do not have the resources to carry out hand pollination.

3 Yield improvement

A wide gap still exists between actual yield per hectare and the theoretical potential yield of oil palm. The current yield stands at about 20 tonnes fresh fruit bunch (FFB) per hectare per year, although some plantations can easily get 30 tonnes while 40–46 tonnes are achievable with well-managed practices. Closing this yield gap will boost production significantly without any impact on surrounding protected area.

Oil palm is a perennial crop with a productive lifespan of 20 years or more. The hybridisation of the *pisifera* and *dura* (*D x P*) resulted in the *tenera* (*T*), the oil palm type with the better value in terms of oil yield. Fruit set starts as early as 25 months or less after field planting with peak yields seen in palms after four years. The quality of seedlings planted will determine the yield and health of oil palms in plantations and therefore good nursery practices are critical factors to consider in oil palm cultivation. Good planting materials can only be obtained from certified selected seeds and it follows that the selection of seeds and the implementation of good nursery practices are key factors in the assurance of high-yielding palms.

3.1 Good agricultural practices

To ensure the availability of good planting materials for more than five million hectares of oil palm planted area in Malaysia (Malaysian Oil Palm Statistics, 2015), in the 1980s the Palm Oil Research Institute of Malaysia (now the MPOB) established a certification scheme, the Oil Palm Nursery Competency Certification. This scheme was later incorporated into the MPOB Code of Good Nursery Practice for Oil Palm Nurseries (Choo et al., 2015).

3.2 Selective breeding

In 1941, Beirnaert and Vandederweyen found that the monogenic inheritance of the oil palm shell of *dura*, *tenera* and *pisifera* was in the ratio of 1:2:1. This was an important discovery as the *tenera* (D x P) hybrid has a thicker mesocarp compared to *dura* and *pisifera* and is therefore the preferred commercial planting material. Since the *pisifera* is partially fertile, it is used for breeding by crossing the male *pisifera* with the *dura* to generate the *tenera* hybrid. Knowledge of the single-gene inheritance of shell thickness in oil palm has led to oil palm breeding for *tenera* planting material.

As the success of breeding programmes depends largely on the availability of genetic variety for improvement, PORIM/MPOB made several prospecting trips to Africa and South America to collect additional oil palm germplasm (Rajanaidu and Rao, 1988; Rajanaidu, 1994). The oil palm breeding programme (Rajanaidu et al., 2000) was initiated with the objectives of:

- increasing oil palm yield, for example, from current 4 t/ha/yr to 9 t/ha/yr;
- selecting palms with low vegetative growth, for example, short palms;
- enhancing of oil quality and composition, for example, high vitamin E content, high carotene content and high oleic acid content;
- increasing resistance to diseases, for example, *Ganoderma* tolerant;
- characterising physiological traits such as bunch index (BI), total dry matter (TDM) and bunch dry matter (BDM);
- exploiting genotype and environmental interaction.

Conventionally, BI (BDM/TDM ratio) has been used to screen for physiologically efficient progenies for high FFB yield. However, the harvesting index (HI) also needs to be taken into account because this would indicate economic yield as the HI is a ratio of economic yield to biological yield.

The breeding programme in MPOB has successfully produced planting materials with slow height increment (PS1), high iodine value (PS2), high kernel content (PS3), high carotene content (PS4), higher oil content thin-shell *teneras* (PS5) and large fruit *duras* (PS6), high BI (PS7) and high vitamin E content (PS8), long stalk palms for easier harvesting

(PS10), high carotene *E. guineensis* (PS11), high oleic palm (PS12) and low lipase palm for lower free fatty acid and high-quality palm oil (PS13).

3.3 Tissue culture for high-yielding clones

The vegetative propagation of oil palm, because of its botanical characteristics, can only be done through tissue culture. The first tissue culture attempt was initiated in the 1960s with the successful production of plantlets in the 1970s (Jones, 1974; Rabéchault and Martin, 1976). MPOB (then PORIM) followed up on this with research on refining *in vitro* propagation. However, this achievement was later challenged because of the need to overcome the incidences of abnormal clones with mantled fruits and other vegetation aberrations (Corley et al., 1986).

Despite the issue of abnormal clones, research in tissue culture to enhance clonal efficiency went ahead with the focus on improving the tissue culture protocol to meet the demand for ramets. Progress made to improve culture media, protocols for cloning and field trials, and efficient bioreactor systems for multiplication of oil palm cultures are some of the solutions to supply ramets to the industry.

Besides the concern on abnormality, another issue is the low embryogenesis rate. To address the latter, gene expression (Ong-Abdullah and Ooi, 2007) and gene mapping (Ting et al., 2013) studies, focusing on the development of biomarkers for tissue culture amenity and abnormality, were successful in identifying putative markers for embryogenesis. The abnormality issue was a problem but improvements in tissue culture processes and stringent culling in many oil palm tissue culture laboratories resulted in more manageable levels of fruit mantling. Jaligot et al. (2014) were able to shed some new insights into the epigenetic origin of the problem; then research work by Ong-Abdullah et al. (2016) later identified the actual cause of mantling, which was due to epigenetic change in the Karma transposon in the intron of the homeotic gene *DEFICIENS*.

4 Oil palm genomics and genetic engineering

4.1 Oil palm genomics

The genome is the genetic material in a chromosome, and unravelling the oil palm genome was one of the major achievements of R&D for the improvement of oil palm cultivation. In 2004, MPOB initiated a small genome sequencing project on the oil palm genome to look at hypomethylated and gene-rich regions which encode for the genes and their regulatory regions. Since then, two private companies including MPOB (Singh et al., 2013a) have sequenced the oil palm genome.

MPOB is also the host and developer of a database available at the Genomsawit portal (https://genomsawit.mpob.gov.my) to manage and share oil palm DNA sequences and other relevant information on oil palm genomic resources. The sharing of data is vital in expediting research particularly for molecular breeding or marker-assisted breeding of the oil palm. Molecular markers are becoming crucial in breeding programmes. These markers could help in reducing long and tedious phenotypic selection in the conventional breeding programme and greatly accelerating the production of improved cultivars. Simple sequence repeat (SSR) markers are the preferred molecular markers for mapping, genetic analyses and market-assisted plant improvement programmes. Various workers

(Singh et al., 2008; Low et al., 2008, 2014; Seng et al., 2011) have described the mining of SSR from available oil palm sequences. SSR markers have also been effectively used to analyse oil palm germplasm (Singh et al., 2008; Zulkifli et al., 2012; Bakoume et al, 2015) and mapping populations (Billotte et al., 2005, 2010; Singh et al., 2009; Ting et al., 2013, 2014; Montoya et al., 2014; Jeennor et al., 2014).

The availability of the oil palm gene sequence also allows for the mining of markers for important agronomic traits such as fruit colour (Singh et al., 2014), presence or absence of shell and oil quality. Indeed the successful identification of a single gene, called Shell, can help to increase the oil palm's yield by 30% (Singh et al., 2013b). In the past, selective breeding was the technique to ensure planting of elite *tenera* palms. However, this technique is not fail-save and about 10% of plantings can still be contaminated with the low-yielding dura because of uncontrolled wind and insect pollination. Confirming whether a palm will be of the desired shell type, that is, *tenera*, will need at least three years when the palm starts to bear fruits, at which time it would be uneconomical to uproot the palm for replanting. The identification of the shell gene enabled the developing of a molecular method on seedlings so that cultivation of *tenera* palms are ensured, thereby raising the yield of oil palm plantations.

As to the issue of abnormal clones from tissue culture, the gene responsible for mantled or shrivelled oil barren fruits was identified by MPOB and its collaborator, the United States-based biotechnology company Orion Genomics in 2015 (Ong Abdullah et al., 2015). The latter has been licensed by MPOB to develop a diagnostic test based on a simple leaf-based assay to identify clones with the mantle gene, thereby circumventing the planting of mantled palms (Ong-Abdullah et al., 2016).

Several oil palm breeding programmes around the world now include objectives to use genomic data to develop disease-resistant palms and produce value-added products. The annotated sequence of the oil palm genome is also a resource for biomarker discovery and also for the development of diagnostic tools subsequent to identification of genes responsible for both desirable and undesirable agronomical traits.

4.2 Genetic engineering

Gene manipulation through genetic engineering has gained attention because of some limitations of conventional breeding. It allows for specific genes of interest to be introduced directly into a genome and radically reduces the time needed for selective breeding to obtain the desired traits. With this technique, the fatty acid composition of palm oil could be modified to produce either a highly saturated or unsaturated oil or a lycopene instead of carotene-rich palm oil (Parveez et al., 2011).

Research in genetic engineering of the oil palm covers work on the following:

- gene and enzyme isolation in the oil palm fatty acid and carotenoid biosynthetic pathways;
- isolation of promoters for effective targeting of transgenes into specific tissue and to ensure optimal timing of the transgenic expression;
- construction of various transformation vectors carrying combinations of genes;
- promoters and other genetic materials for transforming oil palm; and
- development of transgenic oil palm and production of value-added transgenic oil palm.

There are concerns about the putative effects of transgenic crops on the environment and human health. Potential approaches to address such apprehension include the elimination of marker genes using plasmids (Komari et al., 1996), replacement of antibiotic marker genes with more 'friendly' genes (Joesbo et al., 1998) and self-containment of the transgene through chloroplast transformation (Daniell et al., 1998). Research is also required to tackle the probability of integrating a plasmid or DNA vector, besides the gene of interest, into the oil palm genome. More recently, research on plant transformation approaches to obviate the concerns on genetically modified plant products saw the emergence of gene editing technologies such as the CRISPR/CAS system (Belhaj et al., 2013; Alpeter et al., 2016).

5 Cultivation with minimal environmental impact

The world population is now reaching 7.3 billion, with the global middle class making up 1.7 billion, and it is expected to go up to 9.5–9.7 billion by 2050. Vegetable oil demand is expected to go up to 125–150 million tonnes in the next five years. The drivers for vegetable oil demand will come from population growth and the booming middle class. Palm oil will play a key role in meeting this demand because of its stable supply, availability, affordability and its versatile functionality in both food and non-food applications.

Research in oil palm cultivation had previously been more centred on increasing productivity through selection of quality planting materials obtained from selective breeding or tissue culture work. Though genomics will also be key to both quality and designer palms with high productivity and desired oil composition, this is still work in progress because the science is new and ethical, religious and philosophical concerns still revolve around palms engineered by man.

The defining issue at the moment is climate change due to global warming which in turn is triggered by the increasing level of GHG emissions produced by increasing anthropogenic activities. Deforestation and forest degradation are estimated to account for around 12% of global GHG anthropogenic carbon dioxide (CO_2) emissions (van der Werf et al., 2009), while the remaining 88% are through other anthropogenic activities such as transportation and production of industrial goods and the combustion of fossil fuel. These figures clearly indicate that reducing fossil fuel emissions remains the key element for stabilising atmospheric CO_2 concentrations and consequently global warming which lead to climate change.

5.1 Increasing yield to minimise land use change

The key here is the goal of increasing productivity without expansion onto new land, that is, no land use change. Under this scenario, if the yield of FFB per hectare could be increased from 20 to 25 t/ha, any additional amount of palm oil can be produced from the increased FFB without requiring an additional hectare of new land. This is why yield improvement through the tissue culture and genomic routes are top R&D priorities, together with best management practices.

5.2 Integrated pest management for control of pests and diseases

Two important factors governing the measured yield gap between potential yield of hypothetical oil palm genotypes and the actual yield of palms in plantation are attacks by pests and palms infected by diseases. A major disease which threatens the oil palm in Malaysia is the basal stem rot (BSR) caused by *Ganoderma* spp. fungi (Idris, 2011a). Since BSR is lethal and infected palms stop producing and eventually die, studies were carried out to understand its biology and epidemiology with the aim of (a) establishing a rapid and accurate diagnosis of the disease and (b) discovering solutions for its detection and implementing management strategies for controlling the *Ganoderma*. Proper land preparation at replanting is also a way to control the spread of BSR. Other solutions to BSR are the breeding of oil palm progenies which are resistant to *Ganoderma* and the use of biological control such as fungi, mycorrhiza and bacteria. Studies have also shown that the manipulation of fertiliser nutrients can reduce BSR incidences significantly. Although the ultimate objective is to find a cure for BSR infected palms, these interim solutions are important measures to curb the spread of the disease.

Other important diseases of the palms include vascular wilt disease which is also known as *Fusarium* wilt, bud rot, red ring disease, sudden wilt and leaf spot disease (Idris, 2011b).

Cultivation of oil palm in plantations often stretches over a large geographical zone is a monoculture, and many view this method as the worst possible way to grow crops. Monocultures are usually more susceptible to outbreaks of diseases and pest infestations than crops harbouring genetic polymorphism (Mundt, 2002). Thus, one of the greatest risks of monoculture is the danger of the entire crop being wiped out by a single pest or disease. This risk is exacerbated by the normal practice of planting vast areas with palms from a reduced number of distinct genotypes. Indeed, high genetic diversity could be a safeguard against the spread of pest and disease in monoculture crops (King and Lively, 2012). As oil palm is a perennial crop, diverse crop rotation, that is, polyculture, is not a solution because oil palm takes at least three years to start fruiting and establishment of oil palm plantations is costly and laborious. Thus, further work could be done to weigh the advantages and problems associated with the cultivation of palms from various different genetic backgrounds.

The Malaysian palm oil industry is always on the alert with continuing research for ready preventions/solutions against new pests and diseases. Steps are taken to circumvent disasters such as the bud rot scourge which wiped out some 50 000 ha of oil palm estates in South America. Stakeholders at the global level in oil palm R&D must remain very vigilant on emerging diseases and new pathogens and also follow through with ready action plans.

With the growing concerns on the use of agrochemicals to control pest infestation, alternative methods with less negative impact on the environment are important areas of research in oil palm cultivation. Biological control such as the use of microbial pesticides has great potential in avoiding the adverse impacts associated with the use of synthetic agrochemical insecticides. The bacterium *Bacillus thuringiensis* (*BT*) has proven to be a highly successful microbial insecticide (Rosas-Garcia, 2009). BT is mainly a soil-dwelling bacterium and was first discovered by the Japanese biologist Shigetane Ishiwata in 1901. Research has shown that BT is effective for the control of the bagworm (*Metisa plana*), an oil palm defoliator (Siti Ramlah et al., 2011). Other types of biological control agents include contact fungi and ingested viruses. The other common pest infestation in oil palm plantations is that by the rhinoceros beetle (*Oryctes rhinoceros*) and its larvae. The powder formulations of the fungus *Metarhizium anisopliae* and the *Oryctes* virus have proven to be effective in reducing the population of the larvae and the adult beetles, respectively (Ramle et al., 2005, 2008).

5.3 Soil and nutrient management to increase land yield

Available arable land and soils containing sufficient nutrients form the foundation of the cultivation of high-yielding palms. R&D actions are needed to ensure and enhance soil fertility so that the palm can produce yields which are near to its theoretical yield potential (Carron et al., 2015a,b, 2016). The nutrients in soil only support plant growth while the type and characteristics of the soil will determine the palm's ability to absorb water and nutrients. If the palm is to produce a yield near to its genetic potential, the soil must provide a suitable medium for its growth. Fertilisers can be added to make up for soil nutrient deficiency but these can be lost through leaching and runoffs, especially in tropical countries with characteristic heavy rainfall. The issue of fertiliser eutrophication of surface water and contamination of groundwater must be considered. Consequently, any use of either chemical or organic fertilisers exceeding the optimum amount required is not only a direct economic loss but also increases the risk of pollution and environmental burden.

The microbiome is also an important factor in determining the health of soils. Within the soil, organisms function within an ecological food web where nutrients are cycled through the soil biomass. This soil food web is the basis of healthy, living soil. Vital soil organisms involved in the soil food web include bacteria, fungi, protozoa, nematodes, arthropods and earthworms. Besides reducing the use of pesticides and fertilisers, the soil microbiome improves palm growth and health. A detailed study of the entire genome of soil biota through metagenomics approach is expected to provide important information on complex soil microbial communities for innovation of novel agricultural applications (Mocali and Bendetti, 2010).

Hence, in the matter of maintaining soil health, the holistic management approach is proposed as the basis for further research in oil palm cultivation. The relationship among the soil, oil palm and water must be managed as a 'whole' entity so that the cultivation of oil palm can continue indefinitely where land (soil) and water sustainability is created for future generations of oil palm growers.

5.4 Sustainability-centred cultivation

With the increasing demand on the earth's natural resources to feed and house the growing world population, oil palm cultivation will have to yield more oil using less land in order to protect and conserve human population, natural resources, biodiversity and the environment.

High-yielding palms curb land expansion but great care and governance are also needed to ensure that cultivation does not have a negative impact on the natural environment and its biodiversity.

The issue of oil palm cultivation on peat soils is a long-standing debate between the oil palm industry and environmental/conservation groups. Peatlands are important carbon sinks because they capture approximately one-third of the world's soil carbon and 10% of global freshwater sources reside in them. Oil palm grows best on mineral soils but because of the scarcity of arable land, about 8% of Malaysian oil palms are planted on peatlands. As peatlands must be drained before conversion to oil palm estates, oxidation and loss of organic carbon as CO_2 occur. However, in many instances oil palm is planted only after the peat forest is logged/degraded for timber harvesting. If these areas had not been planted with oil palm, it would have been left degraded, thereby incurring heavy GHG emissions anyway. Thus, the burden of GHG emissions should not be entirely borne by oil palm alone.

Despite the fact that the cultivation of oil palm on peat is both difficult and costly, much research over the years (Gurmit et al., 1988; Mutert et al., 1999; Hasnol et al., 2010, 2011; Haniff et al., 2011) have proven that with the correct land preparation, drainage and water management, optimal planting practices and fertiliser regimes, planting on peat can be carried out with minimal carbon losses and production costs.

While there is consensus among the planters that pristine peat and high biodiversity peat areas should be preserved as natural carbon sinks, opinions vary on the extent to which secondary peat forests and degraded peatlands are to be excluded from agricultural development in a developing country. Certain quarters have even advocated that peatlands should be no-go areas for cultivation of oil palm. One noteworthy point which has been overlooked in the eagerness for this moratorium on peatland development is the non-differentiation of tropical and temperate peat. On the contrary, tropical lowland peats have a distinctly different morphology compared to temperate and boreal peats (Veloo et al., 2014). Most publications on tropical peat refer to generic peat without any further characterisation.

Paramananthan and Wahid (2015), based on the mapping of over 700 000 ha of organic soils, suggested that tropical peats, especially in the surface tier (0–50 cm), are mostly sapric with low GHG emissions. In addition, hardwood is often present, again implying low GHG emissions.

Tropical peat soils are quite different from temperate peats not only because of the formation under different climate conditions but also because temperate peats are mainly derived from low growing plants while tropical peats are from forest species. Paramanthan and Wahid (2015) reported that the organic materials making up the peat vertical profile range from highly decomposed sapric material at the surface to a partly decomposed hemic material and to an undecomposed fibric material.

Thus, it is believed that a proper quantification of peat emissions should take note of their inherent differences not only in peat from different climatic regions but also within the same region because of the variations in peat vertical profile (sapric, hemic and fibric). This will require development of system models for different peat ecosystems including characterisation of peat according to an appropriate peat classification system for tropical peat, carbon exchange between peat ecosystems and the atmosphere at different spatial (different regions, countries, etc.) and temporal scales (different seasons, e.g. wet and dry season in the tropics). The unified peat classification (UPCS), as worked out by Paramanathan and Wahid (2015), will be useful for identifying the site characteristics of a peat area (Paramanathan, 1998, 2010a,b; Paramananthan and Wahid, 2008) and whether these could be both economically viable and environmentally acceptable for cultivation of oil palm (Paramanantan, 2014).

Answering to the question on whether development can be done on peat because of the issue of GHG emissions from drainage of peatlands, confirmation can only be obtained from investigation after the measurement of emissions from different peat types as grouped under the UPCS. This work is currently in progress.

6 Conclusion

Clones from tissue culture are the second wave of improvement following high-yielding materials derived from selective breeding. The unravelling of the oil palm genome heralds exciting advances through mining of traits to produce designer palms with high yield; disease, pest and drought resistance; quality; desired fatty acid; and phytochemical composition and which are short for easier harvesting. Such science-based innovations address the pressing needs for the cultivation of oil palm to produce more oil using less inputs and arable land, and with minimal adverse social, environmental and economic impacts.

However, a precautionary approach must be taken in the production of genetically modified palms because of the uncertainty on the effects of genetically modified organisms on human health and the environment and the variable acceptance of such technologies in different societies, especially in the developed world. The complete sequencing, assembling and annotation of the oil palm genome could also be invaluable in reducing the carbon emissions associated with the cultivation of oil palm.

Post-genomic research could provide information to drive the development of designer palms with sustainable high yields, disease resistance and ease of harvesting FFB. Additionally, new techniques for genetic crop improvement including advances in the use of genome editing could address some of the concerns associated with genetically modified crops as the process is claimed to mimic that in nature albeit at an accelerated rate.

With growing attention on the effectiveness of the best cultivation methods, it is no longer just about producing more with less, that is, eco-efficiency. Neither is it about conservation alone because sustainability is now the key word driving research in oil palm cultivation. Be that as it may, science-based approaches will remain the key in ensuring improvement of yield as well as good cultivation practices that protect both the environment and people. In this context, the best approach for research is to integrate a systems or holistic approach to address the three elements of sustainability, namely social equity, environmental protection and economic progress.

7 Where to look for further information

Further information on development of the oil palm in Malaysia can be obtained from:

- Ministry of Plantation Industries and Commodities (MPIC), Malaysia, www.kppk.gov.my
- Malaysian Palm Oil Board (MPOB), www.mpob.gpv.my

8 References

Altpeter, F., Springer, N. M., Bartley, L. E., Blechl, A. E., Brutnell, T. P., Citovsky, V. and et al. (2016). Advancing crop transformation in the era of genome editing. *Plant Cell* 28, 1510–20. doi:10.1105/tpc.16.0019.

Bakoume, C., Wickneswari, R., Siju, S., Rajanaidu, N., Kushairi, A. and Billotte, N. (2015). Genetic diversity of the world's largest oil palm (*Elaeis guineensis* Jacq.) field genebank accessions using microsatellite markers. *Genet. Resour. Crop Evol.* 62, 349–60.

Beirnaert, A. and Vandederweyen, R. (1941). *Contribution a L'etude genetique er biometrique des varieties d'Elaeis guineensis, Jacq*. Publication INEAC, Serie Scientifique 27.

Belhaj, K., Chaparro-Garcia, A., Kamoun, A. and Nekrasov, V. (2013). Plant genome editing made easy: Targeted mutagenesis in model and crop plants using the CRISPR/Cas system. *Plant Methods* 9, 39.

Billotte, N., Jourjon, M. F., Marseillac, N., Berger, A., Flori, A., Asmady, H., Adon, B., Singh, R., Nouy, B., Potier, F., Cheah, S. C., Rodhe, W., Ritter, E., Courtois, B., Charrier, A. and Mangin, B. (2010). QTL detection by multi-parent linkage mapping in oil palm (*Elaeis guineensis* Jacq.). *Theor. Appl. Genet.* 120, 1673–87. DOI:10.1007/s00122-010-1284-y.

Carron, M. P., Pierrat, M., Snoeck, D., Villenave, C., Ribeyre, F., Marichal, R. and Caliman, J. P. (2015a). Temporal variability in soil quality after organic residue application in mature oil palm plantations. *Soil Res.* 53(2), 205–15. https://doi.org/10.1071/sr14249

Carron, M. P., Auriac, Q., Snoeck, D., Villenave, C., Blanchart, E., Ribeyre, F. and Caliman, J. P. (2015b). Spatial heterogeneity of soil quality around mature oil palms receiving mineral fertilization. *Eur. J. Soil Biol.* 66, 24–31.

Carron, M. P., Auriac, Q., Snoeck, D., Villenave, C., Blanchart, E., Ribeyre, F. and Caliman, J. P. (2016). Do the impact of organic residues on soil quality extend beyond the deposition area under oil palm? *Eur. J. Soil Biol.* 75, 54–61.

Choo, Y. M., Tan, Y. A., Ramadasan, K., Ainie, K., Mohd Mokmin, B., Nik Mohd Aznizan, N. I., Puah, C. W. and Fatiha, I. (2015). *Malaysian Oil Palm Sustainability Manual.* Published by the Malaysian Palm Oil Board, p. 141. ISBN: 978-967-961-211-0.

Corley, R. H. V., Lee, C. H., Law, I. H. and Wong, C. Y. (1986). Abnormal flower development in oil palm clones. *The Planter* 62, 233–40.

Daniell, H., Datta, R., Varma, S., Grey, S. and Lee, S. B. (1998). Containment of herbicide resistance through genetic engineering of the chloroplast genome. *Nat. Biotechnol.* 16, 345–8.

Gurmit, S., Tan, Y. P., Rajah Padman, C. V. and Lee, F. W. (1986). Experiences on the cultivation and management of oil palms on deep peat in United Plantations Berhad. 31 2nd International Soil Management Workshop: Classification, Characterization and Utilization of Peat land, Haadyai, Thailand: Preprint.

Haniff, H., Kushairi, A., Tarmizi, A., Wahid, O., Hasnol, O., Farawahida, M. D. and Maisarah, J. (2011). *Guidelines or the Development of a Standard Operating Procedure for Oil Palm Cultivation on Peat.* Published by the Malaysian Palm Oil Board, Kajang, Selangor, Malaysia.

Hasnol, O., Tarmizi, A., Farawahida, M., Mohd Haniff H. and Mohd Pilus, Z. (2011). Best management Practices for Oil Palm Cultivation on Peat Ground water-table maintenance in relation to peat subsidence and estimation of CO_2 emissions at Sessang, Sarawak. *J. Oil Palm Res.* 23, 1078–86.

Hasol, O., Ahmad Tarmizi, M., Mohd Haniff, H., Farawaihida, M. D. and Hasimah, M. (2010). Best management practices for oil palm planting on peat: Optimum groundwater table, MPOB Information Series No. 528, p. 7. http://www.merdekaaward.my/Recipients/By-Category/Outstanding-Contribution-to-the-people-of-Malaysia/Datuk-Leslie-Davidson.aspx (accessed 12 September).

Idris, A. S. (2011a). Biology, detection and control of *Ganoderma* in oil palm. In *Further Advances in Oil Palm Research* (2000–2010), Vol. 1, eds M. B. Wahid, Y. M. Choo and K. W. Chan, Published by the Malaysian Palm Oil Board, Kajang, Selangor, Malaysia, pp. 485–521.

Idris, A. S. (2011b). Other devastating diseases of oil palm. In *Further Advances in Oil Palm Research* (2000–2010), Vol. 1, eds M. B. Wahid, Y. M. Choo and K. W. Chan, Published by the Malaysian Palm Oil Board, Kajang, Selangor, Malaysia, pp. 522–42.

Jaligot, E., Hooi, W. Y., Debladis, E., Richaud, F., Beulé, T., Collin, M., Agbessi, M. D. T., Sabot, F., Garsmeur, O., D'Hont, A., Sharifah Shahrul, R. S. A. and Rival, A. (2014). DNA Methylation and Expression of the *EgDEF1* Gene and Neighboring Retrotransposons in mantledi Somaclonal Variants of Oil Palm. *PLoS One*, 9(3), e91896.

Jeennor, S. and Volkaert, H. (2014). Mapping of quantitative trait loci (QTLs) for oil yield using SSRs and gene-based markers in African oil palm (*Elaeis guineensis* Jacq.). *Tree Genet. Genomes*, 10, 1–14.

Joersbo, M., Donaldson, I., Kreiberg, J., Petersen, S. G., Brunstedt, J. and Okkels, F. T. (1998). Analysis of mannose selection used for transformation of sugar beet. *Mol. Breed.* 19, 798–803.

Jones, L. H. (1974). Propagation of clonal oil palms by tissue culture. *Oil Palm News*, 17, 1–8.

King, K. C. and Lively, C. M. (2012). Does genetic diversity limit disease spread in natural host populations? *Heredity (Edinb).* 109(4), 199–203.

Komari, T., Hiei, Y., Saito, Y., Murai, N. and Kumashiro, T. (1996). Vectors carrying two separate T-DNAs for co-transformation of higher plants mediated by *Agrobacterium tumefaciens* and segregation of transformants free from selection markers. *Plant J.*, 10, 165–74.

Low, E. T., Alias, H., Boon, S. H., Shariff, E. M., Tan, C. Y. A., Ooi, L. C. L., Cheah, S. C., Raha, A. R., Wan, K. L. and Singh, R. (2008). Oil palm (*Elaeis guineensis Jacq.*) tissue culture ESRs: Identifying genes associated with callogenesis and embryogenesis. *BMC Plant Biol.* 8, 62.

Low, E. T., Rozana, R, Nagappan, J., Ab Halim, M. A., Norazah, A., Chan, K. L., Nauman J. M., Maclean, P., Brauning, R., McCulloch, A., Moraga, R., Ong-Abdullah, M. and Singh, R. (2014). Analyses of hypomethylated oil palm gene space. *PLoS One* 9(1), e86728.

Malaysian Oil Palm Statistics, 2015, 35th Edition. Published by the Malaysian Palm Oil Board, Kajang, Selangor, Malaysia.

Mocali, S. and Bendetti, A. (2010). Exploring research frontiers in microbiology: The challenge of metagenomics in soil microbiology. *Res. Microbiol.* 161(6), 497–505.

Montoya, C., Cochard, B., Flori, A., Cros, D., Lopes, R., Cuellar, T., Espeout, S., Syaputra, I., Villeneuve, P., Pina, M., Ritter, E., Leroy, T. and Billotte, N. (2014). Genetic architecture of palm oil fatty acid composition in cultivated oil palm (*Elaeis guineensis* Jacq.) compared to its wild relative *E. oleifera* (H. B.K), Cortés. *PLoS One* 9, e95412.

Mundt, C. C. (2002). Use of multiline cultivars and cultivar mixtures for disease management. *Ann. Rev. Phytopathol.* 40, 381–410.

Mutert, E., Fairhurst T. H. and von Uexküll, H. R. (1999). Agronomic management of oil palms on deep peat. *Better Crops Int.* 13(1).

Ninth Malaysian Plan, www.bnm.gov.my/files/publication/sme/en/2005/chap_8.pdf.

Ong Abdullah, M., Ordway, J. M., Jiang, N., Ooi, S. E., Kok, S. Y., Norashikin, S., Nuraziyah, A., Ahmad Tamizi, H., Zamzuri, I., Samsul K. R., Fadila, A. M., Nor Azwani, A. B., Marhalil, M., Norziha, A., Zulkifli, Y., Mohd Din, A., Rajanaidu, N., Rajinder, S., Low, B., Muhammad, A. B., Van Brunt, A. and et al. (2015). Loss of Karma transposon methylation underlies the mantled somaclonal variant of oil palm. *Nature* 525, 533–53.

Ong-Abdullah, M., Ordway, J. M., Jiang, N., Ooi, S. E., Azizah, M., Rajanaidu. N., Rajinder, S., Low, E. T. L., Mohit, S., Smith, S. W., Lakey, N., Ravigadevi, S. and Martienssen, R. A. (2016). SureSawitTM KARMA – A diagnostic assay for clonal conformity. MPOB Information Series, ISSN 1511–7871, June 2016, MPOB TS No. 156.

Ong-Abdullah, M. and Ooi, S. E. (2007). Biomarkers: Finding a niche in oil palm tissue culture, Part 2 – Targeting the transcriptome. Oil Palm Bulletin No. 54:68–74.

Paramananthan, S. (2014). Tropical lowland peats: To conserve or develop them? In *Selected Papers on Soil Science – Problem Soils.* Paper 2, Published by Agricultural Crop Trust (ACT) and Param Agricultural Soil Surveys (PASS), Petaling Jaya, Selangor, Malaysia, ISBN 978-983-43384-3-5, pp . 15–49.

Paramananthan, S. and Wahid, O. (2008). Classification of organic soils in Malaysia. Paper presented at the Malaysia-Netherlands Joint Committee Seminar on Carbon Emission, 7 July 2008, Palace of Golden Horses, Seri Kembangan, Selangor.

Paramananthan, S. and Wahid, O. (2015). *Soils of the Lower and Middle Baram River Basin, Miri Division, Sarawak, Malaysia.* Published by the Malaysia Palm Oil Board, Kajang, Selangor, Malaysia, ISBN: 978–967-961-205-9.

Paramananthan, S. (1998). *Malaysian Soil Taxonomy (Second Approximation): A Proposal for the Classification of Malaysian Soils, Malaysian Society of Soil Science, Malaysia.* Param Agricultural and Soil Surveys (M) Sdn. Bhd.: Kuala Lumpur, Malaysia.

Paramananthan, S. (2010a). *Malaysian Soil Taxonomy – Revised*, second edition. Param Agricultural Soil Surveys (M) Sdn. Bhd.: Petaling Jaya, Selangor, Malaysia.

Paramananthan, S. (2010b). *Keys to the Identification of Malaysian Soils Using Parent Materials.* Param Agricultural Soil Surveys (M) Sdn. Bhd.: Petaling Jaya, Selangor, Malaysia.

Parveez, G. K., Rasid, O. A. and Sambanthamurthi, R. (2011). Genetic engineering of oil palm. In *Further Advances in Oil Palm Research* (2000–2010), 17. Idris, A. S. (2011a). Biology, detection and control of *Ganoderma* in oil palm. In *Further Advances in Oil Palm Research* (2000–2010), Vol. 1, eds M. B. Wahid, Y. M. Choo and K. W. Chan, Published by the Malaysian Palm Oil Board, Kajang, Selangor, Malaysia, pp. 141–201.

Rabéchault, H. and Martin, J.-P. (1976), Multiplication végétative du palmier à huile (Elaeis guineensis Jacq.) à l'aide de cultures de tissus foliaires. *C. R. Acad. Sci. Paris* 238, 1735–7.

Rajanaidu, N. (1994). *PORIM Oil Palm Genebank*. PORIM: Bangi, p. 19.

Rajanaidu, N. and Rao, V. (1988). Oil palm genetic collections: Their performance and use to the industry. Proceedings of the 1987 International Palm Oil Conference – Agriculture, pp. 59–85.

Rajanaidu, N., Kushairi, A., Rafii, M., Mohd Din, A., Maizura, I. and Jalani, B. (2000). Oil palm breeding and genetic resources. In *Advances in Oil palm Research*, Vol. 1, eds Y. Basiron, B.S. Jalani and K.W. Chan, Published by the Malaysian Palm Oil Board, Kajang, Selangor, Malaysia, pp. 171–237.

Ramle, M., Mohd Basri, W., Norman, K., Siti Ramlah, A. A. and Noor Hisham H. (2006). Research into the commercialization of Metarhizium anisopliae (Hyphomycetes) for biocontrol of the rhinoceros beetle, Oryctes Rhinoceros (Scarabaeidae), in oil palm. *J. Oil Palm Res.*, Vol. Special Issue 2006 Apr, 37–49.

Ramle, M., Norman, K., Siti Ramlah, A. A. and Mohd Basri, W. (2008). Metarhizium granules for control of rhinoceros beetle. MPOB Information Series, ISSN 1511-7871, June 2008 MPOB TT No. 404.

Rosas-Garcia, N. M. (2009). Biopesticide production from Bacillus thuringiensis: An environmentally friendly alternative. *Recent Pat. Biotechnol.* 3, 28–36.

Seng, T. Y., Mohamed Saad, S. H., Chin, C. W., Ting, N. C., Singh, R. S. H., Zaman, F. Q., Tan, S. G. and Syed Alwee, S. S. R. (2011). Genetic linkage map of a high yielding FELDA Deli x Yangambi oil palm cross. *PLos One*, 6, e265293. https://doi.org/10.1371/journal.pone.0026593

Singh, R., Zaki, N. M., Ting, N. C., Rosli, R., Tan, S.-G., Low, E.-T. L., Ithnin, M. and Suan-Choo, C. (2008). Exploiting an oil palm EST database for the development of gene-derived SSR markers and their exploitation for assessment of genetic diversity. *Biologia* 63(2), 227–35.

Singh, R., Low, E. T., Ooi, L. C. L., Ong Abdullah, M., Rajanaidu, N., Ting, N. C., Marjuni, M., Chan, P. L., Maizura, I., Mohd Arif, A. M., Nagappan, J., Chan, K. L., Rozana, R., MRaohd Amin, H., Norazah, A., Mohammad, A. B., Lakey, N., Bacher, B., Van Brunt, A., Wang, C., Ordway, J. M., Ravigadevi, S. and Martienssen, R. A. (2014). The oil palm *VIRESCENS* gene controls fruit colour and encodes a R2R3-MYB. *Nat. Comm.* 5, Article number:4106, DOI:10.1038/ncomms5106 , 1–8.

Singh, R., Ong-Abdullah, M., Low, E. T. L., Mohamad Arif, A. M., Rozana, R., Rajanaidu, N., Ooi, L. C. L., Ooi, S. E., Chan, K. L., Mohd Amin, H., Norazah, A., Jayanthi, N., Bacher, B., Lakey, N., Smith, S. W., He, D., Hogan, M., Muhammad, A. B.,. Lee, E. K., DeSalle, R., Kudrna, D., Goicoechea, J. L., Wing, R. A., Wilson, R. K., Fulton, R. S., Ordway, J. M., Martienssen, R. A. and Ravigadevi, S. (2013a). Oil palm genome sequence reveals divergence of interfertile species in Old and New worlds. *Nature* 500, 335–9.

Singh, R., Low, E. T., Ooi, L. C., Ong-Abdullah, M., Ting, N. C., Nagappan, J., Rajanaidu, N., Amiruddin, M. D., Rosli, R., Manaf, M. A., Chan, K. L., Halim, M. A., Azizi, N., Lakey, N., Smith, S. W., Budiman, M. A., Hogan, M., Bacher, B., Van Brunt, A., Wang, C., Ordway, J. M., Ravigadevi, S. and Martienssen, R. A. (2013b). The oil palm SHELL gene controls oil yield and encodes a homologue of SEEDSTICK. *Nature* 500, 340–4.

Singh, R., Noorhariza, M. Z., Ting, N. C., Rozana, R., Tan, S. G., Low, E. T., Maizura, I. and Cheah, S. C. (2008). Exploiting an oil palm EST database for the development of gene-derived SSR markers and their exploitation for assessment of genetic diversity. *Biologia* 63: 227–35.

Siti Ramlah, A. A., Ramle, M., Norman, K. and Mohd Basri, W. (2011). Microbial approach in pest control. In *Further Advances in Oil Palm Research* (2000–2010), vol 1, eds M. B. Wahid, Y. M. Choo and K. W. Chan, Published by the Malaysian Palm Oil Board, Kajang, Selangor, Malaysia, pp.435–84.

Ting, N. C., Jansen, J., Nagappan, J., Ishak, Z., Chin, C. W., Tan, S. G., Cheah, S. C. and Singh, R. (2013), Identification of QTLs associated with callogenesis and embryogenesis in oil palm using genetic linkage maps improved with SSR markers, *PLoS ONE*, 8, e53076

Ting, N. C., Jansen, J., Mayes, S., Massawe, F., Sambanthamurthi, R., Ooi, L. C. L., Chin, C. W., Arulandoo, X., Seng, T. Y., Alwee, S. S., Ithinin, M., Singh, R. (2014). High density SNP and SSR-based genetics maps of two independent oil palm hybrids. *BMC Genomics*, 14, 309.

Van der Werf, G. R., Morton, D. C., DeFries, R. S., Olivier, J. G. J., Kasibhatla, P. S., Jackson, R. B., Collatz, G. J. and Randerson, J. T. (2009). CO_2 emissions from forest loss. *Nat. Geosci.* 2, 737–8.

Veloo, R., Paramananthan, S. and Van Ranst, E. (2014). Classification of tropical lowland peats revisited: The case of Sarawak. *Catena*, 118, 179–85.
Wong, L. C. Y. (2007). Development of Malaysia's Agricultural Sector: Agriculture as an Engine of Growth? Presented at the ISEAS Conference on the Malaysian Economy: Development and Challenges, 25–26 January 2007, ISEAS Singapore.
Zulkifli,Y., Maizura, I., and Singh, R. (2012). Evaluation of MPOB Oil Palm (*Elaeis guineensis*) Germplasm Using EST-SSR. *J. Oil Palm Res.* 4:1368–77.

Chapter 3

Sustainability pathways in oil palm cultivation: a comparison of Indonesia, Colombia and Cameroon

Ahmad Dermawan, Center for International Forestry Research (CIFOR), Indonesia; and Otto Hospes, Wageningen University, The Netherlands

1 Introduction

Oil palm (*Elaeis guineensis*) is a versatile crop with increasing multiple uses, especially for food and energy (Alonso-Frajedas et al. 2016). Palm oil has a number of comparative advantages over other vegetable oils. The oil palm has the highest oil yield, which could reach as high as ten times of the yield of soybean, rapeseed and canola oils. It also has the lowest land requirement to produce one ton of oil, and the lowest production cost per hectare (Corley and Tinker 2015). While covering 6% of the total global agricultural land, palm oil production contributes to one-third of global vegetable oil production. Oil palm also has a long lifespan, as it can provide continuous harvest in about 25–30 years.

Oil palm can grow on many types of soil, provided the general physical characteristics are not extreme and the climate is suitable. The ideal latitude for its growth is between 15°N and 15°S (Corley and Tinker 2015). An adequate combination of rainfall and sunshine is important for the growth of oil palm. A minimum of 2,000 mm per year of annual rainfall that is evenly distributed, minimum temperature of 20°C and maximum between 28 and 34°C, and at least 1800 hours of sunshine per year are needed to allow the crop to reach its potential (Ngando-Ebongue et al. 2012, Corley and Tinker 2015).

Palm oil, the processed oil from oil palm, has a variety of derivative products. The demand for palm oil has been increasing since the last three decades. It has surpassed soybean

http://dx.doi.org/10.19103/AS.2017.0018.03

as the most consumed and traded oilseeds globally (Ngando-Ebongue et al. 2012, FAO 2014). China, India and the European Union are the largest importing countries for palm oil.

Despite being very versatile, oil palm is currently one of the most debated crops. For the key producing countries, oil palm has become an important contributor to the economy. Cultivation of oil palm absorbs a vast amount of labour for large-scale estates, as well as millions of smallholders who cultivate and process oil palm. Oil palm also has a vast amount of derivative industries that serve many sectors.

Together with the economic benefits that have been provided through the development and expansion of oil palm, there are a number of significant associated environmental and social costs. A number of studies have highlighted deforestation and biodiversity loss as the main environmental costs (Linder and Palkovich 2016). Koh and Wilcove (2008a, 2008b) show that about half of oil palm expansion in Malaysia and Indonesia has gone through prior deforestation. Subsequent studies on the sources of deforestation from 2000s also confirm the contribution of oil palm as one of the key drivers of deforestation (Abood et al. 2014).

Fayle et al. (2010) found that the total number of ant species in oil palm plantations was significantly lower than under forest cover. Conversion to oil palm dramatically reduces species richness, with significantly fewer primary-forest species than found on logged forests, notably for birds, leaf-litter ants, beetles, aerial hymenopterans, flies and true bugs (Edwards et al. 2014b). Earlier studies in Indonesia also found similar results: oil palm plantations support much fewer species than do forests and often also fewer than other tree crops (Fitzherbert et al. 2008).

In terms of social costs, dispossession of smallholders' land by corporations and violent groups has become an important social impact (e.g. Budidarsono et al. 2013, Linder and Palkovitz 2016). Most of the countries with dominant role of corporations in driving the oil palm production are characterized by poor social impacts (IFC 2013, Buitron 2001). Specifically in Colombia, the social cost of oil palm has involved cases of dispossession of land by violent groups (Garcia-Ulloa et al. 2012, Maher 2015).

A number of governance initiatives have been launched in order to minimize the negative ecological and social impacts as well as to enhance positive impacts of oil palm cultivation. Initiatives from public agencies, private multi-stakeholder process as well as corporate self-regulation to develop standards, best management practices, certification and corporate commitments have emerged with the aim of achieving sustainable and equitable oil palm development. Each country experiences different pathways in moving towards sustainable production of oil palm.

The present chapter aims at identifying the pathways towards sustainable oil palm production in three countries: Indonesia, Colombia and Cameroon. Indonesia and Colombia are the leading oil palm countries in Southeast Asia and Latin America, respectively. While Cameroon is not the leading palm oil producer in Africa, its stage of development is similar to the one of other countries in the region. The three countries reflect different levels of complexities of interactions among actors, different governance challenges, as well as different routes and achievements towards sustainable oil palm production.

2 Conceptualizing sustainable pathways

Increasing palm oil production in a sustainable way is a huge challenge for governments, private-sector actors and smallholders. On the production side, actors from both government and private sectors set the production ambitions or targets for palm oil. On the consumer side, there are demands that oil palm is produced in a sustainable manner, that is, not destructive to the environment as well as respectful to the rights of the communities. Though considerations on sustainability are important, the way in which these considerations are translated into specific policies and are combined with production ambitions and targets varies across places (Wittmayer et al. 2014). This leads to a variety of pathways to sustainable palm oil production.

We conceptualize sustainable pathways as trajectories that connect technical, environmental and governance practices that reinforce each other, and consisting of actors at different levels that regulate, manage, implement and monitor these practices towards sustainable production (Hospes et al. 2017; Leach et al. 2010). This conceptualization has three components: sustainable production as the aspiration, as a variety of practices that support each other, and governed by actors at different levels with different roles.

Sustainable production can be seen as a desired situation, a vision or an ambition. Generally, this includes not being destructive to the environment as well as respect for social conditions. It is important to note that sustainable production itself is both a context-specific and contested concept. Both different contexts and different frames of problems and solutions for sustainable production explain different steps, and ideas on steps, towards sustainable production.

Sustainable pathways involve the combination of technical, environmental, regulatory and governance practices that reinforce each other in contributing towards sustainable production. An example in the oil palm context is the combination of best management practices using the best available planting material and agronomic techniques to realize optimal yields, and regulatory practices that set limits to expansion of palm oil production into forest areas and peat land. Regulatory practices can consist of public regulation and regulation by multi-stakeholder initiatives but also corporate self-regulation. An example within the field of governance and regulation is the combination of integrated land-use planning at national and regional level, and procedural requirements on mutual consultation and agreement between companies and communities at local level.

Sustainable pathways involve different actors with different roles (Hospes et al. 2017; Leach et al. 2010). The government is generally seen as a key regulator agency. However, different bodies and levels of government may execute their regulatory role in different ways, for instance, through the development of a national standard for sustainable palm oil, organizing forest moratoria, licensing palm oil production or monitoring the private sector. Next to complying with government regulations, private-sector actors may also set their own regulations, manage resources and make decisions on land use. Smallholders can be seen as private-sector actors from the entrepreneurship point of view, but also as those who need empowerment as they have much more limited resources to produce in sustainable manners.

There is no single pathway towards sustainable oil palm production (Hospes et al. 2017; Leach et al. 2010; Lindahl et al. 2015). Pathways are historically contingent and context-specific. The historical development of the palm oil sector, changes of the political regime, and political frames of the issue together define the scope for sustainable pathways in a country. Different forms and combinations of regulation exist in different countries, as the involvement of different actors in the palm oil sector in, for instance, Indonesia is not necessarily the same as the one in Columbia and Cameroon. In Indonesia, public, private and corporate self-regulation exists. In Colombia, private regulation dominates. Meanwhile, perhaps sustainability has not yet become an important policy agenda in Cameroon.

3 Oil palm production in Indonesia, Colombia and Cameroon

3.1 Indonesia

From four plants brought to Bogor Botanical Garden in 1848, oil palm has gone through a long history to become one of the most economically important crops in Indonesia. Commercial oil palm during the Dutch colonial period began in 1911, where a Belgian company opened plantations in Pulau Raja (Asahan) and Sungai Lipoet (Aceh). The oil palm sector then grew faster through the development of the first palm oil factory in 1919, and by 1937 Indonesia took over Nigeria as the largest palm oil exporter. However, the oil palm sector fell dramatically during World War II, and the trend continued until the late 1960s (PASPI 2014).

During the New Order period that started when Suharto seized power in 1967, the development of the oil palm sector was facilitated by the enactment of Law 1/1967 on foreign investment. The Government of Indonesia, with the assistance of international donors, developed a variety of programmes to boost the development of the oil palm sector. The introduction of programmes for smallholders dramatically increased the participation of smallholders. Another turning point came in late 1990s, when the economic Indonesian crisis opened doors for a significant increase of palm oil exports. By 2006, Indonesia regained the position as the world's leading exporter of palm oil, surpassing Malaysia. Indonesia is not only the largest exporter but also one of the largest consumers of palm oil in the world. The government has also declared cooking oil, that is mainly palm based, as one of the so-called 'nine essential products'. This is yet another reason why the government has supported the increase of palm oil production.

One of the key drivers that has facilitated the expansion of oil palm in Indonesia has been supportive government policies. The Government of Indonesia has formulated policies to promote private-sector investment as well as the involvement of smallholders into the palm oil business through a number of schemes since the 1970s. Indonesia became famous with the introduction of a number of schemes, such as nucleus-plasma and cooperative scheme. Throughout history, the Government of Indonesia has used different financial schemes to achieve economic growth by developing palm oil (Pramudya et al. in press). A number of additional policy measures supported investments in the palm-oil-processing industries. More recently, the national energy policy has provided room for palm oil-based biodiesel to flourish with the biodiesel mandates. These policies are designed to reach a mid-term target of achieving 40 million tons of CPO production by 2020.

Indonesia has also formulated policies to support the development of advanced palm-oil-processing industries. The Government of Indonesia supports biodiesel development through the establishment of biodiesel mandatory targets. Indonesia has been moving up and down with the blending targets, and is aiming to reach 20% of biodiesel by 2016 and 30% by 2020 (Ministry of Energy and Mineral Resources 2016). However, the biofuel mandate that has been in place since 2006 has been hampered by the fact that at the same time the country also heavily subsidized the fossil fuels (Dermawan et al. 2012). The lower the oil price, the higher the amount of subsidies to be allocated for biodiesel.

Currently, the Indonesian palm oil sector is dominated by large-scale private actors. Some 11.5 million hectares in Indonesia are planted palm oil areas, which is about 5% of the country's land area. More than half of this amount, that is 5.9 million hectares, are owned by large-scale private companies. State-owned enterprises, which dominated the sector during 1960s–1970s (see Pramudya et al. in press), currently control 0.8 million hectares. In addition, there are 1601 palm-oil-processing mills throughout Indonesia. The remaining 4.7 million hectares are controlled by smallholders. Smallholders account for 40% of the total planted area. Indonesia is the country with the largest number of smallholders in the world (Central Statistical Agency of Indonesia 2015).

In Indonesia, palm oil yield varies across different business models (smallholders versus private companies), or even within the same business model, for example between independent smallholders and smallholders under partnership with companies (called plasma smallholders). Independent smallholders carry out oil palm cultivation by themselves, often using planting material of unreliable origin. Meanwhile, smallholders under plasma scheme have their cultivation controlled by the nucleus companies. Such smallholders can obtain higher fresh fruit bunches yield by 15% when compared to independent smallholders (IFC 2013). The best estates can reach yield levels of six tons of palm oil per hectare and beyond.

3.2 Colombia

Oil palm was brought to Colombia in 1932 (Potter 2015). It is called the African oil palm as there is another oil palm species *Elaeis oleifera* that naturally grows in the Amazonian region. Currently, oil palm development is concentrated in 16 states in four production zones: 1) the Western Zone, at the south of western Colombia on the Pacific coast; 2) the Northern Zone, in the northeastern part of the country, near the Atlantic coast; 3) the Central Zone, an inter-Andean valley of the Magdalena river system; and 4) the Eastern Zone, at the foothills of the eastern chain of Andes range (Gomez et al. 2011).

In the beginning, there were a small number of companies and local growers that tried to plant oil palm, mainly to supply domestic markets. Expansion took place rapidly since early 2000s as the government provided incentives to increase palm oil production for exports and to meet biodiesel blending targets of 5% by 2008 (Pacheco 2012). Currently, Colombia cultivates the largest oil palm area in South America. The development of oil palm in Colombia is driven by large-scale actors (although in terms of scale they are still far below those in Indonesia and Malaysia). Currently, about 33% of planted areas are between 200 and 1000 hectares, and another 35% are over 1000 hectares (Potter 2015). The National Federation of Oil Palm Growers, or Fedepalma, was formed in 1962 with the aim of organizing the growers and ensuring the progress of oil palm development. Smallholders formed 'Strategic and Productive Alliances', where an association of smallholders forms a contract with the source of funding, usually large-scale plantations (Potter 2015). In 2008, 55 mills were operating in Colombia, about half of which were relatively small (less than 15 tons fresh fruit bunches per hour). Thirteen mills had the capacity of more than 25 tons per hour, of which only two mills had a capacity of more than 60 tons per hour (Pacheco 2012).

Being the largest producer in South America, Colombia has reached over 1.1 million tons of palm oil in 2015 (Index Mundi 2016). However, the oil palm production in the country has to deal with a number of limiting factors, such as topographic, climatic (seasonal dry periods), and less suitable soil conditions, as well as the presence of pests and diseases (Henson 2011, Pacheco 2012, Torres et al. 2016). Waves of cool temperature and the bud rot disease have also hampered the oil palm production in the country in the last few years, with some significant social impacts (Potter 2015). Still, in 2015 the country recorded a national average yield of 3.2 tons of CPO per hectare, which is comparable to the performance of the Southeast Asia (Index Mundi).

The Colombian government has a target of establishing a total three million hectares of oil palm by 2020. In addition, the government also aims to reach a 20% biodiesel blending by the same year. A number of policies to reach the targets have been issued, for example, a policy on tax holidays, implementation of free tax zones, tax reduction from investments in productive assets and credits for establishing and maintaining plantations (Pacheco 2012). Despite these policies, some have argued that the target is overly ambitious. Reaching three million hectares means increasing the current planted area by approximately six times in five years (Garcia-Ulloa et al. 2012, Castiblanco et al. 2013, Pinto et al. 2014).

3.3 Cameroon

Cameroon has been traditionally using a variety of products from oil palm: the red oil from the mesocarp, the oil contained in the kernel, and the sap that ferments to generate palm wine (Nkongho et al. 2015). Local populations harvested oil palm for subsistence and trade. Oil palms were harvested in the wild groves and were introduced on farmland as a mixed crop with other food and cash crops. After the arrival of the German and British colonial powers, large-scale oil palm sector began to emerge.

Five of Cameroon's ten regions are suitable for oil palm cultivation: the Southwest, Littoral, South, Centre and East regions. These regions are deemed suitable for oil palm cultivation as they meet biophysical requirements in terms of temperature, sunshine, precipitation, soil type and altitude (Hoyle and Levang 2012). These regions have become attractive for investments in the oil palm sector.

In 2015, Cameroon had 130 000 hectares of oil palm producing 270 000 tons of palm oil (Index Mundi 2016). Although about three quarters of oil palm areas are under the smallholders, they have very low yield, with the average of 0.8 ton oil per hectare. This is lower than the yield of agro industries, which could reach 2.3 tons per hectare (Hoyle and Levang 2012, Nkongho et al. 2014). Cameroon has a target of reaching 450 000 tons of palm oil production by 2020 (Holye and Levang 2012).

Cameroon is a net importer of palm oil. The country has become an interesting place for oil palm investment to serve domestic markets, regional markets and demand from Europe. The availability of cheap land, political support from the government and the governmental plans to develop agricultural sector are also factors that make investing in Cameroon more interesting (Hoyle and Levang 2012).

Similar to the independent smallholders in Indonesia, the agronomic performance of smallholders in Cameroon is also poor. A study by Nkongho et al. (2014) found that only 35% of smallholders use certified planting material, while the rest use either uncertified Tenera or Dura planting material (Nkongho et al. 2014). Fertilizer application follows a similar trend, with only 1.1% of smallholders applying fertilizer timely and 68% of smallholders not using fertilizers at all. The yields are quite low: smallholders produce only seven tons of fresh fruit bunches per hectare.

This section has highlighted that each of the three selected countries has set targets for oil palm production. Each of the targets often implies a significant amount of expansion. Such expansion means either opening new land from forests or changing existing land use to give way to oil palm. The next two sections discuss sustainability issues that are emerging in each of the three countries and how actors are responding to these issues.

4 Sustainable pathways: challenges and initiatives

4.1 Sustainability challenges

With the economic benefits provided through the development and expansion of oil palm, there are a number of significant associated environmental and social costs. Some of the environmental impacts of the oil palm expansion are deforestation, biodiversity loss, carbon stock losses and greenhouse gas emissions (Linder and Palkovitz 2016). Koh and Wilcove (2008a,b) show that about half of oil palm expansion in Indonesia has gone through prior deforestation. Subsequent studies on the sources of deforestation from the 2000s also confirm the contribution of oil palm to deforestation (Margono et al. 2012). However, there are disagreements on how much oil palm has contributed to deforestation. For example, the study by Gunarso et al. (2013) shows that undisturbed forests are only about 5% of the land converted to oil palm, which is much smaller than concluded by other studies (Miettinen et al. 2011, Margono et al. 2012, Busch et al. 2015, Vijay et al. 2016). In Colombia about half of the oil palm plantations established in 2002–2008 were previously classified as pastures, and less than 15% of the oil palm replaced natural vegetation (forests, savannah).

With regard to biodiversity, Fayle et al. (2010) concluded that total ant species in Sabah, Malaysia, decreased from 309 to 110 due to the conversion of forests to oil palm plantations. In general, conversion of habitat has decreased the number and richness of species, with significantly fewer primary-forest species than found on logged forests, notably for birds, leaf-litter ants, beetles, aerial hymenopterans, flies and true bugs (Edwards et al. 2014b). Other studies in Indonesia found similar results: oil palm plantations support much fewer species than do forests and often also fewer than other tree crops (Fitzherbert et al. 2008, Kurz et al. 2016).

The associated greenhouse gas emissions from the expansion of oil palm, especially those that come through forest clearing using fires, are also significant. Establishment of oil palm reduces soil organic carbon (van Straaten et al. 2015). Van Straaten et al. (2015) also found that the higher the initial soil organic carbon, the higher the losses. Carbon losses from forest peat conversion to oil palm could reach 405 tonnes in one planting cycle (25 years) (Murdiyarso et al. 2009, Schrier-Uijl et al. 2013). The greenhouse gas emissions depend on the previous land use. In Colombia, the differential production of greenhouse gas depends on whether the land use that precedes oil palm is forest or pasture. In the case of pasture, the greenhouse gas intensity is lower (Castanheira et al. 2014).

In terms of social impact, dispossession of smallholders' land by corporations or (para) military forces is one of the key social impacts (e.g. Budidarsono et al. 2013, Maher 2015). Most of the countries with a dominant role of corporations in driving the oil palm production are characterized by poor social impacts of oil palm (Obidzinski et al. 2012, IFC 2013, Li 2015). Examples are lack of smallholder involvement in large-scale oil palm projects and poor working conditions of farmer-labourers (Hoyle and Levang 2012). In Colombia, the social cost of oil palm involves dispossession of land by (para)military groups (Garcia-Ulloa et al. 2012, Maher 2015). Oil palm companies are also found to show lack of respect to traditional rights (Obidzinski et al. 2012).

4.2 Taking steps towards sustainability: certification and beyond

A number of initiatives have emerged to deal with the sustainability issues in the palm oil sector. In 2001, the World Wildlife Fund (WWF) and a number of European food manufacturers explored the idea of establishing a Roundtable for Sustainable Palm Oil (RSPO), which was eventually established in 2004. Through multi-stakeholder consultation, RSPO developed principles and criteria for sustainable palm oil. The vision of the RSPO is to transform markets and to make sustainable palm oil the norm. Specific objectives of the RSPO are to promote the production, procurement, finance and use of sustainable palm oil products; to develop, implement, verify, assure and periodically review credible global standards for the entire supply chain of sustainable palm oil; to monitor and evaluate the economic, environmental and social impacts of the uptake of sustainable palm oil in the market; and to engage and commit all stakeholders throughout the supply chain, including governments and consumers. Membership of the RSPO is voluntary for palm oil companies as well as for the estates that wish to have their operations certified.

The RSPO has agreed on eight principles: commitment to transparency, compliance with applicable laws and regulations, commitment to long-term economic and financial viability, use of appropriate best practices by growers and millers, environmental responsibility and conservation of natural resources and biodiversity, responsible consideration of employees and of individuals and communities affected by growers and mills, responsible development of new plantings, and commitment to continuous improvement in key areas of activity (RSPO 2013). For each of these principles specific criteria and indicators have been formulated against which a palm oil operation is evaluated. The principles and criteria are generic. Countries have to develop their own national interpretation in order to ensure that the principles and criteria are relevant under national contexts. Colombia and Indonesia have their own national interpretations of the RSPO principles and criteria.

Both the governments of Indonesia and Malaysia have also established official principles and criteria for sustainable palm oil. Indonesia launched a national sustainability standard in 2011 that was updated in 2015: the Indonesian Sustainable Palm Oil (ISPO) standard. The Government of Indonesia emphasized that the plantation sector in Indonesia has to be developed in accordance with a number of principles, such as sovereignty, sustainability, efficiency, fairness and environmental integrity. ISPO is aimed to ensure that oil palm planters and processing companies apply the principles and criteria correctly and consistently to produce sustainable palm oil (Ministry of Agriculture Regulation 11 of 2015). Since the ISPO is established as a regulation by ministerial decree, it is mandatory for processing companies and large-scale plantations. Plasma and independent smallholders as well as eligible oil palm companies that produce palm oil to serve the renewable energy are exempted.

Similar to RSPO, ISPO also consists of a number of principles, criteria and indicators. The principles and criteria are differentiated for different types of plantations: plantations that are integrated with the processing facilities, plantations that are not integrated with processing facilities, processing facilities that are not integrated with plantations, and plantations for biodiesel. Though not mandatory for plasma and independent smallholders, the ISPO does specify the principles and criteria for these smallholders. These principles include business legality, plantation management, protection to use primary forests and peatland, environmental management and monitoring, responsibilities to the workers, social responsibilities and community economic empowerment, and continuous improvements (Ministry of Agriculture Regulation 11 of 2015).

The Malaysian Sustainable Palm Oil (MSPO) standard was launched in 2013. It was officially implemented in January 2015. Unlike ISPO, MSPO is voluntary. The government has the ambition to get all actors certified by 2020. Given this ambition, several observers expect that the MSPO may turn into a mandatory standard in the future. MSPO is applicable to independent smallholders, oil palm plantations and organized smallholders and palm oil mills. MSPO has eight principles related to management commitment and responsibility; transparency; compliance to legal requirements; social responsibility, health, safety and employment conditions; environment, natural resources, biodiversity and ecosystem services; best practices; and development of new planting.[1]

The third kind of initiatives towards producing sustainable palm oil comes from corporate self-regulation. The most recent one of this kind is the pledge of six giant palm oil companies to zero-deforestation in Indonesia under the Indonesian Palm Oil Pledge (IPOP) initiative.[2] These six companies are Golden Agri Resources, Wilmar International Limited, Cargill, Asian Agri, Musim Mas and Astra Agro Lestari, with support from the Indonesian Chamber of Commerce. IPOP has a vision to advance Indonesia's sustainable palm oil business practices by collaborating with the government and all stakeholders to attain a sustainable palm oil sector. The giant palm oil companies have to work in a collaborative way to develop sustainable palm oil production that delivers stakeholder value, is deforestation free and respects human and community rights.

Indonesia is a perfect example of how the three regulations – RSPO, ISPO and self-regulation – are coexisting. By May 2016, there were approximately 115 members of the RSPO in Indonesia. Thirty-five companies have a number of their estates and mills certified, covering 1.6 million hectares that produce 6.6 million tonnes of certified sustainable palm oil (CSPO) from approximately 24 million tonnes of fresh fruit bunches. By the same period, there were 148 estates or mills that were certified under ISPO. The Government of Indonesia originally targeted to certify all actors by 2014, but the level of compliance to date remains low. It was reported that by 2014, approximately 200 out of the 881 companies that are eligible to pursue ISPO certification, had registered. However, only 67 were certified.[3] Next to seeking compliance with ISPO, six giant palm oil companies in Indonesia established IPOP. Although the size of their plantations and the capacity of their mills are not known publicly, it is estimated that these companies control at least 90% of the CPO intake in Indonesia (AgroIndonesia 2015).

In Colombia and Cameroon, the discussion on sustainability has centred on RSPO compliance, although Colombia is more advanced than Cameroon in terms of engagement with RSPO. A national interpretation of the RSPO principles and criteria exists for Colombia and it is under revision to align with the 2013 version of RSPO principles and criteria. There are 28 RSPO members in Colombia, where five of them produce 107 000 tons of CSPO from 39 500 hectares of total certified areas. Fedepalma has a desire to implement the principles and criteria of RSPO and move its members gradually into certification (Potter 2015). With Colombia's CPO export at approximately 30% of the production, and European countries being the main export destination, it could be expected that Colombia will move towards being fully compliant with RSPO certification.

As of 2016, there are no RSPO members in Cameroon, although some efforts are geared towards moving to RSPO (RSPO n.d., TFA2020 2015). Both Colombia and Cameroon have laws or regulation on the practices of oil palm plantations and processing units, although they might not be structured in the same way as the RSPO or ISPO.

1. Hospes, O. (2014) explains that the ISPO and RSPO standards are look-alikes but differ on critical points. These differences are also listed by Efeca (2016) in its comparative study of RSPO, ISPO and MSPO.
2. It is important to note that the members of IPOP have ended IPOP as a group by end of June 2016. However, each individual company will continue to implement zero-deforestation commitment individually.
3. Wibowo, A. D. 2014. Baru 7% Perusahaan Sawit Miliki Sertifikat ISPO. http://ekonomi.metrotvnews.com/read/2014/10/05/300849/baru-7-perusahaan-sawit-miliki-sertifikat-ispo.

For example, WWF and Greenpeace reported that one large company in Cameroon did not follow the requirement (in the form of presidential decree) to establish the oil palm concession in terms of avoiding forest clearing. Allegedly, the same company also violated the court order to suspend the operation after complaints by local people (Greenpeace 2013).

5 The dynamics of sustainability: actors, regulations and practices

Decision-making processes in the oil palm sector take place in different policy arenas at different levels. Three policy arenas can be distinguished. First is land-use planning at national, regional and local level. The specific issue is whether or not to allocate land for oil palm, and if so, in what areas. Second is the issuance of permits, that is, decision making on whether a company or smallholder can be given a licence to cultivate palm oil in the areas designated for palm oil cultivation. Third is planning of palm oil production at the estate level, which takes place after the permit has been issued by the government and obtained by the permit holder (Paoli et al. 2015).

Decision-making processes in the oil palm sector are complex because of three issues. The first issue is that the different arenas involve a diversity of public authorities with different, often overlapping, mandates on palm oil issues. In Indonesia, the Ministry of Agriculture has issued the ISPO standard but is not mandated to take decisions on land-use planning, which is a delicate and longstanding issue in Indonesia (Brockhaus et al. 2012, Bettinger 2015, McCarthy and Robinson 2016, Setiawan et al. 2016). Spatial planning is not clear or contested. Different ministries use different maps to define and designate areas for agriculture, mining and forestry. As a result, (unclear) spatial planning is generally seen as one of the most important problems in the palm oil sector in Indonesia (Gaveau et al. 2016). Decentralization processes have not made things easier or more transparent. Decentralization has led not only to more regulations (at provincial and district level) but also to lack of clarity on how decisions on land allocation, issuance of permits for oil palm concession and operational practices are made in Indonesia. This lack of clear rules of the game has enabled local authorities to abuse their public power and to enrich themselves personally. Similarly, Colombia is faced with competing authorities, competing land uses and politicization of palm oil production. One of the drivers that facilitated the expansion of oil palm in Colombia was the political support given during the president Uribe administration with the notion that the country could reach three million hectares of oil palm. As a result of this political support and compared to other commodities, oil palm got extraordinary support through the Rural Capital Incentive programme (Potter 2015).

The second issue is that different public and private regulatory systems are used or referred to in the different policy arenas. These systems may complement, but also compete with, each other. RSPO and ISPO generally serve the same purpose. Also, they have a number of similar principles and criteria. However, beyond their respective voluntary/mandatory status, there are major differences between RSPO and ISPO in many key areas, such as the treatment on areas called high conservation value and the implementation of free and prior informed consent (Efeca 2016, Hospes 2014, Ministry of Agriculture and RSPO Secretariat 2016). This is not (yet) the case in Colombia and

Cameroon, as these countries do not have any mandatory national public sustainability standards, although these countries have legislations that regulate how oil palm must be cultivated and processed.

The complexity has become even more intense with the pledge of six giant palm oil groups to establish IPOP to pledge for 'zero deforestation, no peat land and no exploitation' practices in the oil palm production. These companies have their estates and mills both RSPO and ISPO certified. Although IPOP is currently only directed at palm oil production in Indonesia, the pledge could have implications for other countries. Depending on how each member company defines and translates the commitment, each of these companies may move their expansion plans to other countries. Without improving their practices, environmental and social problems of palm oil production in Indonesia and Malaysia could be 'transported' to Africa (Linder 2013, Wich et al. 2014, Linder and Palkovitz 2016). Since IPOP has been disbanded as a group and each member carries out its own zero-deforestation commitment individually, it remains to be seen how each company defines, translates, implements and monitors its commitment to zero-deforestation.

Interestingly, the Government of Indonesia and Malaysia established the Council of Palm Oil Producing Countries (CPOPC) in 2015. The objective of the Council is 'to promote, develop and strengthen cooperation in the oil palm cultivation and industry among the Member Countries as well as to ensure long term benefits of such palm oil endeavours to the economic development and well-being of the people of the Member Countries' (President of the Republic of Indonesia 2016, p.2). The Council invites other producer countries to participate as members of CPOPC. Whereas the CPOPC is an intergovernmental initiative, the council could play a role in harmonizing or orchestrating RSPO, ISPO and MSPO standards to generate a unique and recognized standard for the sustainable production of palm oil. Of course, there is also the risk of more confusion and controversy. The council could turn into a political tool or arena to challenge private initiatives, like the RSPO and IPOP.

The third issue is about the translation of different regulatory systems into technical, land use and governance practices. While each practice may well be within the scope of a specific regulatory system, it may conflict with other regulatory system. The conflicts may be more severe under the conditions of unclear land use and tenure. Land-use planning and how it is enforced has been a contentious issue in each of the countries (Hoyle and Levang 2012, Njoh 2012, Garcia and Slunge 2015, McCarthy and Robinson 2016). For example, in Indonesia, the Plantation Law 39 of 2014 still requires that all land under concession is planted, including those with high conservation values. Meeting production and biodiesel targets may require additional forests to make room for oil palm. Large-scale companies may be able to implement best management practices as a means for intensification to increase the yield, although under unclear land tenure they may also expand their plantations outside their concessions (Gaveau et al. 2016). However, implementing best management practices might still be a dream for independent smallholders, as most of them are trapped into a vicious circle of low inputs, low production and low income. To increase income, smallholders often consider clearing of new land for oil palm as the only option. This only aggravates the struggle and problems of smallholders to meet the requirements of both public and private regulations. Even worse, the largest companies that committed themselves as a group and then individually to zero-deforestation will not accept their products if and when these smallholders establish their plantations through deforestation.

6 Conclusions: creating sustainable pathways

The variety of sustainability initiatives of public and private actors shows that all actors in the palm oil industry acknowledge environmental and social impacts of palm oil expansion and the need to minimize these. However, there is no agreement on what actor or regulatory system should be leading in addressing these impacts. Given the many different actors and increasing number of regulatory systems in the palm oil sector, we consider organizing deliberation and collaborative interaction between different actors as critical for defining and achieving context-specific sustainable pathways for palm oil production.

Taking lessons from Indonesia where all regulatory systems are coexisting, it is possible to see how pathways towards sustainability may emerge and to distinguish different steps. The first step is to agree on a common vision and to discuss this vision with a view to identify shared principles, acknowledging that the definition of sustainability is not rigid and is open to multiple interpretations. When there is a common vision, the second step is to specifically define the pathways or the actor-practice connections that reinforce the technical, land use and governance practices. This framework allows for context-specific combinations. The third step could be to improve one standard by learning from other standards, using the principle of continuous improvement (that is common for both RSPO and ISPO). The fourth step could be to acknowledge diversity of authorities and regulatory systems and at the same time seek mutual recognition. Mutual recognition is a principle and mechanism to cope with plural legal order (Berman 2009).

The establishment of the RSPO, the rise of national standards for palm oil in Indonesia and Malaysia, and, more recently, the emergence of private self-regulation and commitment to zero-deforestation show that sustainable palm oil has been tabled by all policymakers and stakeholders in the field of palm oil. The challenge is now to define mutual responsibility and shared sovereignty, such as to prepare a concerted effort and to create context-specific sustainable pathways for palm oil production in Indonesia, Columbia, Cameroon and all other oil-palm-producing countries of the world.

7 Where to look for further information

Wageningen University, the Netherlands, has embarked on various interdisciplinary, multi-level and multi-PhD research programmes on global value chains and sustainable food and agriculture. These include the INREF-funded SUSPENSE programme on the identification of sustainable pathways for the production, processing and governance of palm oil, and the NWO-funded programme on the Next Generation of Global Value Chains. Both programmes are focused on sustainable palm oil. The research consortia of these programmes are led by Katrien Termeer and Otto Hospes of the Public Administration and Policy Group of Wageningen University.

8 References

Abood, S. A., Lee, J. S. H., Burivalova, Z., Garcia-Ulloa, J. and Koh, L. P. 2015. Relative Contributions of the logging, fiber, oil palm, and mining industries to forest loss in Indonesia. *Conservation Letters* 8(1): 58–67.

AgroIndonesia. 2015. *IPOP langgar kedaulatan RI. Jakarta, Indonesia*: AgroIndonesia.

Alfonso-Frajedas, A., Liu, J., Salerno, T. and Xu, Y. 2016. Inquiring into the political economy of oil palm as a global flex crop. *Journal of Peasant Studies* 43(1): 141–65.

Berman, P. S. 2009. The new legal pluralism. *Annual Review of Law and Social Sciences* 5: 225–42.

Bettinger, K. A. 2015. Political contestation, resource control and conservation in an era of decentralisation at Indonesia's Kerinci Seblat National Park. *Asia Pacific Viewpoint* 56(2): 252–66.

Brockhaus, M., Obidziski, K., Dermawan, A., Laumonier, Y. and Luttrell, C. 2012. An overview of forest and land allocation policies in Indonesia: Is the current framework sufficient to meet the needs of REDD+? *Forest Policy and Economics* 18: 30–7.

Budidarsono, S., Susanti, A. and Zoomers, A. 2013. Oil Palm Plantations in Indonesia: The implications for migration, settlement/resettlement and local economic development. In Fang, Z. (ed.), *Biofuels – Economy, Environment and Sustainability*. Intech. http://www.intechopen.com/books/biofuels-economy-environment-and-sustainability. Accessed 5 September 2016.

Busch, J., Ferretti-Gallon, K., Engelmann, J., Wright, M., Austin, K. G., Stolle, F., Turubanova, S., Potapov, P. V., Margono, B., Hansen, M. C. and Baccini, A. 2015. Reductions in emissions from deforestation from Indonesia's moratorium on new oil palm, timber, and logging concessions. *Proceedings of the National Academy of Sciences* 112(5): 1328–33.

Castanheira, E. G., Acevedo, H. and Freire, F. 2014. Greenhouse gas intensity of palm oil produced in Colombia addressing alternative land use change and fertilization scenarios. *Applied Energy* 114: 958–67.

Castiblanco, C., Etter, A. and Aide, T. M. 2013. Oil plantations in Colombia: A model of future expansion. *Environmental Science and Policy* 27: 172–83.

Central Statistical Agency of Indonesia. 2015. *Indonesian oil palm statistics 2014*. Jakarta: Indonesia: CSA.

Corley, R. H. V. and Tinker, P. B. 2015. *Oil palm*. 5th ed. London: Wiley-Blackwell.

Dermawan, A., Obidzinski, K. and Komarudin, H. 2012. Withering before full bloom?: Bioenergy development in Southeast Asia. *CIFOR Working Paper 94*. Bogor, Indonesia: CIFOR.

Edwards, D. P., Magrath, A., Woodcock, P., Ji, Y., Lim, N. T. L., Edwards, F. A., Larsen, T. H., Hsu, W. W., Benedick, S., Khen, C. Y., Chung, A. Y. C., Reynolds, G., Fisher, B., Laurance, W. F., Wilcove, D. S., Hamer, K. C. and Yu, D. W. 2014b. Selective-logging and oil palm: multitaxon impacts, biodiversity indicators, and trade-offs for conservation planning. *Ecological Applications* 24(8): 2029–49.

Efeca. 2016. Comparison of the ISPO, MSPO and RSPO Standards. http://www.sustainablepalmoil.org/wp-content/uploads/sites/2/2015/09/Efeca_PO-Standards-Comparison.pdf. Accessed 5 September 2016.

FAO (Food and Agricultural Organization of the United Nations). 2014. *Oil crops – Market Summaries*. Food Outlook May 2014. Rome: FAO.

Fayle, T. M., Turner, E. C., Snaddon, J. L., Chey, V. K., Chung, A. Y. C., Eggleton, P. and Foster, W. A. 2010. Oil palm expansion into rainforest greatly reduces ant biodiversity in canopy, epiphytes and leaf-litter. *Basic and Applied Ecology* 11: 337–45.

Fitzherbert, E. B., Struebig, M. J., Morel, A., Danielsen, F., Bruhl, C. A. and Donald, P. 2008. How will oil palm expansion affect biodiversity? *Trends of Ecology and Evolution* 23: 538–45.

Garcia, J. J. and Slunge, D. 2015. *Environment and climate change management: perspectives for post-conflict Colombia*. Gotheburg and Uppsala, Sweden: Sida's Helpdesk for Environment and Climate Change.

Garcia-Ulloa, J., Sloan, S., Pacheco, P., Ghazoul, J. and Koh, L. P. 2012 Lowering environmental costs of oil-palm expansion in Colombia. *Conservation Letters* 5: 366–75.

Gaveau, D. L. A., Pirard, R., Salim, M. A., Tonoto, P., Yaen, H., Parks, S. A. and Carmenta, R. 2016. Overlapping land claims limit the use of satellites to monitor no-deforestation commitments and no-burning compliance. *Conservation Letters* May 2016: 1–8. doi: 10.1111/conl.12256.

Gomez, P. L. Mosquera, M. and Castilla, C. 2005. Oil palm: a sustainable agro-industry in Colombia. *Oilseeds and fats Crops and Lipids* 12(2): 121–4.

Greenpeace. 2013. *Herakles Farms in Cameroon: A showcase in bad palm oil production*. Washington, D.C.: Greenpeace.

Gunarso, P., Hartoyo, M. E., Agus, F. and Killeen, T. J. 2013. *Oil palm and land use change in Indonesia, Malaysia and Papua New Guinea*. Kuala Lumpur, Malaysia: RSPO.

Henson, I. E., Romero, R. R. and Romero, H. M. 2011. The growth of the oil palm industry in Colombia. *Journal of Oil Palm Research* 23: 1121–8.

Hospes, O. 2014. Marking the success or end of global multi-stakeholder governance? The rise of national sustainability standards in Indonesia and Brazil for palm oil and soy. *Agriculture and Human Values* 31: 425–37.

Hospes, O., Kroeze, C., Oosterveer, P., Schouten, G. and Slingerland, M. 2017. New generation of knowledge: Towards an inter- and transdisciplinary framework for sustainable pathways of palm oil production. *NJAS - Wageningen Journal of Life Sciences*. http://dx.doi.org/10.1016/j.njas.2017.01.001. Accessed 20 January 2017.

Hoyle, D. and Levang, P. 2012. *Oil palm development in Cameroon*. Yaounde, Cameroon: WWF, IRD and CIFOR.

IFC (International Finance Corporation). 2013. *Diagnostic study on Indonesian oil palm smallholders: Developing a better understanding of their performance and potential*. Jakarta, Indonesia: IFC.

Index Mundi. 2016. Palm Oil Production by Country in 1000 MT. http://www.indexmundi.com/agriculture/?commodity=palm-oil&graph=production. Accessed 5 September 2016.

Koh, L. P. and Wilcove, D. S. 2008a. Oil palm: disinformation enables deforestation. *Trends of Ecology and Evolution* 23: 67–8.

Koh, L. P. and Wilcove, D. S. 2008b. Is oil palm agriculture really destroying tropical biodiversity? *Conservation Letter* 2008(1): 60–4.

Kurz, D. J. , Turner, E. C., Aryawan, A. A., Barkley, H. C., Caliman, J.-P., Konopik, O., Sudharto, Ps. and Foster, W. A. 2016. Replanting reduces frog diversity in oil palm. *Biotropica* 48: 483–90.

Leach, M., Scoones, I. and Stirling, A. 2010. *Dynamic Sustainabilities: Technology, Environment, Social Justice*. London: Routledge.

Li, T. M. 2015. Social impacts of oil palm in Indonesia: A gendered perspective from West Kalimantan. *Occasional Paper 124*. Bogor, Indonesia: CIFOR.

Lindahl, K. B., Baker, S., Rist, L. and Zachrisson, A. 2015. Theorising pathways to sustainability. *International Journal of Sustainable Development and World Ecology*. http://dx.doi.org/10.1080/13504509.2015.1128492

Linder, J. M. 2013. African primate diversity threatened by 'new wave' of industrial oil palm expansion. *African Primates* 8: 25–38.

Linder, J. M. and Palkovitz, R. E. 2016. The threat of industrial oil palm expansion to primates and their habitats. In Waller, M. T. (ed.) *Ethnoprimatology – Primate conservation in the 21st century*. Switzerland: Springer.

Maher, D. 2015. Rooted in violence: Civil war, international trade and the expansion of palm oil in Colombia. *New Political Economy* 20(2): 299–330.

Margono, B. A., Potapov, P. V., Turubanova, S., Stolle, F. and Hansen, M. C. 2014. Primary forest cover loss in Indonesia over 2000–2012. *Nature Climate Change* 4: 730–5.

McCarthy, J. and Robinson, K. 2016. *Land and development in Indonesia: searching for the people's sovereignty*. Singapore: ISEAS Publishing.

Miettinen, J., Shi, C. and Liew, S. C. 2011. Deforestation rates in insular Southeast Asia between 2000 and 2010. *Global Change Biology* 17: 2261–70.

Ministry of Agriculture and RSPO Secretariat. 2016. *Joint study on the similarities and differences of the ISPO and the RSPO certification systems*. Jakarta, Indonesia: Ministry of Agriculture and RSPO Secretariat.

Ministry of Energy and Mineral Resources. 2014. *Industri Biodiesel Perlu Investor Baru*. Jakarta, Indonesia: MEMR. http://ebtke.esdm.go.id/post/2014/12/05/734/industri.biodiesel.perlu.investor.baru. Accessed 5 September 2016.

Murdiyarso, D., Hergoulalc'h, K. and Verchot, L. V. 2010. Opportunities for reducing greenhouse gas emissions in tropical peatlands. *Proceedings of the National Academy of Sciences* 107(46): 19655–60.

Ngando-Ebongue, G. F., Ajambang, W. N., Koona, P., Firman, B. L. and Arondel, V. 2012. Oil palm. In: Gupta, S. K. (ed.), *Technological innovations in major world oil crops*, Volume 1 breeding. London: Springer.

Njoh, A. J., 2012. Equity, fairness and justice implications of land tenure formalization in Cameroon. *International Journal of Urban and Regional Research* 37(2): 750–68.

Nkongho, R. N., Feintrenie, L. and Levang, P. 2014. The non-industrial palm oil sector in Cameroon. *CIFOR Working Paper 139*. Bogor, Indonesia: CIFOR

Nkongho. R. N., Ndjogui, T. E. and Levang, P. 2015. History of partnership between agro-industries and oil palm smallholders in Cameroon. *Oilseeds and fats Crops and Lipids* 22(3): A301.

Obidzinski, K., Andriani, R., Komarudin, H. and Andrianto, A. 2012. Environmental and social impacts of oil palm plantations and their implications for biofuel production in Indonesia. *Ecology and Society* 17(1): 25–43.

Pacheco, P. 2012. Soybean and oil palm expansion in South America A review of main trends and implications. *CIFOR Working Paper 90*. Bogor, Indonesia: CIFOR.

Paoli, G. D., Gillespie, P., Wells, P., Hovani, L., Sileuw, A., Franklin, N. and Schweithelm, J., 2015. *Oil palm in Indonesia: Governance, decision making, and implications for sustainable development – summary for policy makers and practitioners*. Jakarta: The Nature Conservancy Indonesia Program.

PASPI (Palm Oil Agribusiness Strategic Policy Institute). 2014. *The Sustainability of Indonesian Palm Oil Industry*. Bogor, Indonesia: PASPI.

Pinto, A. D., Haruna, A., Creamer, B., Hyman, G., Li, M., Kwon, H. Y., Garcia, J. B. V., Coca, A, C., Tapasco, J., Martinez, J. D. and Hoyos, J. D. 2014. *Low emission development strategies for agriculture and other land uses: The case of Colombia*. Washington, D. C.: IFPRI.

Potter, L. 2015. Managing oil palm landscapes: A seven-country survey of the modern palm oil industry in Southeast Asia, Latin America and West Africa. *Occasional Paper* 122. Bogor, Indonesia: CIFOR.

Pramudya, E., Hospes, O. and Termeer, K. (in press). Governing the palm oil sector through finance: the changing roles of the Indonesian state. *Bulletin of Indonesian Economic Studies*.

President of the Republic of Indonesia. 2016. *Presidential Regulation 42 of 2016 on the Ratification of the Charter of the Establishment of the Council of Palm Oil Producing Countries (CPOPC)*.4 May 2016. Jakarta, Indonesia. http://www.bphn.go.id/data/documents/16pr042.pdf. Accessed 5 September 2016.

RSPO (Roundtable on Sustainable Palm Oil). 2013. *Principles and criteria for the production of sustainable palm oil 2013*. Kuala Lumpur: RSPO.

RSPO. N.d. Report on capacity building and awareness-raising event in Cameroon. http://www.ifc.org/wps/wcm/connect/1316c2804300dc85b7d6b732ece34d2d/Proforest_Del18__Roadshow_Cameroon.pdf?MOD=AJPERES. Accessed 5 September 2016.

Schrier-Uijl, A. P., Silvius, M., Parish, F., Lim, K. H., Soeharto, R. and Anshari, G. 2013. *Environmental and social impacts of oil palm cultivation on tropical peat – a scientific review*. Kuala Lumpur: RSPO.

Setiawan, E. N., Maryudi, A., Purwanto, A. H. and Lele, G. 2016. Opposing interests in the legalization of non-procedural forest conversion to oil palm in Central Kalimantan, Indonesia. *Land Use Policy* 58: 472–81.

TFA2020. 2015. TFA 2020 action plan on oil palm development in Africa. Briefing Note Three. May 2015. https://www.tfa2020.org/wp-content/uploads/2014/07/3_Briefing_Note_Three_Africa_Initiative_Public_TFA_2020_Final.pdf. Accessed 9 September 2016).

Torres, G. A., Sarria, G. A., Martinez, G., Varon, F., Drenth, A., and Guest, D. I. 2016. Bud Rot Caused by Phytophthora palmivora: A Destructive Emerging Disease of Oil Palm. *Phytopathology* 106(4): 320–9.

Van Straaten, O., Corre, M. D., Wolf, K., Tchienkoua, M., Cuellar, E., Matthews, R. and Veldkamp, E. 2015. Conversion of lowland tropical forests to tree cash crop plantations loses up to one-half of stored soil organic carbon. *Proceedings of the National Academy of Sciences* 112(32): 9956–60.

Vijay, V., Pimm, S. L., Jenkins, C. N., and Smith, S. J. 2016. The Impacts of Oil Palm on Recent Deforestation and Biodiversity Loss. *PloS one*, 11(7), e0159668.

Wich, S. A., Garcia-Ulloa, J., Kuhl, H. S., Humle, T., Lee, J. S. H. and Koh, L. P. 2014. Will oil palm's homecoming spell doom for Africa's great apes? *Current Biology* 24: 1659–63.

Wittmayer, J. M., Schäpke, N., van Steenbergen, F. and Omann, I. 2014. Making sense of sustainability transitions locally: how action research contributes to addressing societal challenges. *Critical Policy Studies* 8(4): 465–85.

Chapter 4

The palm oil governance complex: progress, problems and gaps

Pablo Pacheco, Center for International Forestry Research (CIFOR), Indonesia; Patrice Levang, Center for International Forestry Research (CIFOR), Indonesia and Research Institute for Development (IRD), France; Ahmad Dermawan, Center for International Forestry Research (CIFOR), Indonesia; and George Schoneveld, Center for International Forestry Research (CIFOR), Kenya

1 Introduction

Oil palm is one of the most controversial global commodity crops, which has undergone one of the highest rates of expansion in comparison with other crops in the tropical world. It is a controversial crop since the conditions under which it expands as well as their social and environmental impacts are ambiguous. On one hand, oil palm expansion has delivered important economic development in host countries, including indirect benefits for local infrastructure development and rural poverty reduction, and multiplier effects for the national economies. On the other hand, its development has often come at the expense of basic human and customary rights and of biodiverse, carbon-rich tropical forests, as local communities have been evicted from their lands and precious primary forest and peatland ecosystems have been destroyed (Sayer et al. 2012; Pacheco et al. 2017; Sheil et al. 2009).

While the crop originated in West Africa, much of its industrial expansion under large-scale monocrop plantation systems has occurred in Southeast Asia, notably in Malaysia and Indonesia (Rival and Levang 2014). Governments in palm-oil-producing countries have seen its rural development potential and therefore supported its expansion

http://dx.doi.org/10.19103/AS.2017.0018.31

through different incentive and tenure policies that attracted private investments in the plantations development. However, the growing environmental impacts associated with oil palm expansion, along with the increasing concerns of consumer markets in developed economies, have led to the emergence of institutional architectures and arrangements for supporting the adoption of more sustainable socio-environmental practices (Hospes 2014). In addition, following consumer pressures, some major companies have adopted 'No deforestation, no peat and no exploitation' commitments among traders and producers (Climate Focus 2015). Those concerns, along with growing pressure from civil society groups in oil-palm-producing countries, are also prompting the adoption of policies with the potential to guide a more sustainable expansion of the palm oil sector (Daemeter Consulting 2015).

The palm oil value chain has increased in complexity over time; while the main palm-oil-producing countries are Malaysia and Indonesia, they fill markets all around the globe and their products have multiple uses ranging from cooking oil to ingredients in industrial products (e.g. processed foods, detergents and cosmetics), as well as biodiesel (Lai et al. 2012). The palm oil global value chain comprises a wide range of stakeholders, from producers of all sizes to processors, traders, consumer goods manufacturers (CGMs) and retailers. Despite being dominated by a handful of companies at the refining and international trading stages, production involves a wide range of suppliers from companies to smallholders and manufacturing involves a wide range of CGMs in a market that is diversifying. The main markets consist of developed economies, mainly the European Union, and emerging markets, particularly China and India. This global complexity makes the palm oil value chain harder to govern for social and environmental outcomes.

The palm oil sector faces some critical performance issues. The first is persistent conflicts over land and benefit flows linked to industrial plantation expansion. For example, in Indonesia, several unresolved land disputes persist since local customary rights have often been neglected in plantations development, either associated with companies or immigrants (Colchester and Chao 2013). The second is the large yield gap obtained by independent smallholders with regard to those of smallholders in partnership schemes with companies, which can be even higher than those obtained by well-managed large-scale plantations (Molenaar et al. 2013). The third is the detrimental environmental impacts since palm oil tends to place pressure on forests and agroforestry lands, resulting in biodiversity loss and water pollution (Sheil et al. 2009), and relatively large carbon emissions, particularly when it expands in peatlands (Khasanah et al. 2012). The governance of palm oil production is to a large extent aimed at tackling these three issues, which are embedded in wider institutional contexts in the landscapes where oil palm is expanding. Thereby, the governance of palm oil intersects with key challenges associated with enhancing the governance of landscapes.

The palm oil governance conundrum expresses how multiple agents' interests can be met while addressing short-term economic goals under long-term environmental objectives. The main questions that this chapter wants to address are: Why is that in spite of evolving governance architectures and arrangements still no major changes are observed in the performance of the palm oil sector? What governance factors prevent progress in achieving a faster and widespread transition towards a more sustainable palm oil production in the tropics?

Answering these questions requires an in-depth understanding of the palm oil governance complex. This chapter aims to unpack it under three different dimensions. The first comprises the constituent parts of the governance systems (i.e. action arenas, actors, architectures and arrangements), and their interactions. The second dimension embraces the various governance problems encountered in the palm oil sector, which require a resolution in order to transition towards more sustainable development scenarios. These problems are related to allocation and access to resources favouring some social groups over others, power distribution across actors in decision-making that allows some rules to be imposed in detriment to others, upward and downward accountability and institutional checks-and-balances, and transparency in monitoring compliance and verifying progress towards achievement of targets and commitments to the wider society. The third dimension relates to the major gaps associated with the process of governing the palm oil sector, aimed at addressing some of the main social and environmental performance issues. This chapter provides a thorough analysis of these three dimensions (i.e. constituent parts, governance problems and gaps) shaping the governance complex when applied to oil palm development in the tropics.

2 Conceptual framework: the palm oil governance complex

The palm oil governance complex is constituted by the multiple interactions that different stakeholders establish along the production, processing and trade in the value chains, and the interactions unfolding in the landscapes where those value chains are embedded. The complex reflects how different institutional architectures and arrangements, linked to mandatory state regulations and voluntary private standards, work in conjunction to regulate, provide incentives and improve the economic performance of the palm oil sector, while addressing the main social and environmental impacts and trade-offs emerging from their rapid development, which affect its overall performance. More complex transnational policy regimes are emerging for governing natural resources (Overdevest and Zeitlin 2012).

We propose here a framework comprising three dimensions that help to understand the performance of the palm oil governance complex and to identify corrective measures for potential public and private policy action (Fig. 1). The first dimension has to do with the constituent parts that characterize the palm oil governance complex per se. These are 1) action arenas, which configure the space where roles and responsibilities are exerted; 2) actors and agency, which contribute to reshaping the rules and regulations that apply in the social, political and economic systems; 3) the evolving institutional architectures, in both the public and private domains; and 4) the agreements and arrangements established by social actors. The second dimension looks at the major governance problems that have an influence on shaping the performance of the palm oil governance complex. The main problems relate to the allocation and access to resources, issues of unequal power distribution among actors, weak accountability and institutional checks-and-balance systems, and opaque transparency in decision-making and reporting of both progress and failures in the accomplishment of targets. The third dimension relates to the gaps observed when assessing the functioning of the governance complex, in the light of the different problems faced by the institutional system for governing both supply chains and landscapes. We argue that both the problems and gaps have to be evaluated while considering how they contribute in addressing the socio-environmental performance and trade-offs of the sector as a whole. Table 1 provides more detail on the specific elements or features identified in each of these dimensions.

The first dimension in our framework has to do with the constituent parts characterizing the governance of the palm oil complex. These parts constitute the material configurations and networks shaping the performance of the palm oil sector (e.g. differentiated yields among producers, unequal benefit sharing, land conflicts and detrimental environmental

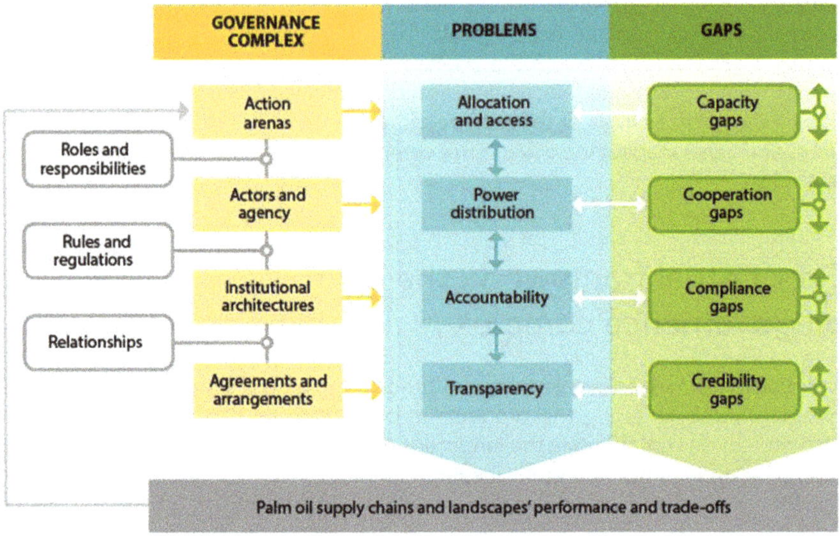

Figure 1 A framework to assess the performance of the palm oil governance complex.

Table 1 The main dimensions configuring the palm oil governance complex

Governance system	Problems	Gaps
Action arenas involving global supply chains and local territorialities	Allocation of resources and benefits associated with access to diverse economic and institutional rents	Capacity gaps in linking effectively supply chain and landscape management institutional initiatives
Multiplicity of actors involving public, private and civil society organizations	Power distribution across different actors with effect on decision making and imposition of rules	Cooperation gaps across different government levels and within private organizational structures
Evolving architectures involving public administrative systems and voluntary standards	Upward and downward accountability mechanisms and the associated existing checks-and-balances	Compliance of voluntary self-regulations and pledges, and enforcement of public mandatory regulations
Multi-level agreements and arrangements involving competition and collaboration	Transparency in the monitoring and verification of compliance of public and private commitments	Credibility and legitimacy gaps of public regulations and private standards

impacts), under which they evolve and are deployed in the different actions and initiatives for addressing the challenges of palm oil development in both the supply chains and landscapes hosting oil palm plantations expansion. Below is a description of these constituent parts.

- *Action arenas* refer to the social space where individuals and actors exchange goods and services and establish social interactions to solve problems or dominate one another, among other things (Ostrom et al. 1994). Two action arenas are involved in palm oil development: first, the global palm oil supply chain bounded by a network of market transactions and flows along the chain; and second, the landscapes hosting oil palm plantations within specific territorial boundaries, which commonly are jurisdictional administrative units at the sub-national level with defined government systems.
- *Actors* are individuals and collectives exercising *agency*. The social actors in the palm oil sector are multiple. They include government officials, with different positions in the structure of government, and a diverse range of private actors involving consumer organizations, transnational and national palm oil companies. They also include diverse types of producers which constitute second- or third-party suppliers in the value chain, as well as indigenous communities, service providers and civil society organizations (Oosterveer 2015). All of these actors have a stake either in the value chains, or in the territories where oil palm is expanding.
- *Architectures* are comprised of institutional structures, systems and mechanisms defined by clusters of norms and principles aimed at regulating the behaviour of social actors and controlling social processes, in ways that comply with certain objectives embraced by those systems (Biermann et al. 2009). Governance architectures in the palm oil sector are composed by a mix of public regulations and private standards (Hospes 2014).
- *Agreements and arrangements* refer to the formal and informal interactions established by different individual or collective actors. While the formal interactions are established within formal governance architectures, by making use of the existing institutional systems and mechanisms, there are others established more informally, that is outside of those formalized institutionalized social spaces (Williamson 2009). Both types of formal and informal interactions take place in the value chains and the landscape's action arenas.

These constituent parts establish several interactions among themselves. For example, the action arenas may contain specific type of interactions among actors struggling to defend and/or advance their own positions in the value chains, which in turn leads to the evolution of different architectures and institutional arrangements, something that also has implications for specific landscape management challenges. The latter may, in turn, lead to redefining the architectures and institutional arrangements linked to disparate state and non-state actors' perspectives and interests in the value chains, which may have different implications across landscape contexts.

The constituent parts of palm oil governance define different types of governance problems or challenges. We follow Biermann (2014) in the selection of some key governance problems that merit to be analysed in order to understand the politics of palm oil governance when addressing the key performance issues facing the palm oil sector with regard to differential yields among different types of producers and unequal economic benefits distribution, land and social conflict, and negative environmental impacts. Major problems addressed here are:

- *Allocation* of material resources (e.g. land and inputs) and benefits associated with the differential *access* that social groups have to diverse economic rents associated from the productive use of those resources and institutional rents linked to state incentives.
- *Power* that different social groups hold and exercise in the devising of rules as part of their role and influence in decision-making processes, and the means that they have at their disposal to impose those rules and affect the socio-environmental impacts.
- *Accountability* refers to the upward and downward accountability mechanisms, and the related checks-and-balances that are in place in order for social actors to become accountable for their actions and the associated outcomes. In addition, it implies to also have in place the mechanisms to punish deviant behaviours, and ways to appeal inappropriate decisions.
- *Transparency* in the monitoring and verification of individual or collective compliance of state regulations and/or self-regulations expressed in specific individual commitments or adherence to voluntary goals stated in collective processes or pledges.

Achieving progress in dealing with these different governance problems, while trying to improve the performance issues facing the palm oil sector, becomes critical for the governance architectures and frameworks – in ways that ensure equitable allocation and access to resources, fair access to decision-making processes by all stakeholders, and effective accountability on the outcomes of those decisions. Yet, this does not always work as expected, and different gaps can be found in the implementation process. Following Boström and colleagues (2015), we select four critical gaps that affect the governance system's performance. Below is a description of these gaps.

- *Capacity gaps* resulting from issues associated with articulating better supply chains and landscape management with broader consumers' expectations on sustainability and the capacities and resources held by upstream producers to uptake improved practices and standards. Capacities are required not only for the supply chain actors to effectively organize production, processing and marketing, but also for actors with territorial management authority to formulate and implement territorial planning, and secure tenure rights, among others.
- *Cooperation gaps* result from the lack of coordination between policy design and implementation across different levels of government from national to sub-national. These gaps also occur within decision-making in the corporate side, specifically between the parent company and their subsidiary groups, and third-party suppliers. Cooperation is prevented by increasingly decentralized administrative systems with no incentives to coordinate and it is challenged by the complexity and fragmentation along a commodity chain and across different jurisdictional levels.
- *Compliance gaps* linked to how able private actors are to follow their own self-regulations and pledges, as well as to comply with the mandatory public regulations enforced by the state in ways that are empirically verified, with the outcomes

disseminated in transparent ways to a wider society. Many factors may prevent their effective implementation and lead to failures, and even to undesired direct outcomes, as well as indirect ones such as exclusion of poor smallholders or environmental impacts due to leakage effects.
- *Credibility gaps* associated with the effectiveness of public regulations in comparison to private standards, and of emerging institutional arrangements to address the problems facing the supply chains and landscapes due to palm oil expansion. This gap also refers to how widespread and durable are the solutions associated with these different policy instruments, regulations and mechanisms, and their benefits for the wider society.

The gaps observed in the governance complex, to some extent, reflects the inability of the actors' agency, and institutional architectures and arrangements to overcome the performance issues facing the palm oil sector, contributing to enlarge, or at least maintain, the governance challenges preventing the sector from embracing enhanced sustainability and inclusivity. In line with this framework, understanding the interactions between the three dimensions (i.e. constituent parts, governance challenges and observed gaps) that define the politics of palm oil governance is therefore critical. When related to supply chains and landscapes, a major step is to explain what factors prevent change, and what further public and private action is needed in order to either change the governance's constituent parts or minimize the observed gaps.

3 Constituent parts of the palm oil governance complex

3.1 Action arenas

The two action arenas shaping the development, and thus governance challenges for palm oil, are the global palm oil supply chain and the landscapes hosting the development of oil palm plantations. While the first action arena is related to a network bounded for a multiplicity of transactions of goods and services for oil palm development, the second is bounded for specific territorial boundaries. The two arenas host specific actors and institutional architectures and arrangements that ultimately shape the development and impacts of oil palm expansion.

The global palm oil supply chain has been growing in complexity over time. The palm oil sector is dominated by a handful of conglomerates involved in production, processing and trade (i.e. Wilmar, Musim Mas, Golden Agri-Resources (GAR), Cargill and Asian Agri in Indonesia and IOI, Sime Darby and FELDA in Malaysia). These groups source palm oil from their own plantations as well as from a large number of second- and third-party suppliers. According to the Malaysian Palm Oil Board (MPOB), in Malaysia there are 445 fresh fruit bunches (FFBs) mills, 44 palm kernel crushers, 52 refineries and 19 oleochemical plants (MPOB 2016). In Indonesia, according to information released by the five major corporate groups, these groups would control about 40 refineries, and source from about 850 mills and 1600 plantations. The official statistics present lower estimates (Statistik Indonesia 2015). According to MPOB (2016), in 2015, about 5.6 million ha of oil palm had been planted in Malaysia, of which 2.6 million ha were located in Peninsular Malaysia.

In Indonesia, 10.4 million ha of oil palm were planted by 2013, out of which about 42% were cultivated by smallholders (Directorate General of Estates 2014).

The main companies processing and trading crude palm oil (CPO) supply to a diversified number of end users, including a wide range of CGMs and retailers delivering a range of products in the food, chemical, pharmaceutical and cosmetic industries. While much processing and refining of CPO and palm kernel oil (PKO) take place in Indonesia, Malaysia and Singapore, most manufacturing takes place in the countries of consumption and in China, where transnational corporations manufacture products for consumers around the world. The palm oil processing industry is well developed in Malaysia, where companies maintain higher comparative advantage based on their efficiency (Abdullah et al. 2015), yet over time Indonesian companies have been able to slowly expand their palm oil refining capacity.

The landscapes where oil palm plantations tend to expand consist of places relatively well connected to ports, where refineries have been established, and with road infrastructure that makes possible the transport of CPO from the plantations to the mills. A key feature for the development of plantations is the establishment of palm oil mills, since production areas must be relatively closer to the processing facilities. In recent years, the number of mills and refineries has grown in Indonesia, which is contributing to greater sector integration and therefore improved efficiency. FFBs are transported from plantations to mills, where they are processed into CPO and KPO. Transportation is limited to an area of 50–100 km, due to the rapid deterioration of the fruit quality after harvesting. Once the FFB is milled, the resulting CPO is transported to the refineries, which are often located in the main export ports. Social exclusion of local indigenous groups has consistently accompanied the expansion of large-scale plantations in frontier lands, facilitating the dispossession of native populations that held customary rights over lands taken over by companies (Colchester et al. 2011; Li 2015; Gellert 2015). In addition, these plantations have attracted immigrant workers for both permanent and casual labour in the plantations, who may also look for land once they make enough savings. The latter were promoted by colonization programmes for securing the supply of workers to plantations such as FELDA in Malaysia and Transmigration in Indonesia (Potter 2015).

3.2 Actors and agency

CGMs and retailers in Western markets have been key drivers in shaping the governance of the palm oil sector by introducing environmental concerns in the sourcing of palm oil. In addition, environmental non-governmental organizations (NGOs) and advocacy groups have targeted consumer brands such as Nestlé, Unilever, Ferrero, Krispy Kreme and Dunkin' Donuts to leverage change among their suppliers (Pretty et al. 2008; Dauvergne and Lister 2013; Bregman 2015). The challenge with the palm oil industry is that the uptake of palm oil is highly fragmented. One of the biggest consumers is Unilever, which uses 4% of the world's supply of palm oil (Unilever 2014). This means that individual CGMs and retailers have limited influence and leverage on the supply chain and the sustainability standards of production. Nonetheless, some major platforms, such as the Consumer Goods Forum (CGF), embracing a large number of CGMs begun to adopt policies promoting the adoption of sustainable sourcing (CGF 2013). Imposing standards on producers is particularly challenging, as a large proportion of palm oil is manufactured and sold in India and China, countries whose consumers are – so far – more price sensitive and less concerned with sustainability.

Major plantation, processing and trading corporations responded to the pressures from CGMs and retailers by making public some collective commitments. In 2013, a wave of sustainability commitments began to emerge in the palm oil sector. The 'No Deforestation, No Peat, No Exploitation' movement was driven by a handful of international advocacy and civil society groups. Their message was simple and aimed to achieve what the Roundtable on Sustainable Palm Oil (RSPO) had failed to do – stop the deforestation of biodiverse and carbon-rich primary and secondary forests and peatlands. Campaigns targeted major oil palm traders such as Wilmar, GAR, Musim Mas, Cargill, Asian Agri and Astra Agro. These commitments differed from past sustainability policies in that they were applied to not only their operations but also their third-party suppliers. As such, commitments moved down the supply chain and were imposed on producers who had, until then, faced limited exposure to global sustainability demands.

The Government of Indonesia has made visible an important agency in the debates on palm oil, by eroding initiatives from the private sector, which could affect government's regulatory and control power. As a way to reinforce the role of the government, it has made the national regulations prevalent, around the Indonesian Sustainable Palm Oil (ISPO) for regulating oil palm plantations development (Hospes 2014). Nonetheless, as a way to respond to international debates on climate change, and due to the detrimental effects of palm oil development in peatlands, the Indonesian government has also played an important role in supporting greater control of oil palm development in peatlands, and has supported an agenda of peatlands restoration. The government, however, is not a monolithic body, and it embraces antagonistic positions of different state agencies. In addition, the sub-national governments have also established some agencies and initiatives to formulate local regulations on sustainable plantation sector including palm oil. Some of the provincial governors such as those of Kalimantan also embraced sustainability initiatives such as the one promoted worldwide by the Governors' Climate and Forests Task Force (GCF), and the West and Central Kalimantan governments were signatories of the New York Declaration on Forests.

A much lower capacity of organization and agency is seen on the smallholder palm oil producers and customary local populations affected by the development of oil palm plantations (Brandi et al. 2015). This is likely due to the pressure exerted by the government while supporting the private sector, and the strong political patronage when the plantations were developed (Varkkey 2012). While there are smallholder associations in Indonesia – the Oil Palm Smallholders Union (SPKS) and the Oil Palm Smallholders Association (APKASINDO), they have a relatively small membership base, whose agency power is very limited. In addition, the indigenous organizations, notably the Indigenous Peoples' Alliance of the Archipelago (AMAN), are mostly focused on demands around recognition of customary tenure claims. Grassroots smallholder and indigenous organizations, therefore, have a very limited capacity to build a movement able to reach national scale with some capacity to contest the power of the private sector. In addition, national producers, mostly medium-sized palm oil growers, are organized in the Indonesian Palm Oil Association (GAPKI), which has a significant lobbying power over the government.

3.3 Architectures

The institutional architecture underpinning the palm oil sector is structured by three major elements. The first are the national policies and regulations in the oil-palm-producing countries to incentivize or control the development of palm oil production, processing and trade. The second are the import policies from consumer countries increasingly interested in sourcing some verified sustainable supply. The third are a different range of private standards, and self-regulatory commitments endorsed by the corporate sector to delink their operations from supply associated with deforestation, and increasingly with violations of workers' rights. These types of regulatory initiatives are creating a relatively complex institutional architecture.

With regard to the public policy in Indonesia, in 2011 the government launched its ISPO standard, a mandatory, third-party-audited, verification system, based on the existing Indonesian regulations. The standard was introduced to ensure the adherence of oil palm plantations to government laws and policies. Meeting the ISPO standard is mandatory for all plantations and palm oil mills and voluntary for smallholders; so far, the standard has fallen short in a number of respects since only 226 companies hold ISPO certificates by March 2017 (ISPO 2017). ISPO has gone through a number of revisions in relation to its timeframe for compliance, and in March 2015 it was rebranded as the ISPO Certification System (Ministry of Agriculture et al. 2015). Currently, ISPO is under the process of strengthening led by the Coordinating Ministry of Economic Affairs. The Malaysian version, the Malaysian Sustainable Palm Oil (MSPO) standard, was introduced in 2013. In contrast to ISPO, it is still voluntary, although a government statement on February 2017 has announced a timeline for the mandatory implementation of MSPO across the country by 2019 (MPOCC 2017).

The Government of Indonesia has also issued a number of policies to address Indonesia's growing contributions to global greenhouse gas (GHG) emissions and climate change. The previous administration installed policies to protect primary forests and peatlands, notably the moratorium on issuing new permits on primary natural forest and peatland (Government of Indonesia 2013), which has been called into question (Busch et al. 2015), along with other regulations related to the land use planning, the protection and restoration of peatlands, and the One Map initiative, which aims to develop a unified map of land use agreed upon by all ministries. Most of these initiatives, while benefiting from the support of the central government, have yet to materialize. Furthermore, in reaction to the fire and haze crisis of 2015, the government issued a presidential instruction banning the clearance and exploitation of peatlands, and new planting in burned areas, as well as setting up a peatland restoration agency (BRG) with the goal of restoring about two million hectare of damaged peatlands in the next five years (Jong 2016). More recently, the Indonesian Government announced an additional moratorium on concessions for oil palm development, yet no decree was issued.

The new administration's commitments to empower rural communities and indigenous peoples' rights over the management of their land reflect the Constitutional Court's Decision no. 35/PUU-X/2012. The court declared that customary forests are no longer state forests, effectively returning the stewardship of these forests to indigenous peoples. But the Ministry of Forestry has been slow to implement the necessary changes (AMAN n.d.). In addition, a major issue constraining the governance of the palm oil sector is that a large but still unknown portion of land under palm oil cultivation is occupied illegally by smallholders, as these lands are still classified as state forests. The government has made some efforts to regularize these settlements, but these endeavours are still in their infancy (Sirait 2015).

One of the most significant regulations adopted by consumer countries is the European Union Renewable Energy Directive (EU-RED) (Directive 2009/28/EC), which requires all biofuel feedstocks, including biodiesel derived from palm oil, to meet the defined standards. As such, a number of palm oil producers in Malaysia and Indonesia, mainly those that are part of the largest palm oil conglomerates, certify their oil under the International Sustainability and Carbon Certification (ISCC) standard. Additionally, 11 European countries have adopted some form of sustainable palm oil commitment at the national level, and three more will potentially follow this trend (Esselink and van der Wekken 2015). In October 2012, a number of UK sector associations with significant interests in the supply or use of palm oil made a public statement of their various commitments to sourcing sustainable palm oil. In March 2016, France introduced a tax on noncertified sustainable palm oil imports as part of a national biodiversity bill, yet it still has to be reviewed and ratified in the upper house. Recently, a call from the European Parliament recommended to phase out the use of palm oil for biodiesel by 2020 within the European Union (European Parliament 2016), a move that has been challenged by the Indonesian and Malaysian governments.

Biodiesel companies and product manufacturers in Europe have established the European Sustainable Palm Oil (ESPO) initiative with a commitment toward 100% sustainable palm oil in 2020. The agreement is supported by an alliance of refineries, and food and feed manufacturers and retailers in the Netherlands, Denmark, France, Belgium, Germany, the UK, Italy and Sweden. It is facilitated by three European sustainability organizations, that is, Caobisco, FEDIOL and IMACE. The parties confirmed their commitment by signing the Amsterdam Palm Oil Declaration on 7 December 2015, whose mission and objectives are to support the uptake of more sustainable palm oil in Europe by working in close collaboration with the national initiatives, RSPO and EU associations. It will do this by encouraging the involvement of non-member companies, sectors and countries and the synchronization of activities. It is still not clear how influential ESPO will be in the global palm oil market.

The third type of initiatives is related to the emergence of private standards. The most prominent is the RSPO, which was formally established in 2004. It emerged through a multistakeholder process, which included both private sector and civil society organizations. Since its inception, the RSPO has seen a slow but steady uptake of the standard, which now certifies roughly 21% of the total global supply (RSPO 2016). In spite of its coverage of the palm oil market, the RSPO has received criticism regarding the stringency of its principles and criteria and its ability to enforce companies' compliance to standards on the ground. As a number of its corporate members have adopted additional 'No Deforestation' commitments, the RSPO recently launched RSPO Next. This is the second initiative to emerge from RSPO member companies and goes beyond the principles and criteria of the original RSPO. The Palm Oil Innovation Group (POIG) was another initiative promoted by Greenpeace and a few companies to demonstrate companies' ability to not only mitigate negative environmental impacts, but also create positive ones.

In September 2014, the growing number of 'No Deforestation' commitments spurred the New York Declaration on Forests. This declaration, which includes national and sub-national governments (in Indonesia) as well as civil society and private sector organizations, aims to halve the rate of deforestation by the end of 2020. These commitments fostered regional- and local-level alliances such as the Malaysian-dominated Sustainable Palm Oil Manifesto (SPOM), while in Indonesia this took the form of the Indonesian Palm Oil Pledge (IPOP), an association of five companies working toward the same goal. Interestingly, since the inception of IPOP, the Indonesian government has strongly opposed it, branding it a cartel, in violation of Indonesia's competition laws and ultimately threatening the government's sovereignty, and also argued its negative impacts on smallholders (Anderson et al. 2005). IPOP was disbanded in July 2016, while companies kept their individual commitments to sustainable and deforestation-free palm oil production.

3.4 Arrangements

Despite the animosity over sovereign rights in relation to private standards in Indonesia, there have been attempts to find common ground between public and private standards. A recent joint study between ISPO and RSPO identifies a number of similarities and differences between the two standards. While there is a significant scope for alignment, particularly in the auditing process, key differences remain in the treatment and definition of high conservation value (HCV) areas within concessions, and the rules to follow for developing new plantations, including the Free Prior Informed Consent (FPIC), which are more stringent in RSPO vis-à-vis ISPO criteria (Ministry of Agriculture et al. 2015). By today, however, it is not clear what steps will be taken by the two parties in order to promote greater alignment of these two standards. Nonetheless, as mentioned, the Indonesian government is conducting efforts to strengthen the ISPO standards in ways that could close the existing gaps.

The commitments to deforestation-free supply chains led to the development of the High Carbon Stock Approach (HCSA) by The Forest Trust (TFT), GAR and Greenpeace in order to identify no-go areas for plantation development (Greenpeace 2013). This approach somehow wanted to provide more clarity on which type of forests should be preserved under more simple methodologies than those provided by the HCV methods, still based on objective and verifiable criteria (Greenpeace 2014). In 2014, a separate group of major palm oil producers known as the Manifesto group announced a voluntary moratorium on clearance of HCS areas, based on empirically valid thresholds of carbon stocks, under the so-called HCS+ approach, which deviated from the Greenpeace-initiated HCSA. Since having in place two different criteria added some confusion, a process was initiated in 2015 to harmonize HCSA and HCS+, and in late 2016, a single set of principles was agreed upon by the involved stakeholders. In 2017, a HCSA toolkit that merges the two approaches will be finalized, which will be expected to be adopted by all major corporate stakeholders (HCSA 2016).

The challenge to implementing corporate and public sector sustainability commitments with new methods in place to comply with and verify progress has resulted in a range of organizations and associations emerging in the institutional landscape to help develop the methodologies and structures needed. One of the most prominent organizations is TFT, an NGO that works with companies to implement their zero deforestation commitments. A number of organizations are also working to develop new business models and value chain structures to support the inclusion of smallholders in the value chains, while supporting the uptake of improved management practices. These include development organizations such as the Sustainable Trade Initiative (IDH) and the Netherlands Development Organisation (SNV), multilateral banks such as the International Finance

Corporation (IFC), and private sector associations and working groups such as PISAgro and the SMART working group. Many of these models seek to increase transparency and traceability in the supply chain and aggregate smallholders in order to access the financial investments needed to replant and increase yields, as well as meet internationally recognized sustainability standards.

In addition, private sector actors, NGOs, donors and development organizations are making progress towards the adoption of arrangements at the sub-national level inspired by approaches that embrace sustainability goals at the jurisdictional level. These include efforts to map smallholders, and enable district-level monitoring, reporting and certification (Wolosin 2016). Provincial-level regulations are also emerging in support of market-based mechanisms for sustainability, such as the commitment by the South Sumatra Provincial Government to turn South Sumatra into a sustainable province (IDH 2016) and the provincial regulation issued in 2011 by the Provincial Government of Central Kalimantan, acknowledging the concept of HCV and allowing palm oil companies to retain and protect areas within their concessions (Gnych and Wells 2014). Many such governments hope that by supporting market-based sustainability standards, they will encourage investment from responsible multinational corporations in their provinces or districts, which will deliver economic development. Such jurisdictional-based view has been championed by Unilever and Marks & Spencer, who in late 2015 launched the 'Produce and Protect' approach through which they commit to prioritize their commodity procurements from areas that are implementing jurisdictional forest and climate initiatives (CGF 2015).

Finally, it is important to highlight some arrangements that are being prompted by financial actors, both in the international and national arenas. For example, the Banking Environment Initiative (BEI), along with the CGF, has developed the Soft Commodities Compact (CISL 2016), which consists of a set of technical guidelines to help the banking industry transform soft commodity supply chains, including the elimination of deforestation, through their clients. The main sustainability criteria for the palm oil sector rely on the framework developed by RSPO principles and indicators. A complementary initiative unfolding at a national level consists of the progress made by Indonesia's eight largest commercial banks to adopting responsible lending practices across all their portfolios, in the context of a Sustainable Financial Roadmap developed by the Indonesian Financial Services Authority (OJK) (OJK 2014). A few international banks are supporting Indonesian banks to develop the internal procedures and build internal capacities to implement these responsible lending policies.

4 Critical problems affecting the palm oil governance complex

The governance of the palm oil sector, in spite of new regulations embracing the objectives of environmental sustainability and peatland restoration, as well as wider recognition of the potential of private standards to foster the uptake of sustainability practices, still faces some critical problems. These problems are related with issues around uneven allocation and access to productive resources and finance, unequal distribution of power among different stakeholders involved in the sector, relatively poor accountability mechanisms, and opaque decision-making and sharing of information. These are explained in more detail below.

4.1 Allocation and access

Allocation of resources and access is relatively uneven when related to the palm oil sector since a few transnational corporate groups control a major portion of the markets and capital flows, yet over time there is an increasing number of small- and medium-scale landholders, and other local investors, struggling to get access to land and other resources, such as finance and inputs, to make a profit from FFB production. In Indonesia, there is a strong legacy of policies, mainly on land allocation and fiscal incentives that privileged private investments for developing large-scale industrial plantations under wider discourses of agricultural modernization. For example, public policies have favoured the allocation of land to industrial plantations, while promoting production and services delivery agreements under different partnership schemes, between companies and local villagers and/or immigrants (Cramb and Curry 2012). The effects of these policies are contradictory, depending on the contexts.

Some local villagers suffered from land dispossession, while others were able to enter into agreements with companies, thereby benefiting from local employment and access to complementary sources of income (Cahyadi and Waibel 2015; Gellert 2015). Access to land in contexts where there is significant development of oil palm tends to also facilitate access to finance – formal or informal – and to some subsidized inputs, such as fertilizers (Rist et al. 2010; Diegues 1992). Nonetheless, still a major portion of smallholders cannot benefit from secure access to land and access to affordable finance, and thus are unable to make proper use of fertilizers to sustain their yields. There are emerging efforts to channel finance to smallholders for improving the uptake of management practices and increasing the yields; while a few pilot projects are emerging, it is unlikely that these initiatives will get impact at scale in order to foster substantive changes (Rainforests Alliance 2016).

4.2 Power distribution

Power is unevenly distributed among stakeholders in the value chain and within landscapes where oil palm tends to expand. A few corporate groups have enormous influence on the supply of FFB through contractual agreements with second- and third-party suppliers. Influential individuals form palm oil consortiums that control oil palm production, marketing and distribution. These consortiums are using that power to impose the adoption of environmental standards through the supply chain, particularly those related to zero deforestation, in part responding to the pressure from traders, CGMs and retailers (Dauvergne and Lister 2013; Pichler 2013). The power of major corporate groups, however, has been contested by the Government of Indonesia, which in an attempt to reinforce national sovereignty is imposing the use of the ISPO standards over the international RSPO standards, or specific company standards (Hospes and Kentin 2014), and is making use of fiscal incentives, such as the ones linked to the CPO fund, to ratify its power in guiding the sector development, under some specific views on sustainability. This latter development has been accompanied by lobby from GAPKI, which is deploying initiatives in the Indonesian Parliament to safeguard the interest of national palm oil producers.

The power and authority in the landscapes, in a similar way, are shared between company actors, who have an important role in local economic development, and government officials from the provincial and district levels, who gained an important weight on the decisions about allocation of concession permits for oil palm development, and thereby on deciding who are the groups that can capture the economic rents originated from logging and forest conversion (Setiawan et al. 2016). In some cases, the power concentrated by local elites at the district and provincial level tends to be used to reinforce their privileged position, while in other cases it can also be used, as part of patronage systems, to grant land and other public resources to local constituencies in exchange of votes as a way to secure their power (Gellert 2015). In Indonesia, local politics tend to largely be dominated by practices of patrimonial manipulation (Fukuoka 2013), which gives limited influence to organized local village or associations and cooperatives in local decision-making. Their participation, in some cases, tends to be regulated, for example, when defining FFB prices.

4.3 Accountability

Different accountability problems do persist in the governance of the palm oil sector. Most are associated with the difficulties encountered by national government in decentralized contexts to be accountable to provincial and district governments, as well as for the district and provincial governments to make upper-level governments more accountable (Setiawan et al. 2016). Major issues refer to very opaque decision-making process for land use planning, granting licenses to palm oil companies, taxes collection from activities associated with timber extraction from land clearing and palm oil production, and the use of such public resources. In most of these issues, tensions exist among different levels and sectors of government due to intertwined interests of local politicians and investors especially related to the local government election, uneven law enforcement and government's desire to collect locally generated income (Brockhaus et al. 2012). This leads to clashes between local and national interests around oil palm development (Sahide and Giessen 2015), yet under very poor accountability.

In Indonesia, major government revenues from palm oil, such as value-added, corporate income and export taxes, are channelled through the central government in Jakarta, so they can be redistributed among provinces. In addition, many of the national earnings from oil palm are captured through export taxes, which are collected centrally and only a small fraction flows back to production provinces (Falconer et al. 2015). Other resources or benefits are transferred through social programmes. It is often the case that local politicians and elites find ways to retain institutional rents at the district or province level. Much of these rents are generated through land allocation. There is a strong argument to see more of the fiscal revenues from palm oil retained in the areas in which it is produced in order to support the mitigation of environmental impacts as well as local infrastructure development. But the retention of profits is often distorted and corrupted by local political economies and self-interest, in scenarios where local communities rarely benefit (McCarthy 2012). The institutional systems to make government officials accountable for their decisions and public money are weak, and many interests exist to prevent it from happening.

4.4 Transparency

Oil palm plantations have developed through consortiums involving political elites, senior bureaucrats and businessmen ranging from the district to the national levels, thus connecting private and public actors in complex and opaque processes of decision-making (Varkkey 2012, 2013). The system of political patronage under which oil palm has expanded in Indonesia explains how many companies benefited from the weak policies and corrupted processes of land allocation. This has made vast swathes of forested land available for plantations at low cost (Bakker and Moniaga 2010). To some extent, this opacity in land allocation prevents the government and companies to disclose information regarding who holds what amount of land, how much land has been developed, what is the magnitude of land banks held by corporate groups and how much of those have been granted in peatlands. There have been attempts by environmental advocacy groups to systematize this information, so to make publicly available the current land banks in the hands of companies. In addition, some legal clause impedes the Ministry of Agriculture and the National Land Agency to disclose this information, particularly on company ownership. Civil society organizations have requested this information to be disclosed as part of debates of zero deforestation commitments. While some corporate groups (e.g. Wilmar and GAR) have made progress in disclosing information on their supply chain from mills to refineries, additional efforts are needed to trace upstream suppliers including third-party suppliers, and independent smallholders, and make that information open to the public.

5 Major gaps in the governance of the palm oil sector

The problems analysed above indicate the presence of several structural aspects that affect not only the overall social, economic and environmental performance of the sector, but also the governance architectures and arrangements to foster concrete changes in the material conditions and institutional structures shaping the sector's performance. This section explores four types of gaps that are observable in the governance of the palm oil sector, specifically capacity, cooperation, compliance and credibility gaps. These gaps have to be analysed in order to identify the public and private interventions with potential to reshape the action arenas, strengthen the agency of the most disempowered social groups, and improve the public and private rules and the arrangements to accommodate the diverse interests of the palm oil sector's stakeholders.

5.1 Capacity gaps

The main capacity gaps result from uneven access to information and knowledge, as well as the technical skills to embrace and put in place mechanisms and processes that articulate better initiatives to improve supply chain management with others for enhancing landscape management. Capacity gaps are important, since they influence the possibility for linking more effectively the value chain interventions to territorial approaches to sustainability, as well as may contribute to linking better monitoring of progress towards sustainability in specific jurisdictions with expectations from CGMs and retailers on addressing sustainability issues in accordance with certain agreed standards. In addition, there is new demand for knowledge and technologies linked to the compliance with sustainability commitments such as, for example, the identification and delimitation of HCV and HCS areas, the implementation of traceability systems from upstream suppliers to processing facilities, and the assessment of social and environmental risks, among others. Currently, such capacity gaps are fulfilled by consultancy groups (e.g. Daemeter, TFT and

Aidenvironment) and environmental organizations (e.g. The Nature Conservancy (TNC) and the World Wide Fund for Nature (WWF)), which are making use of new technologies and tools for remote sensing analysis, mapping, geo-traceability, data storage and analysis. Yet several specialized IT service providers are growing in number in order to fulfil this emerging demand. One of the most critical aspects is who is filling the needs of capacity building for local stakeholders including not only producer organizations but also district and provincial governments in order to use emerging knowledge for informing public-driven processes, which are essential to support sustainability at the jurisdictional level such as land use and spatial planning. Another set of capacities that are lacking are those related to enforcement capacities by state agencies in order to put in place the mechanisms to control and sanction regulations more effectively.

5.2 Cooperation gaps

The cooperation gaps in the palm oil governance complex are several, which in part depend on the incentives and motivations of stakeholder involved in the sector. There are issues of cooperation within different ministries influencing the palm oil sector governance, such as those between the ministries of agriculture, environment and forestry, and finance. In addition, there are significant coordination gaps between the central government and the provincial levels of government with regard to the implementation of specific policies, mainly those related to concession permits and spatial planning, and fiscal resources allocation. Furthermore, important coordination gaps exist between the public and the private sectors mainly with regard to the adoption of norms and procedures for supporting conservation of high-carbon and biodiversity-rich forests within state forestlands granted as oil palm concessions. In this regard, it is widely known that the government forced IPOP's disbandment as a way to reinforce its position of power in rules design and implementation, which was eroded by the major palm oil corporations. Finally, another cooperation gap is the one existing within the palm oil value chains, for example, those between manager of parent companies and plantation managers on the ground. It is not necessarily that managers along the value chain share the same perspectives of sustainability and ways to safeguard social and environmental risks, or implement existing policies and standards. In light of these different cooperation gaps, there is wide scope for improving collaboration not only within state agencies, but also across different levels of government, as well as between the government and the private sector, and along management operations in the supply chain.

5.3 Compliance gaps

Compliance gaps relate, on the one hand, to the capacity of companies to comply with their own pledges, and the rules and procedures in place to accomplish their commitments and, on the other hand, with their compliance with public regulatory frameworks in ways that can be empirically verified. In regard to the former, the aggressive move of companies to delink their supply chains from deforestation, accompanied by efforts to put in place traceability systems to monitor sources of supply as well as measures to avoid the exclusion of smallholders from their supply chains, raises doubts about effective compliance even more since companies have not defined clear targets and timelines. There are several obstacles that prevent effective compliance such as the difficulty for companies to put in place traceability systems that capture all their suppliers – mainly when smallholders are involved

– and to design expansion plans that do not place additional pressure on peatlands. This is associated with the importance of putting in place monitoring systems that are able to trace the progress of compliance, while alerting on potential social and environmental risks. In some cases, however, compliance to these commitments requires complementarities with public regulations, particularly in regard to tenure legality and spatial planning. In addition, company commitments can produce leakage effects that can only be addressed with effective law enforcement. While likely large-scale companies are more prepared to make progress towards their commitments, and thus place pressure on their third-party suppliers, it is likely that the latter do not have the resources or the motivation to comply with those commitments. This raises the issue of the costs of compliance and who pays for it.

5.4 Credibility gaps

There are important credibility and legitimacy gaps, which refer to how effective private self-regulations can be vis-à-vis public regulations in achieving goals of zero deforestation, no expansion in peatlands and respect to local people's rights. Currently, the lack of transparency, in the absence of independent verification of commitments, puts into question the credibility of companies. Reducing credibility gaps is going to require an increasing participation of independent civil society organizations in scrutinizing the progress of companies. A similar situation occurs with regard to public interventions in reversing the governance challenges in the palm oil sector, since the public sector is largely discredited due to its inability to solve issues related to encroachment of state lands, illegal tenure and opaque management of public money. The anti-corruption efforts made by the government have contributed to address credibility gaps, yet more has to be done. It will be important in the future that major palm oil corporate groups disclose information regarding progress in implementing their own commitments (e.g. supplier and production zones), and the approaches and methods used to verify them, as well as that the state make public information that is critical to assess compliance, such as the status of concession permits as well as land banks held by companies. Additional information that has to be shared with the society comprises the flows of fiscal resources related to palm oil taxes, subsidies and incentives, and who are benefiting from them.

6 Conclusions

The present chapter was aimed at deciphering some aspects of the palm oil governance complex, with strong emphasis on Indonesia. The major conundrum facing the palm oil sector refers to how multiple agents' interests can be harmonized while addressing short-term development goals but embracing long-term environmental objectives. Recent years have witnessed important changes in the policy frameworks regulating the palm oil sector as well as a more active intervention of the private sector in embracing private standards and commitments. This has led to a reconfiguration of the action arenas for sustainable palm oil. Both value chains and landscapes have become spaces where public regulations and private standards complement and/or conflict, as well as where public and private sector actors, and other stakeholders, cooperate and/or compete when looking for the sector's sustainability. This chapter tackles the issue of why is that in spite of evolving governance architectures and arrangements there are still no major changes observed in the performance of the palm oil sector. This analysis does not look at the performance of the palm oil sector but rather at the performance of the palm oil governance complex made of action arenas, actors, architectures and arrangements that govern the sector.

We conclude that in spite of significant changes to improve the institutional architecture as well as the agreements and arrangements established between public and private actors, the governance complex faces several structural constraints that constitute governance challenges. These constraints are related to issues of unequal allocation of and access to resources; uneven power distribution across different players along the palm oil supply chain as well as in the landscapes where oil palm expansion takes place; poor upward and downward accountability; and limited transparency on how resources are allocated, who benefit from those resources, and how those benefits are distributed. As result of this analysis, we have identified some major gaps across different dimensions, specifically gaps in capacities, cooperation, compliance and credibility. Addressing the specific challenges associated with these gaps may lead to improved performance of the palm oil governance complex in order to build more sustainable and inclusive oil palm supply chains and landscapes.

7 Where to look for further information

Additional information can be found in the following sources:

- Carbon networks
  - HCV network, https://www.hcvnetwork.org/
- Service providers
  - Daemeter, http://daemeter.org/en/home
  - TFT, http://www.tft-earth.org/
- Research organizations
  - CIFOR, http://www.cifor.org/
  - CIRAD, http://www.cirad.fr/en
  - ICRAF, http://www.worldagroforestry.org/
- Palm oil platforms
  - ZSL, https://www.zsl.org/
- Sustainability initiatives
  - ISPO, http://www.ispo-org.or.id/
  - InPOP, http://www.foksbi.id/en/home/
- Private standards
  - RSPO, http://www.rspo.org/
- Monitoring of commitments
  - Supply Change, http://supply-change.org/
- Financial initiative
- Banking Environment Initiative, http://www.cisl.cam.ac.uk/business-action/sustainable-finance/banking-environment-initiative
- Corporate group commitments
  - Musim Mas, http://www.musimmas.co.id/
  - Wilmar, http://www.wilmar-international.com/
  - GAR, http://goldenagri.com.sg/

8 References

Abdullah, I., W. H. W. Mahmood, M. Fauadi, M. N. Ab Rahman and F. A. A. Jali (2015) Sustainability in Malaysian palm oil: A review on manufacturing perspective. *Polish Journal of Environmental Studies*, 24, 1463–75.

AMAN (n.d.) *Petition for MK No.35 Ruling and Indigenous Peoples Bill*. Jakarta, Indonesia: Indigenous Peoples Alliance of the Archipelago.

Anderson, L. O., Y. E. Shimabukuro, R. S. Defries and D. Morton (2005) Assessment of deforestation in near real time over the Brazilian amazon using multitemporal fraction images derived from terra MODIS. *IEEE Geoscience and Remote Sensing Letters*, 2, 315–18.

Bakker, L. and S. Moniaga (2010) The space between: Land claims and the law in Indonesia. *Asian Journal of Social Science*, 38, 187–203.

Biermann, F. (2014) *Earth System Governance: World Politics in the Anthropocene*. Massachusetts, USA: The MIT press.

Biermann, F., P. Pattberg, H. van Asselt and F. Zelli (2009) The fragmentation of global governance architectures: A framework for analysis. *Global Environmental Politics*, 9, 14–40.

Boström, M., A. M. Jönsson, S. Lockie, A. P. J. Mol and P. Oosterveer (2015) Sustainable and responsible supply chain governance: Challenges and opportunities. *Journal of Cleaner Production*, 107, 1–7.

Brandi, C., T. Cabani, C. Hosang, S. Schirmbeck, L. Westermann and H. Wiese (2015) Sustainability standards for palm oil: Challenges for smallholder certification under the RSPO. *The Journal of Environment & Development*, 24, 292–314.

Bregman, T. (2015) *Achieving Zero (Net) Deforestation Commitments: What it Means and How to Get There*. Oxford, UK: Global Canopy Programme.

Brockhaus, M., K. Obidzinski, A. Dermawan, Y. Laumonier and C. Luttrell (2012) An overview of forest and land allocation policies in Indonesia: Is the current framework sufficient to meet the needs of REDD+? *Forest Policy and Economics*, 18, 30–7.

Busch, J., K. Ferretti-Gallon, J. Engelmann, M. Wright, K. G. Austin, F. Stolle, S. Turubanova, P. V. Potapov, B. Margono, M. C. Hansen and et al. (2015) Reductions in emissions from deforestation from Indonesia's moratorium on new oil palm, timber, and logging concessions. *Proceedings of the National Academy of Sciences*, 112, 1328–33.

Cahyadi, E. R. and H. Waibel (2015) Contract farming and vulnerability to poverty among oil palm smallholders in Indonesia. *The Journal of Development Studies*, 52, 681–95.

CGF (2013) *Sustainability Activation Toolkit*. Issy-les-Moulineaux, France: The Consumer Goods Forum.

CGF (2015) *Statement from Consumer Goods Forum Co-chairs, Acting Individually: Production Protection 1*. France: The Consumer Goods Forum.

CISL (2016) The Banking Environment Initiative (BEI) and Consumer Goods Forum (CGF)'s 'Soft Commodities' Compact. Cambridge Institute for Sustainability Leadership. http://www.cisl.cam.ac.uk/business-action/sustainable-finance/banking-environment-initiative/programme/soft-commodities (last accessed13 March 2016).

Climate Focus. 2015. Progress on the New York Declaration on Forests: An Assessment Framework and Initial Report. Climate Focus, in collaboration with Environmental Defense Fund, Forest Trends, The Global Alliance for Clean Cookstoves, The Global Canopy Program. Amsterdam, Netherlands: Climate Focus.

Colchester, M., S. Chao, J. Dallinger, H. E. P. Sokhannaro, V. T. Dan and J. Villanueva (2011) *Oil Palm Expansion in South East Asia: Trends and Implications for Local Communities and Indigenous Peoples*. Moreton-in-Marsh, UK: Forests People Program, Perkumpulan Sawit Watch.

Colchester, M. and S. Chao (2013) *Conflict or Consent? The Oil Palm Sector at a Crossroads*. Moreton-in-Marsh, UK: Forest Peoples Programme (FPP), Sawit Watch and TUK INDONESIA, p. 422.

Cramb, R. and G. N. Curry (2012) Oil palm and rural livelihoods in the Asia-Pacific region: An overview. *Asia Pacific Viewpoint*, 53, 223–39.

Daemeter Consulting (2015) *Indonesia's Evolving Governance Framework for Palm Oil: Implications for a No Deforestation, No Peat Palm Oil Sector*. Bogor, Indonesia: Daemeter Consulting.

Dauvergne, P. and J. Lister (2013) *Eco-business: A Big-brand Takeover of Sustainability*. Cambridge, MA.: MIT Press.

Diegues, A. C. (1992) The social dynamics of deforestation in the Brazilian Amazon: an overview. *Discussion Paper – United Nations Research Institute for Social Development*, 36.

Directorate General of Estates (2014) Tree Crop Estate Statistics of Indonesia 2012–2014: Oil Palm. Jakarta, Indonesia: Ministry of Agriculture of Republic of Indonesia.

Esselink, E. and D. van der Wekken (2015) Sustainable palm oil in Europe: The national initiatives approach. In *European Palm oil conference*. Milan, Italy.

European Parliament (2016) Draft Opinion of the Committee on International Trade for the Committee on the Environment, Public Health and Food Safety on palm oil and deforestation of rainforests (2016/2222(INI)). *European Parlimanet*. http://www.europarl.europa.eu/sides/getDoc.do?pubRef=-%2f%2fEP%2f%2fNONSGML%2bCOMPARL%2bPE-592.301%2b01%2bDOC%2bPDF%2bV0%2f%2fEN (last accessed 18 March 2017).

Falconer, A., T. Mafira and G. Sutiyono (2015) *Improving Land Productivity Through Fiscal Policy: Early Insights on Taxation in the Palm Oil Supply Chain*. San Francisco, CA: Climate Policy Initiative.

Fukuoka, Y. (2013) Local politics in Indonesia: Pathways to power. *Journal of Contemporary Asia*, 43, 186–8.

Gellert, P. K. (2015), Palm Oil Expansion in Indonesia: Land Grabbing as Accumulation by Dispossession, in Jon Shefner (ed.) *States and Citizens: Accommodation, Facilitation and Resistance to Globalization* (Current Perspectives in Social Theory, Volume 34) Emerald Group Publishing Limited, pp. 65–99.

Gnych, S. and P. Wells (2014) *Land Use in Central Kalimantan: Combining Development and Sustainability Goals for Land Optimization*. Bogor, Indonesia: CIFOR.

Government of Indonesia (2013) Presidential Instruction No. 6 Year 2013 regarding moratorium on the issuance of new licenses and improvement of governance of primary forests and peatlands. Jakarta, Indonesia: Sekretariat Kabinet Republik.

Greenpeace (2013) *Identifying High Carbon Stock (HCS) Forest for Protection: Towards Defining Natural Forests and Degraded Lands (formerly forest) in the Tropics*. Amsterdam, The Netherlands: Greenpeace.

Greenpeace (2014) *The HCS Approach: No Deforestation in Practice*. Amsterdam, The Netherlands: Greenpeace.

HCSA (2016) Agreement on unified approach to implementing no deforestation commitments. *High Carbon Stock Approach*.

Hospes, O. (2014) Marking the success or end of global multi-stakeholder governance? The rise of national sustainability standards in Indonesia and Brazil for palm oil and soy. *Agriculture and Human Values*, 31, 425–37.

Hospes, O. and Kentin, A. (2014) Tensions Between Global-Scale and National-Scale Governance: The Strategic Use of Scale Frames to Promote Sustainable Palm Oil Production in Indonesia, in *Scale-sensitive Governance of the Environment* (eds F. Padt, P. Opdam, N. Polman and C. Termeer), John Wiley & Sons, Ltd, Oxford, UK. doi: 10.1002/9781118567135.ch13.

IDH (2016) *The Sustainable Trade Initiative. 2016. Green Growth Plan*. https://www.idhsustainabletrade.com/project/green-growth-plan-south-sumatra/(last accessed 13 March 2017).

ISPO (2017) *Sertifikasi ISPO*. Indonesian Sustainable Palm Oil. http://www.ispo-org.or.id/index.php?option=com_content&view=article&id=79&Itemid=225&lang=ina (last accessed 13 March 2017).

Jong, H. N. (2016) *Peatland Agency Gears Up for Haze Crisis*. http://www.thejakartapost.com/news/2016/03/10/peatland-agency-gears-haze-crisis.html (last accessed 10 March 2016).

Khasanah, N., M. van Noordwijk, A. Ekadinata, S. Dewi, S. Rahayu, H. Ningsih, A. Setiawan, E. Dwiyanti and R. Octaviani (2012) *The Carbon Footprint of Indonesian Palm Oil Production*. Technical brief No. 25. Bogor, Indonesia: World Agroforestry Center (ICRAF).

Lai, O.-M., C.-P. Tan and C. C. Akoh (2012) *Palm Oil: Production, Processing, Characterization, and Uses*. Urbana, IL: AOCS Press.

Li, T. M. 2015. Social impacts of oil palm in Indonesia: A gendered perspective from West Kalimantan. Occasional Paper 124. Bogor, Indonesia: CIFOR.

McCarthy, J. F. (2012) Certifying in contested spaces: private regulation in Indonesian forestry and palm oil. *Third World Quarterly*, 33, 1871–88.

Ministry of Agriculture, RSPO, ISPO, Mutuagung Lestari & UNDP (2015) Joint study on the simmilarities and diffferences on the ISPO and the RSPO certification systems. Bogor, Indonesia.

Molenaar, J. W., M. Persch-Orth, S. Lord, C. Taylor and J. Harms (2013) *Diagnostic Study on Indonesian Oil Palm Smallholders*. Jakarta, Indonesia: International Finance Corporation.

MPOB (2016) *Sectoral Status*. Malaysian Palm Oil Board. http://bepi.mpob.gov.my/index.php/statistics/sectoral-status/137-sectoral-status-2015/752-number-a-capacities-of-palm-oil-sectors-2015.html (last accessed [Accessed on 28 May 2016]).

MPOCC (2017) *Way Forward for MSPO as a Mandatory Certification Scheme*. Malaysian Palm Oil Certification Council. http://www.mpocc.org.my/single-post/Way-Forward-for-MSPO-as-a-Mandatory-Certification-Scheme (last accessed 6 February 2017).

OJK (2014) *Siaran Pers Bersama. Peluncuran "Arah Jalan Keuangan Berkelanjutan dan Buku Pedoman"*. Otoritas Jasa Keuangan. http://www.ojk.go.id/id/berita-dan-kegiatan/siaran-pers/Documents/SIARANPERSOJKKLH_rev_KLHK4Des_1418003025.pdf (last accessed 13 March 2017).

Oosterveer, P. (2015) Promoting sustainable palm oil: Viewed from a global networks and flows perspective. *Journal of Cleaner Production*, 107, 146–53.

Ostrom, E., R. Gardner and J. Walker (1994) *Rules, Games, and Common-Pool Resources*. Ann Arbor, Michigan: The University of Michigan Press.

Overdevest, C. and J. Zeitlin (2012) Assembling an experimentalist regime: Transnational governance interactions in the forest sector. *Regulation & Governance* 8, 22–48.

Pacheco, P., S. Gnych, A. Dermawan, H. Komarudin and B. Okarda (2017) The palm oil global value chain: Implications for economic growth and social and environmental sustainability. Working Paper 220. Bogor, Indonesia: CIFOR.

Pichler, M. (2013) 'People, planet & profit': Consumer-oriented hegemony and power relations in palm oil and agrofuel certification. *The Journal of Environment & Development*, 22, 370–90.

Potter, L. (2015) *Managing Oil Palm Landscapes: A Seven-country Survey of the Modern Palm Oil Industry in Southeast Asia, Latin America and West Africa*. CIFOR Occasional Paper, Volume 122. Bogor, Indonesia: CIFOR.

Pretty, J., G. Smith, K. W. T. Goulding, S. J. Groves, I. Henderson, R. E. Hine, V. King, J. van Oostrum, D. J. Pendlington, J. K. Vis and et al. (2008) Multi-year assessment of Unilever's progress towards agricultural sustainability II: Outcomes for peas (UK), spinach (Germany, Italy), tomatoes (Australia, Brazil, Greece, USA), tea (Kenya, Tanzania, India) and oil palm (Ghana). *International Journal of Agricultural Sustainability*, 6, 63–88.

Rainforests Alliance (2016) Smallholder palm oil financing in Indonesia: Needs and opportunites for sustainable finance interventon. *Rainforests Alliance.*

Rist, L., L. Feintrenie and P. Levang (2010) The livelihood impacts of oil palm: Smallholders in Indonesia. *Biodiversity and Conservation*, 19, 1009–24.

Rival, A. and P. Levang (2014) *Palms of Controversies: Oil Palm and Development Challenges*. Bogor, Indonesia: CIFOR.

RSPO (2016) *Impacts*. Kuala Lumpur, Malaysia: Roundtable of Sustainable Palm Oil.

Sahide, M. A. K. and L. Giessen (2015) The fragmented land use administration in Indonesia – Analysing bureaucratic responsibilities influencing tropical rainforest transformation systems. *Land Use Policy*, 43, 96–110.

Sayer, J., J. Ghazoul, P. Nelson and A. Klintuni Boedhihartono (2012) Oil palm expansion transforms tropical landscapes and livelihoods. *Global Food Security*, 1, 114–19.

Setiawan, E. N., A. Maryudi, R. H. Purwanto and G. Lele (2016) Opposing interests in the legalization of non-procedural forest conversion to oil palm in Central Kalimantan, Indonesia. *Land Use Policy*, 58, 472–81.

Sheil, D., A. Casson, E. Meljaard, M. van Noordwijk, J. Gaskell, J. Sunderland-Groves, K. Wertz and M. Kanninen (2009) The Impacts and Opportunities of Oil Palm in Southeast Asia What Do We Know and What Do We Need to Know? Center for International Forestry Research (CIFOR), Bogor, Indonesia, CIFOR Occasional Paper no. 51, 67p.

Sirait, M. T. (2015) Inclusion, Exclusion and Agrarian Change. Experiences of Forest Land Redistribution in Indonesia. In *CERES, Research School for Resource Studies for Development*. Rotterdam, Netherlands: Erasmus University Rotterdam, p. 241.

Statistik Indonesia. (2015) *Direktori perusahaan perkebunan kelapa sawit 2014*. Jakarta, Indonesia: Central Statistical Agency.

Unilever (2014) Sustainable Palm Oil Progress Report 2014. https://www.unilever.com/Images/uslp-unilever-palm-oil-report-nov14_tcm244-424235_en.pdf (last accessed 17 March 2017).

Varkkey, H. (2012) Patronage politics as a driver of economic regionalisation: The Indonesian oil palm sector and transboundary haze. *Asia Pacific Viewpoint*, 53, 314–29.

Varkkey, H. (2013) Oil palm plantations and transboundary haze: Patronage networks and land licensing in Indonesia's Peatlands. *Wetlands*, 33, 679–90.

Williamson, C. R. (2009) Informal institutions rule: Institutional arrangements and economic performance. *Public Choice*, 139, 371–87.

Wolosin, M. (2016) *Jurisdictional Approaches to Zero Deforestation Commodities*. Washington DC, USA: WWF

Part 2

Plant physiology and breeding

Chapter 5

Advances in understanding oil palm reproductive development

Estelle Jaligot, CIRAD, UMR DIADE (IRD, UM), France

1 Introduction

The African oil palm (*Elaeis guineensis*, Jacq.) is the highest-producing oil crop, with a mere 6% of cultivable lands planted with oil palm yielding one-third of the global vegetable oil market share (FAO, 2015). However, the combination of a strong demographic pressure and increased per capita income in developing countries (chiefly in Asian countries such as China and India) puts a strain on the demand for edible oils, with its consumption predicted to nearly double by 2050 (Corley, 2009; Murphy, 2014). Palm oil production, on the other hand, is constrained by growing international concerns over the negative impacts of plantations on both the environment and human communities. Changes in land use from high conservation value, high carbon stock primary forests or peatlands to oil palm plantations are paralleled by a drop in both biodiversity and carbon sequestration, together with a rise in greenhouse gases emissions. Under the impulse of both state-owned holdings and private companies, the implantation or expansion of large oil palm plantations sometimes occurs at the expense of local populations and cases of illegal land appropriation were reported, especially in Africa (Koh and Wilcove, 2007; Rival and Levang, 2014; Wich et al., 2014). The relative stalling of plantations growth that ensues is becoming a pregnant problem in the two main oil palm producing countries, Indonesia and Malaysia, which together account for nearly 90% of the world's supply of palm oil and which rely on this commodity for their economic growth.

Consequently, strategies aiming at increasing oil yield per surface unit without increasing agricultural inputs should contribute to increasing the sustainability of oil palm cultivation

http://dx.doi.org/10.19103/AS.2017.0018.04

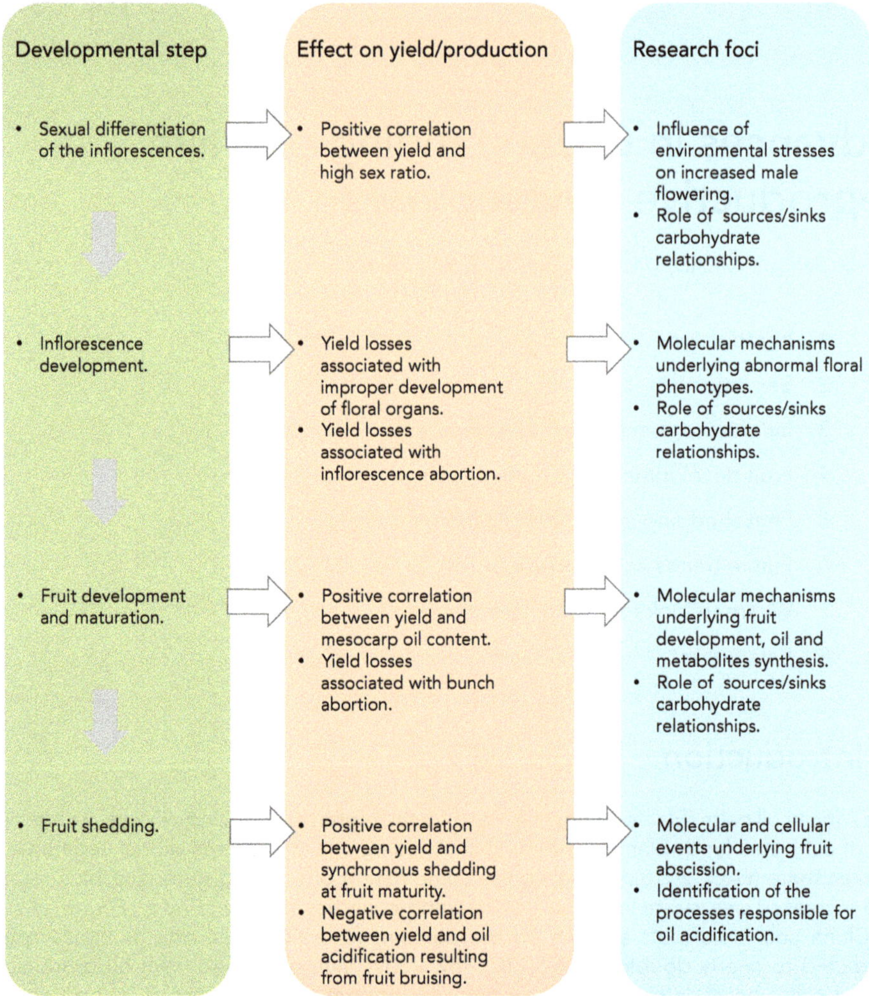

Figure 1 Steps of reproductive development of the oil palm with an effect on yield, and associated research interests.

by alleviating the need for continually expanding planted areas (Murphy, 2014). And there is, indeed, a lot of room for improvement, considering that the average yield in smallholder plantations, which contribute nearly half the palm oil production in Malaysia and Indonesia, is estimated to be around 2–3 tons per hectare and that the average yield in industrial plantations reaches 4–6 t/ha. By comparison, the theoretical yield that can be expected from material resulting from decades of breeding is estimated to be 10–15 t/ha (Barcelos et al., 2015; Murphy, 2014). Moreover, it has been estimated that over 25% of the palms that are planted in Malaysia are approaching the end of their productive lifetime while most of the remainder have reached or will soon reach their peak productivity

(Murphy, 2014). Altogether, these observations suggest that the yield gap between actual and theoretical figures could therefore be partially closed through the replanting of elite, high-yielding material in ageing or low-yielding plantations.

As far as palm oil production is concerned, yield is influenced by a variety of parameters, among which some are connected to the regulation of developmental processes affecting the reproductive organs: the ratio of female to total inflorescences (sex ratio), inflorescence development, fruit development, maturation (during which oil synthesis takes place) and shedding (see Fig. 1). Obviously, a better understanding of such key steps in the reproductive development of oil palm is of paramount importance so that the genetic, molecular, physiological and environmental mechanisms or the stimuli controlling them can be identified and, ultimately, used towards the optimization of oil production through both breeding and agricultural practices.

Beyond what can be achieved within the boundaries set by current planted areas and climate conditions, how climate change may affect oil palm development and oil palm cultivation must be anticipated so that food security can be ensured. In that aspect, it is most urgent to explore how the present environmental conditions may influence the reproductive developmental stages listed above. This way, we can better anticipate the oil palm's responses to future challenges, mitigate their negative impact on production and elaborate appropriately adapted ideotypes.

2 Sex ratio

Oil palm is a monoecious species in which the emergence of male and female inflorescences alternates on the same individual; hence, the term 'temporal dioecy' has been coined to describe this type of reproductive development.

From the time oil palm reaches reproductive maturity, approximately 3–4 years after planting, oil palm inflorescences are continuously produced from the single Shoot Apical Meristem (SAM) of the plant and differentiate from the stem of the palm, at the axil of each new leaf. The sexual determination of oil palm inflorescences occurs about two years prior to flower maturity (anthesis) and 29–32 months before bunch harvest (Adam et al., 2011; Combres et al., 2013; Cros et al., 2013; Durand-Gasselin et al., 1999). Although, in principle, male and female inflorescences alternate so that self-pollination is prevented, various factors come into play and influence the actual outcome. Clues to the involvement of genetic factors can be found: the phenological comparison between different oil palm genotypes grown in the same environment shows substantial variations in the sex ratio (T. Durand-Gasselin, pers. comm.). Also, since fruit production is directly dependent on the rate of female flower emergence, elite oil palm genotypes obtained through breeding schemes tend to produce a higher number of female inflorescences, sometimes at the expense of pollen availability (Cochard et al., 2005; Durand-Gasselin et al., 1999).

In addition to these likely genetic determinants, non-genetic factors have long been known to exert a strong influence on sex differentiation in oil palm. The external application of growth regulators such as gibberellins resulted in increased male flowering (Corley, 1976a). Although the mechanism governing this phenomenon is as yet undetermined, it could be related to the direct or indirect action of growth regulators on the expression of the genes controlling the development of floral organs. A similar hypothesis has been suggested to explain the *in vitro* induction of stamen development in female date palm flowers submitted to variations of the balance between exogenously applied auxins and cytokinins (Masmoudi-Allouche et al., 2009).

Most strikingly, it has been shown that a variety of stress-related conditions may result in modifications of sex ratio in oil palm. For instance, the ablation of either leaves (defoliation), or inflorescences or fruit bunches (pruning) has opposite effects on the sex differentiation of inflorescences since they result in an increased number of male inflorescences and an increased number of female inflorescences, respectively (Corley et al., 1995; Corley and Tinker, 2016; Legros et al., 2009b). Also, it has been noted that the mutual shading of individual palms in areas with high-density planting patterns tends to stimulate the emission of male inflorescences (Durand-Gasselin et al., 1999), although the stress associated with competition for soil-associated resources such as nutrients and water may account partly for this effect. In order to account for these phenomena, it has been proposed that sex differentiation in oil palm was mainly driven by competitive supply-and-demand relationships between organs behaving mainly as producers (or 'sources') (i.e. photosynthetic leaves) or as importers (or 'sinks') of carbohydrates which include, but are not limited to, reproductive structures such as flowers and fruits. Indeed, Cros et al. (2013) showed that sex ratio is negatively correlated with bunch load. Thus, the increase in male inflorescence emission that is obtained in conditions of impaired photosynthesis (resulting from either shading or defoliation) could be attributed to the comparatively higher physiological 'cost' of female inflorescences (Combres et al., 2013). Conversely, the selective removal of sink organs through inflorescence or bunch pruning tends to increase sex ratio because of partial alleviation of the competition for photosynthetic assimilates (Corley and Tinker, 2016). In support to this model, recent publications analysing changes in the inflorescence transcriptome after defoliation treatment (Ajambang et al., 2015, 2016a, 2016b) showed an enrichment of differentially expressed genes with biological functions pertaining to carbohydrate metabolism and abiotic stress response pathways.

In oil palm, the water status of the plant at the time of sex differentiation of the inflorescence plays a significant role in the latter phenomenon. Indeed, sex ratio variability between planting sites under contrasting water regimes has long been reported by oil palm planters and breeders, with African plantations experiencing a marked seasonality, alternating dry and rainy seasons, displaying a higher frequency of male inflorescences than South East Asian locations under important and relatively stable rainfall (Corley, 1976b). Obviously, this difference in sex ratio cannot be explained solely by the use of different genotypes or the stronger prevalence of highly selected planting material in Asia, since it has also been observed with identical progenies planted in sites under distinct climatic conditions (Marcel de Raïssac, pers. comm.). In accordance with these observations, a positive correlation between sex ratio and water balance has been established and inter-cross variability in drought tolerance traits has been hypothesized to account for its variations (Cros et al., 2013). Also, the analysis of a Quantitative Trait Locus (QTL) explaining 11.3% of the phenotypic variation for sex ratio led to the identification of the *EgAKR1* (aldo-keto reductase) gene with putative functions in response to abiotic stresses (Somyong et al., 2014). However, as pointed out by Corley and Tinker (2016), the induction of increased male flowering by drought periods occurring at different times along the inflorescence development process makes it likely that there is more than one period of sensitivity to sex determination-altering processes in oil palm.

3 Inflorescence and flower development

Following sex determination, inflorescence and then floral structures differentiate and develop through a process that culminates with the emergence and maturation of floral organs (Adam et al., 2005). In angiosperms, this chain of events is regulated through the sequential action of genes associated with inflorescence meristem, flower meristem and floral organ identities, respectively, as it has been elucidated through the study of the model plant species *Arabidopsis thaliana* (Fig. 2) (Liu and Mara, 2010; Ó'Maoiléidigh et al., 2014; Wellmer and Riechmann, 2010).

The occurrence of flower abnormalities, which negatively impact oil production, has been a strong incentive to tackle the question of flower development regulation in oil palm. The naturally occurring *Poissoni* phenotype, also known under the vernacular name *diwakkawakka* in Asia, displays an apparent feminization of male floral organs in inflorescences from both sexes (Adam et al., 2005; Hartley, 1988). This morphological variant, which has also been identified in the Latin American relative of oil palm (*Elaeis oleifera*), is assumed to result from a mutation, although this has not been formally demonstrated.

In contrast with these relatively rare occurrences of aberrant flower morphotypes affecting seed-derived palms, the *mantled* variation only arises in clonal progenies of tissue culture-derived oil palms, where it affects 5% of the regenerants on average with unpredictable variations of both its frequency and its severity between genotypes and production batches (Eeuwens et al., 2002). Whereas the initial purpose of the somatic embryogenesis process was to generate a virtually unlimited number of *in vitro*-generated identical copies from high-yielding individuals, the mantled floral phenotype (which is morphologically similar to the *diwakkawakka*) makes the affected palms mostly unproductive due to fertility issues and impaired oil accumulation in the fruits (Corley et al., 1986). Visual detection of mantled palms is only possible from the time they reach the flowering stage, that is, 3–4 years from the time plantlets generated through the somatic embryogenesis protocol (which takes 2–3 years to complete and requires a highly qualified manpower) have been transferred to the field. Therefore, the *mantled* variation is not only costly in terms of oil production but also in human labour and plantation space. As a result, the large-scale use of clonal oil palms in plantations has been significantly hampered by the occurrence of the mantled variation. Obviously, finding markers enabling a screening of this phenotype as early as the *in vitro* culture stages is of high importance to the industry and might greatly facilitate the future adoption of somatic embryogenesis-derived material by end users (Rival et al., 2000).

The slow reversion of the *mantled* phenotype through time was an early indication that its source is not genetic, but more likely epigenetic and thus associated with changes in gene regulation mechanisms since these are commonly altered by tissue culture (Bairu et al., 2011; Kaeppler et al., 2000; Kaeppler and Phillips, 1993; Miguel and Marum, 2011). Indeed, it was soon demonstrated that the genomic DNA of *in vitro* as well as adult *mantled* tissues is significantly depleted in cytosine methylation (Jaligot et al., 2000, 2002, 2004; Matthes et al., 2001). Parallel investigations targeting oil palm genes belonging to the MADS-box family of transcription factors involved in floral organ identity showed that, consistently with the appearance of the *mantled* phenotype, the genes involved in male organs (stamens) formation are downregulated throughout the development of variant flowers (Adam et al., 2005, 2006, 2007). Further work on the regulation of one of these stamen identity genes, namely *EgDEF1*, recently suggested that its expression defect in

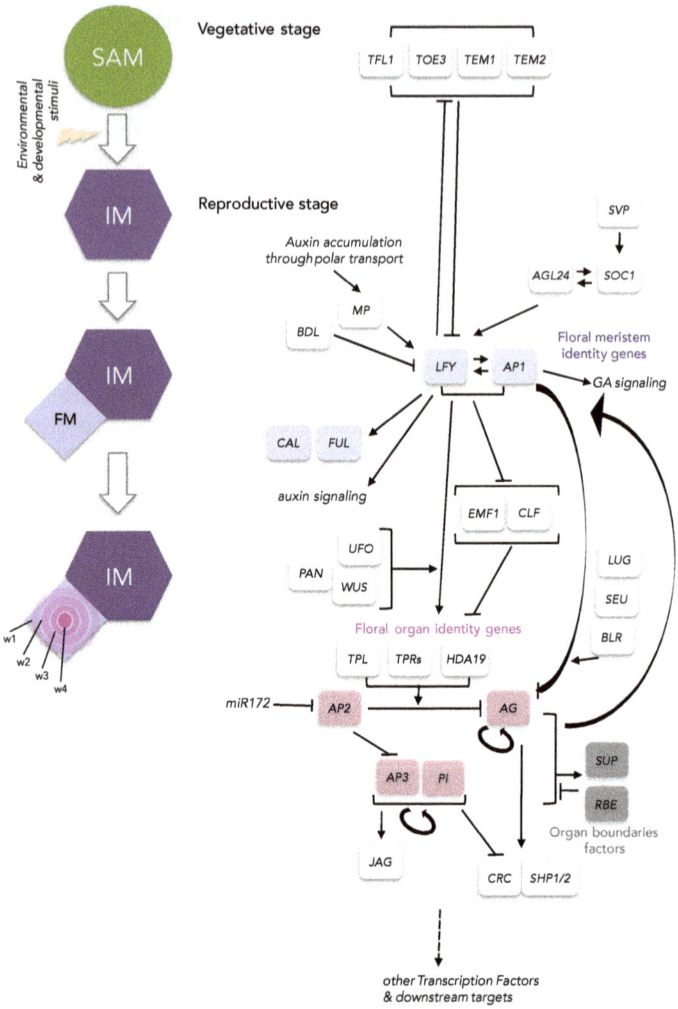

Figure 2 Summary of the flower development process (based on Liu and Mara, 2010; Ó'Maoiléidigh et al., 2014; Wellmer and Riechmann, 2010). Pointed arrowheads: activation; blunt connectors: repression; curved arrows: positive feedback loops. Abbreviations: SAM = shoot apical meristem; IM= inflorescence meristem; FM = flower meristem; w1-w4: floral whorls 1 to 4; *TFL1* = *TERMINAL FLOWER1*; *TOE3* = *TARGET OF EAT3*; *TEM1/2* = *TEMPRANILLO1/2*; *MP* = *MONOPTEROS*; *BDL* = *BODENLOS*; *LFY* = *LEAFY*; *AP1* = *APETALA1*; *AGL24* = *AGAMOUS-LIKE24*; *SVP* = *SHORT VEGETATIVE PHASE*; *SOC1* = *SUPPRESSOR OF OVEREXPRESSION OF CONSTANS1* ; *CAL* = *CAULIFLOWER*; *FUL* = *FRUITFULL*; *EMF1* = *EMBRYONIC FLOWER1*; *CLF* = *CURLY LEAF*; *PAN* = *PERIANTHA*; *UFO* = *UNUSUAL FLORAL ORGANS*; *WUS* = *WUSCHEL*; *TPL/TPRs* = *TOPLESS/TOPLESS-RELATED*; *HDA19* = *HISTONE DEACETYLASE19*; *LUG* = *LEUNIG*; *SEU* = *SEUSS*; *BLR* = *BELLRINGER*; *AP2* = *APETALA2*; *AG* = *AGAMOUS*; *AP3* = *APETALA3*; *PI* = *PISTILLATA*; *SUP* = *SUPERMAN*; *RBE* = *RABBIT EARS*; *NZZ/SPL* = *NOZZLE/SPOROCYTELESS*; *JAG* = *JAGGED*; *CRC* = *CRABS CLAW*; *SHP1/2* = *SHATTERPROOF1/2*.

mantled flowers might result from improper RNA processing caused by methylation loss of a nearby transposable element (TE) (Jaligot et al., 2014; Ong-Abdullah et al., 2015). While these studies propose a compelling molecular model for the flower phenotype in adult clonal oil palms, much remains to be done to understand where, how and (perhaps more importantly) when this epigenetic alteration arises during the tissue culture process so as to provide a reliable basis for early detection tests that could be implemented in commercial production units.

Beyond the specific issue of the *mantled* variation, understanding and reducing the emergence of somaclonal variations is of considerable practical importance. Indeed, because of their very low genetic variability and highly controlled culture conditions, clonal plants are a material of choice in the study of the response to environmental factors and stress adaptation mechanisms (Rai et al., 2011; Us-Camas et al., 2014; Wang and Wang, 2012). Moreover, it is crucial that the output of micropropagation protocols is under control in the prospect of developing efficient transformation and gene editing techniques, as such protocols always include an *in vitro* regeneration step (Etienne et al., 2016).

Studies carried out in both annual and perennial temperate plant species have shown that flowering responds to environmental cues so that the reproductive phase is adjusted to seasonal changes (Andrés and Coupland, 2012; Kudoh, 2015; Shrestha et al., 2014). As a matter of fact, many of the genes that play a role in either floral induction or flower development pathways respond to stimuli such as changes in photoperiod, light quality or temperature and can be affected by a variety of abiotic stresses (Liu and Mara, 2010; Ó'Maoiléidigh et al., 2014; Wellmer and Riechmann, 2010). It is likely that the balance between the different environmental signals required for flowering is different for a plant growing in tropical regions that are characterized by a narrower range in both temperatures and day length between seasons. However, Combres et al. (2013) proposed that the flowering rhythmicity of oil palm is influenced by photoperiod so as to minimize the adverse effects of strong seasonality on flower and fruit development: this effect is therefore more significant in West Africa than it is in South East Asia. In line with the carbohydrate partition model (see previous section), several studies have also concluded that seasonal variations in oil palm production could be explained mainly through the negative effects of environmental stresses such as drought on the availability of photosynthetic assimilates. Such an imbalance would lead to a biased sex ratio, delayed inflorescence development and enhanced inflorescence and fruit bunch abortions (Combres et al., 2013; Cros et al., 2013; Legros et al., 2009a; Pallas et al., 2013). It has also been reported that a strongly reduced carbohydrate supply such as that obtained through the reduction of the number of source organs through severe defoliation might lead to a preferential abortion of female inflorescences, thereby reinforcing the sex ratio bias (Pallas et al., 2013). However, data are scarce and conflicting regarding this phenomenon and, just as with sex differentiation, there might be several distinct stages of inflorescence development that show a critical sensitivity to abortion-promoting signals (Corley and Tinker, 2016). In their recent paper, Somyong et al. (2016) showed that the development of female oil palm inflorescences likely involves ethylene-induced regulations and the action of non-coding regulatory small RNAs. Because of the sensitivity of both this growth regulator and small RNA-based regulations to environmental challenges, this result can be considered as additional evidence regarding an environmental influence over reproductive development of oil palm.

4 Fruit development and maturation and oil composition

Once the female flower has been pollinated, the development of oil palm fruit (or drupe) is set in motion through steps that have been dissected at the transcriptomic and metabolomic levels in two important articles (Bourgis et al., 2011; Tranbarger et al., 2011). The first phase of early fruit development, which runs up to 100 days after pollination (dap), includes a strong cell division and expansion activity and involves the action of the growth regulators such as gibberellins, cytokinins and finally auxins as well as the induction of the corresponding growth regulators-responsive genes. During the subsequent maturation and ripening phases, it is followed by an increase in both ethylene and abscissic acid production until the time when the fruit reaches full ripeness, around 160 dap (Tranbarger et al., 2011).

As demonstrated through the work of Legros et al. (2009b, 2009c), sucrose is exported from mature photosynthetic source leaves and temporarily stored as starch in the stipe (or pseudo-trunk) of oil palm. Sucrose is later imported to fruits where it provides precursors of oleosynthesis through the glycolysis pathway, with carbon allocation to oil palm fruits amounting to about 30% of photosynthetic C fixation (Corley and Tinker, 2016; Dussert et al., 2013; Lamade et al., 2016). Palm oil synthesis then occurs in the course of the maturation and ripening phases through the massive transcriptional activation of genes involved in fatty acids (FAs) biosynthesis, whereas the genes responsible for the assembly of triacylglycerols (TAGs), the core components of the oil which are formed from the esterification of these FAs, are not significantly upregulated (Bourgis et al., 2011; Tranbarger et al., 2011).

A unique case among all the oleaginous plant species, oil palm synthesizes two oils: palm oil, which accumulates in the pulp or mesocarp, and kernel oil, which is extracted from the storage tissue (endosperm) of the stone. Both tissues rely on different members of the WRINKLED (WRI) family of transcription factors to achieve the upregulation of FA synthesis (Dussert et al., 2013). Ultimately, the markedly different regulations of their respective oil synthesis pathways result in distinct oil compositions and global contents in the tissues. While palm oil makes up 80% of the dry weight of the mesocarp and includes a majority of palmitic (C16:0; 41%) and oleic (C18:1; 31%) acids, the endosperm includes 50% of kernel oil made mostly of medium-chain FA such as lauric (C12:0; 49%) and myristic (C14:0; 15%) acids. Finally, the embryo contains 27% of the oil with a higher concentration of unsaturated FA: 25% palmitic acid, 35% oleic acid and 23% linoleic acid (C18:2) (Dussert et al., 2013).

Now that a thorough characterization of the molecular events underlying fruit development and maturation is available, the molecular basis of the yield gap between high oil- and low oil-producing genotypes is being investigated. Teh et al. (2013) have compared high oil- versus low oil-yield genotypes of oil palm for bunch and fruit morphological traits, and the accumulation of primary and secondary metabolites during fruit maturation. These authors conclude that morphological differences in number of bunches, mesocarp weight per bunch and mesocarp oil content – which they assume result from genetic variations between different crosses – account for most of the yield gap, with none of the trade-off between fruit number and size that had been observed in previous works (Cros et al., 2013; Pallas et al., 2013). The authors also observe a differential accumulation of primary metabolites (amino acids, glycolysis intermediates and tricarboxylic acids) in the mesocarp between high-producing and low-producing genotypes, which they suggest may be indicative of differing strengths of carbon partitioning processes in favour of lipid biosynthesis.

On the other hand, Guérin et al. (2016) have explored the regulation of oil synthesis in back-cross population between *E. guineensis* and *E. oleifera*. Indeed, the latter is characterized by a lower oil accumulation in the mesocarp (15–25%) and a distinct composition that includes nearly half as much palmitic acid (25%) and a higher concentration of unsaturated FAs, which make this species a potential source of genes of interest for the improvement of the nutritional value of palm oil (Barcelos et al., 2015; Murphy, 2014). Using co-expression analyses, the authors confirm the negative correlation between the expression of genes involved in starch and FA biosynthesis, and they identify two additional transcription factors, namely NF-YB1 (nuclear factor Y) and ZFP1 (zinc finger) with a significant influence in oil synthesis in different oil palm tissues, alongside *WRI* family genes. In accordance with previous findings (Parveez et al., 2014), they further highlight the central role of transcriptional regulations of the gene encoding the plastidial enzyme β-ketoacyl-ACP synthase II (KASII, responsible for the elongation of C:16 FA to C:18) in palmitic acid accumulation. These different genes provide potential targets for breeding strategies aimed at altering palm oil FA composition through either introgression or biotechnology In parallel, complementary research efforts are focusing on improving the nutritional value of palm oil through its contents in metabolites such as carotenoids and antioxidant compounds, in the aim of reducing vitamins A and E deficiencies in developing countries (Barcelos et al., 2015; Tranbarger et al., 2011).

5 Fruit shedding and oil acidification

Fruit abscission, the process through which the fruit becomes detached from the pedicel, takes place once full maturity has been achieved, that is, about 160–180 dap. In plantations, the presence of detached fruits at the foot of oil palm is used as an indicator that the bunch is ready to be harvested (Corley and Tinker, 2016). However, premature abscission occurring before both fruit ripeness and oil accumulation are optimal, as well as shedding between harvest site and oil mill cause losses in production, prompting the need for a better understanding of the mechanisms governing the abscission process.

In a first study of the cellular and tissular events associated with abscission of oil palm fruit, Roongsattham et al. (2012) illustrated the differentiation of an abscission zone (AZ) between the mesocarp and the pedicel of the fruit and showed the involvement of polygalacturonases (PGs) and pectin depolymerization in the alterations of cell wall composition leading to effective cell separation. They also demonstrated that the sensitivity of this process to the stimulating effect of ethylene is acquired in the course of the maturation process. In a later work, the same authors suggested that inter-cellular exchange of signalling molecules between cell layers involved in AZ formation might trigger the coordinated changes in gene expression that are required in order to achieve cell separation (Roongsattham et al., 2016).

As a result of shedding or other mechanical shocks affecting the fruits over the course of transportation or processing, the activity of a lipase enzyme in the bruised mesocarp triggers the time-dependent release of Free FA (FFA). When their concentration reaches 5%, palm oil is deemed improper for human consumption and trade according to international standards. Fruit heating is used as a post-harvest treatment to inactivate the enzyme but it is unaffordable for many smallholders, especially in Africa, and it does not fully prevent oil acidification. The gene encoding the enzyme responsible for oil acidification has been mapped and characterized, paving the way for its use in marker-assisted breeding schemes (Morcillo et al., 2013). It is expected that the generalization of oil palms with a 'low-lipase' phenotype in plantations will lead to substantial gains in productivity because of their extended ripening time frame and more flexible harvesting period on one hand, and through the decreased post-harvest costs on the other hand.

6 Future trends and conclusion

The ongoing climate change will have multiple consequences on oil palm culture, depending on the region that is considered.

Over the last 65 years, droughts have increased in frequency and length in Western Africa (Spinoni et al., 2014), resulting in increased rainfall seasonality (Feng et al., 2013). Climate models predict that drought intensity in Africa will further increase in the future (Li et al., 2009). Meanwhile, in South East Asia, increased male and hermaphrodite flower production has been recorded in plantations, in conjunction with erratic rainfalls, leading to relative drought events in certain areas, whereas other regions experience increased precipitations and flooding. Studies spanning the last decades show that drought duration and seasonality tend to increase with time in this area (Feng et al., 2013; Li et al., 2009; Spinoni et al., 2014). The occurrence of El Niño–Southern Oscillation (ENSO) events in the Pacific Ocean has long been correlated with decreased precipitations and longer and more intense droughts, bushfires and hazes, and thus a decline in productivity for oil palm plantations in Indonesia and Malaysia (Corley, 2016). Opposite trends regarding the future severity and the frequency of ENSO can be derived from climatic models because of the overall poor predictability of this event (Collins et al., 2010). However, it seems that more recent projection efforts point towards an increased frequency of extreme El Niño as a result of climate change (Cai et al., 2014; Santoso et al., 2013). In any case, the exceptional magnitude of the 2015/2016 event (the strongest recorded in the last 20 years) makes it a chief concern in this key producing region. As for oil palm-growing countries in Latin America, the forecast is highly uncertain due to conflicting data (Corley, 2016).

The predicted rise in temperatures as a result of increased greenhouse gases (among which is CO_2) concentrations in the atmosphere is not, in itself, an issue for oil palm. Up to a point, rising CO_2 concentrations are predicted to result in increased yields due to improved photosynthetic activity (Henson, 2006). However, high temperatures can become an issue when occurring in conjunction with the relative drought conditions triggered by changes in rainfall patterns. Indeed, the drier air around the leaves associated with such conditions is associated with an increased Vapour Pressure Deficit (VPD), a parameter to which the stomates of oil palm leaves are very sensitive and which leads to their closure (Henson, 1991; Smith, 1989). Closed stomates prevent not only excessive water evaporation but also carbon dioxide uptake for photosynthesis, and extended drought periods are therefore detrimental to photosynthesis and carbohydrate biosynthesis. In many crops, it has been suggested that drought-related decreases in carbon allocation to reproductive structures such as flowers and fruits may lead to decreased pollen fertility or flower abortion and reduced grain or fruit filling (Albacete et al., 2014; Luquet et al., 2008; Pinheiro and

Chaves, 2011). Studies focusing on seasonal and inter-annual variations in phenology and production seem to indicate similar trends in oil palm. The trophic competition model between sink organs (Combres et al., 2013; Pallas et al., 2013) show that certain stages of inflorescence and fruit bunch development display higher sensitivity to limiting environmental and trophic conditions. It accounts well for inflorescence abortion under limiting carbohydrate supply at the early stage of rapid inflorescence growth (*ca.* 10 months before harvest), but it does not for earlier inflorescence abortion, nor for later bunch failure occurring after anthesis. By contrast, data presented by Cros et al. (2013) are in disagreement with some of these conclusions, highlighting the complex effects of seasonal water deficit on production and, more specifically, on inflorescence abortion. Analyses of oil palm inflorescence transcriptome have shown that the expression of genes involved in carbohydrate metabolism was central to inflorescence development (Ajambang et al., 2015) and that, conversely, carbohydrate deprivation triggered the expression of cell death-associated genes, hinting at a possible connection with abortion (Ajambang et al., 2016b). Furthermore, if the drought period overlaps with the fruit maturation process during which the synthesis and accumulation of storage compounds is supposed to take place, it might bear a significant negative impact on the final yield, as it has been shown in oil palm clones (Lamade and Setiyo, 1996). The transient storage of non-structural carbohydrates in the stem partially compensates/buffers the negative effects of small-scale changes in environment on source–sink carbon balance and ensures oil synthesis in the fruit (and further, continuous fruit production) (Legros et al., 2009c); however, this mechanism is likely insufficient in true stress conditions (Combres et al., 2013). Water stress can also be anticipated to have a more qualitative influence on the accumulation of lipids, carotenoids and other metabolic compounds in fruits: since their synthesis is, in most plants, an adaptative trait that is strongly influenced by the environment, it is likely that the nature of the produced molecules might be altered (Murphy, 2014).

Although, ideally, the expansion of the areas planted with oil palm should be avoided, the adverse effects of climate change might prompt the need for growing oil palm outside of the region that is currently the most favourable for its culture, that is, within 20° of the equator (Barcelos et al., 2015; Corley and Tinker, 2016). Also, as Murphy (2014) notes, if oil palm is to provide an increasing share of the world's consumption in vegetable oils in the future, it might be desirable, from a strategical, food security perspective, to deconcentrate palm oil production from South East Asia. Oil palm has a threshold temperature for growth of 15°C (Corley and Tinker, 2016) which currently puts a limit to its cultivation in higher altitudes or lower latitudes, but this might change locally due to rising temperatures and lead to the cultivation of oil palm in moderate altitude (Andean plateaus and African Highlands) or outside of the tropics, in areas that are currently under a drier, more seasonal climate such as southern China (Barcelos et al., 2015).

Lately, several groups have set up to unravelling the molecular mechanisms underlying responses to either low temperatures or drought in oil palm (Azzeme et al., 2016; Ebrahimi et al., 2016; Lei et al., 2014; Silva et al., 2015; Xiao et al., 2014). In addition to genetic determinants, as highlighted by the inter-genotype variations that have been observed, there is a considerable phenotypic plasticity in stress response, which is made visible through the contrasting behaviours displayed by plants with equivalent genetic backgrounds. Overall, these studies have shown that the response to such abiotic stresses involves the activation of signalling pathways with key transcription factors (e.g. CBFs for cold stress) triggering the expression of downstream target genes acting for the mitigation of or the adaptation to the physiological effects of the constraint (e.g. the implementation of antioxidant strategies in drought conditions). In general, it is observed that stress-sensitive and -tolerant genotypes or individuals differ by the strength and/or the precocity of this response, or by their ability to recover more promptly and with lesser long-term effects upon return to favourable growth conditions. It must, however, be observed that stress response often occurs at the expense of biomass production and growth. In the case of the breeding for drought tolerance in particular, the traits that are favourable for enhanced water use efficiency (e.g. early stomatal closure) may be partially incompatible with yield maintenance in low water availability conditions (e.g. because of altered carbon allocation to the reproductive organs) (Albacete et al., 2014; Pinheiro and Chaves, 2011). Ultimately, the goal is to identify traits (in terms of plant architecture, physiology, gene expression, etc.) that are associated with optimal production in a specific set of limited resource availability conditions. These will be thereafter combined through new, adaptation-focused breeding schemes in order to generate varieties that are either tailor-made to suit a specific agro-climatic environment or highly plastic ones that will be able to face any further change in conditions without significant yield loss. The involvement of epigenetic regulations of gene expression in the response and adaptation to most environmental constraints (Jaligot and Rival, 2015; Mirouze and Paszkowski, 2011; Richards, 2011) is abundantly documented. It has been suggested that epigenetic variations might explain the phenotypic plasticity of these responses. The inclusion of the epigenetic information 'layer' as part of the multiple 'omics' approaches, fuelling these novel breeding schemes as well as in field trial evaluation protocols, will thus be of paramount importance for the elaboration of future adapted oil palm ideotypes.

7 Where to look for further information

The following list of references will provide the curious reader with a wealth of interesting developments on the subject of oil palm. Corley and Tinker's reference book *The Oil Palm* is a most recommended starting point.

8 References

Adam, H., Collin, M., Richaud, F., Beulé, T., Cros, D., Omoré, A., Nodichao, L., Nouy, B. and Tregear, J. W., 2011. Environmental regulation of sex determination in oil palm: current knowledge and insights from other species. *Ann. Bot.*, 108, 1529–37. doi:10.1093/aob/mcr151.

Adam, H., Jouannic, S., Escoute, J., Duval, Y., Verdeil, J. L. and Tregear, J. W., 2005. Reproductive developmental complexity in the African oil palm (Elaeis guineensis, Arecaceae). *Am. J. Bot.*, 92, 1836–52.

Adam, H., Jouannic, S., Morcillo, F., Richaud, F., Duval, Y. and Tregear, J. W., 2006. MADS box genes in oil palm (Elaeis guineensis): patterns in the evolution of the SQUAMOSA, DEFICIENS, GLOBOSA, AGAMOUS, and SEPALLATA subfamilies. *J. Mol. Evol.*, 62, 15–31. doi:10.1007/s00239-005-0333-7.

Adam, H., Jouannic, S., Orieux, Y., Morcillo, F., Richaud, F., Duval, Y. and Tregear, J. W., 2007. Functional characterization of MADS box genes involved in the determination of oil palm flower structure. *J. Exp. Bot.*, 58, 1245–59. doi:10.1093/jxb/erl263

Ajambang, W., Ardie, S. W., Volkaert, H., Ngando-Ebongue, G. F. and Sudarsono, S., 2015. Comparative expression profiling of three early inflorescence stages of oil palm indicates that vegetative to reproductive phase transition of meristem is regulated by sugar balance. *Funct. Plant Biol.*, 42(6), 589–98.

Ajambang, W., Mondjeli, C., Ntsefong, G. N. and Sudarsono, S., 2016a. RNA-Seq analysis reveals influence of sugar level and photoperiod on seasonality in oil palm (Elaeis guineensis Jacq.) sex-specific inflorescence emergence. *J. Appl. Biol. Biotechnol.*, 4, 16–22.

Ajambang, W., Volkaert, H. and Sudarsono, S., 2016b. Carbohydrate deprivation upsurges the expression of genes responsible for programmed cell death in inflorescence tissues of oil palm (Elaeis guineensis Jacq.). *Turk. J. Biol.*, 40, 1320–7.

Albacete, A. A., Martínez-Andújar, C. and Pérez-Alfocea, F., 2014. Hormonal and metabolic regulation of source–sink relations under salinity and drought: From plant survival to crop yield stability. *Biotechnol. Adv.*, Plant Biotechnology 2013: "Green for Good II", 32, 12–30. doi:10.1016/j.biotechadv.2013.10.005.

Andrés, F. and Coupland, G., 2012. The genetic basis of flowering responses to seasonal cues. *Nat. Rev. Genet.*, 13, 627–39. doi:10.1038/nrg3291.

Azzeme, A. M., Abdullah, S. N. A., Aziz, M. A. and Wahab, P. E. M., 2016. Oil palm leaves and roots differ in physiological response, antioxidant enzyme activities and expression of stress-responsive genes upon exposure to drought stress. *Acta Physiol. Plant.*, 38, 1–12. doi:10.1007/s11738-016-2073-2.

Bairu, M. W., Aremu, A. O. and Van Staden, J., 2011. Somaclonal variation in plants: causes and detection methods. *Plant Growth Regul.*, 63, 147–73. doi:10.1007/s10725-010-9554-x.

Barcelos, E., Rios, S. de A., Cunha, R. N. V., Lopes, R., Motoike, S. Y., Babiychuk, E., Skirycz, A. and Kushnir, S., 2015. Oil palm natural diversity and the potential for yield improvement. *Front. Plant Sci.*, 6, 190. doi:10.3389/fpls.2015.00190

Bourgis, F., Kilaru, A., Cao, X., Ngando-Ebongue, G.-F., Drira, N., Ohlrogge, J. B. and Arondel, V., 2011. Comparative transcriptome and metabolite analysis of oil palm and date palm mesocarp that differ dramatically in carbon partitioning. *Proc. Natl. Acad. Sci.*, 108, 12527–32. doi:10.1073/pnas.1106502108

Cai, W., Borlace, S., Lengaigne, M., van Rensch, P., Collins, M., Vecchi, G., Timmermann, A., Santoso, A., McPhaden, M. J., Wu, L., England, M. H., Wang, G., Guilyardi, E. and Jin, F.-F., 2014. Increasing frequency of extreme El Nino events due to greenhouse warming. *Nat. Clim. Change*, 4, 111–16. doi:10.1038/nclimate2100

Cochard, B., Amblard, P. and Durand-Gasselin, T., 2005. Oil palm genetic improvement and sustainable development. *Ol. Corps Gras Lipides*, 12, 141–7.

Collins, M., An, S.-I., Cai, W., Ganachaud, A., Guilyardi, E., Jin, F.-F., Jochum, M., Lengaigne, M., Power, S., Timmermann, A., Vecchi, G. and Wittenberg, A., 2010. The impact of global warming on the tropical Pacific Ocean and El Niño. *Nat. Geosci.*, 3, 391–7. doi:10.1038/ngeo868

Combres, J.-C., Pallas, B., Rouan, L., Mialet-Serra, I., Caliman, J.-P., Braconnier, S., Soulié, J.-C. and Dingkuhn, M., 2013. Simulation of inflorescence dynamics in oil palm and estimation of environment-sensitive phenological phases: a model based analysis. *Funct. Plant Biol.*, 40, 263–79.

Corley, R. H. V., 1976a. Sex Differentiation in Oil Palm: Effects of Growth Regulators. *J. Exp. Bot.*, 27, 553–8. doi:10.1093/jxb/27.3.553.

Corley, R. H. V., 1976b. Inflorescence abortion and sex differentiation, in, Hardon, J. J. and Woods, B. J. (Eds.), *Developments in Crop Science*, Elsevier, Amsterdam, The Netherlands, 7–21.

Corley, R. H. V., 2009. How much palm oil do we need?. *Environ. Sci. Policy*, 12, 134–9. doi:10.1016/j.envsci.2008.10.011.

Corley, R. H. V., 2016. Climate Change – What Does it Mean for Oil Palm? Presented at the 5th International Conference on Oil Palm and Environment (ICOPE) « Sustainable Palm Oil and Climate Change: The Way Forward Through Mitigation and Adaptation », Bali, Indonesia.

Corley, R. H. V., Lee, C. H., Law, I. M. and Wong, C. Y., 1986. Abnormal flower development in oil palm clones. *Planter*, 62, 233–40.

Corley, R. H. V., Ng, M. and Donough, C. R., 1995. Effects of Defoliation on Sex Differentiation in Oil Palm Clones. *Exp. Agric.*, 31, 177–90. doi:10.1017/S0014479700025266.

Corley, R. H. V. and Tinker, P. B. H., 2016. *The Oil Palm*. Wiley-Blackwell.

Cros, D., Flori, A., Nodichao, L., Omoré, A. and Nouy, B., 2013. Differential Response to Water Balance and Bunch Load Generates Diversity of Bunch Production Profiles Among Oil Palm Crosses (Elaeis guineensis). *Trop. Plant Biol.*, 6, 26–36. doi:10.1007/s12042-013-9116-2

Durand-Gasselin, T., Baudouin, L., Cochard, B., Adon, B. and Cao, T.V., 1999. Oil palm genetic improvement strategies. *Plant. Rech. Dév.*, 6, 344–58.

Dussert, S., Guerin, C., Andersson, M., Joët, T., Tranbarger, T. J., Pizot, M., Sarah, G., Omore, A., Durand-Gasselin, T. and Morcillo, F., 2013. Comparative Transcriptome Analysis of Three Oil Palm Fruit and Seed Tissues That Differ in Oil Content and Fatty Acid Composition. *Plant Physiol.*, 162, 1337–58. doi:10.1104/pp.113.220525

Ebrahimi, M., Abdullah, S. N. A., Abdul Aziz, M. and Namasivayam, P., 2016. Oil palm EgCBF3 conferred stress tolerance in transgenic tomato plants through modulation of the ethylene signaling pathway. *J. Plant Physiol.*, 202, 107–20. doi:10.1016/j.jplph.2016.07.001

Eeuwens, C. J., Lord, S., Donough, C. R., Rao, V., Vallejo, G. and Nelson, S., 2002. Effects of tissue culture conditions during embryoid multiplication on the incidence of 'mantled' flowering in clonally propagated oil palm. *Plant Cell Tissue Organ Cult.*, 70, 311–23.

Etienne, H., Guyot, R., Beulé, T., Breitler, J.-C. and Jaligot, E., 2016. Plant Fidelity in Somatic Embryogenesis-Regenerated Plants, in, Loyola-Vargas, V. M. and Ochoa-Alejo, N. (Eds.), *Somatic Embryogenesis: Fundamental Aspects and Applications*, Springer International Publishing, 121–50. doi:10.1007/978-3-319-33705-0_8

FAO, 2015. FAOSTAT [WWW Document]. URL http://faostat3.fao.org/home/E (accessed 3.24.16).

Feng, X., Porporato, A. and Rodriguez-Iturbe, I., 2013. Changes in rainfall seasonality in the tropics. *Nat. Clim. Change*, 3, 811–15. doi:10.1038/nclimate1907

Guérin, C., Joët, T., Serret, J., Lashermes, P., Vaissayre, V., Agbessi, M. D. T., Beulé, T., Severac, D., Amblard, P., Tregear, J., Durand-Gasselin, T., Morcillo, F. and Dussert, S., 2016. Gene coexpression network analysis of oil biosynthesis in an interspecific backcross of oil palm. *Plant J.*, 87, 423–41. doi:10.1111/tpj.13208

Hartley, C. W. S., 1988. *The Oil Palm (Elaeis guineensis Jacq.)*, 3rd edition, Longman Scientific and Technical Company, Harlow, U.K.

Henson, I. E., 1991. Limitations to gas exchange, growth and yield of young oil palm by soil water supply and atmospheric humidity. *Malays. Soc. Plant Physiol.*, 2, 51–7.

Henson, I. E., 2006. *Modelling the Impact of Climatic and Climate-related Factors on Oil Palm Growth and Productivity*, Malaysian Palm Oil Board, Selangor.

Jaligot, E., Beulé, T. and Rival, A., 2002. Methylation-sensitive RFLPs: characterisation of two oil palm markers showing somaclonal variation-associated polymorphism. *Theor. Appl. Genet.*, 104, 1263–9. doi:10.1007/s00122-002-0906-4.

Jaligot, E., Beulé, T., Baurens, F. C., Billotte, N. and Rival, A., 2004. Search for methylation-sensitive amplification polymorphisms associated with the 'mantled' variant phenotype in oil palm (Elaeis guineensis Jacq.). *Genome*, 47, 224–8.

Jaligot, E. and Rival, A., 2015. Applying Epigenetics in Plant Breeding: Balancing Genome Stability and Phenotypic Plasticity, in, Al-Khayri, J. M., Jain, S. M. and Johnson, D. V. (Eds.), *Advances in Plant Breeding Strategies: Breeding, Biotechnology and Molecular Tools*. Springer International Publishing, 159–92. doi:10.1007/978-3-319-22521-0_6.

Jaligot, E., Hooi, W. Y., Debladis, E., Richaud, F., Beulé, T., Collin, M., Agbessi, M. D. T., Sabot, F., Garsmeur, O., D'Hont, A., Alwee, S. S. R. S. and Rival, A., 2014. DNA Methylation and Expression of the EgDEF1 Gene and Neighboring Retrotransposons in mantled Somaclonal Variants of Oil Palm. *PLoS ONE*, 9, e91896. doi:10.1371/journal.pone.0091896.

Jaligot, E., Rival, A., Beulé, T., Dussert, S. and Verdeil, J.L., 2000. Somaclonal variation in oil palm (Elaeis guineensis Jacq.): the DNA methylation hypothesis. *Plant Cell Rep.*, 19, 684–90.

Kaeppler, S. M., Kaeppler, H. F. and Rhee, Y., 2000. Epigenetic aspects of somaclonal variation in plants. *Plant Mol. Biol.*, 43, 179–88.

Kaeppler, S. M. and Phillips, R. L., 1993. DNA methylation and tissue culture-induced variation in plants. *Vitro Cell. Dev. Biol. - Plant*, 29, 125–30. doi:10.1007/BF02632283.

Koh, L. P. and Wilcove, D. S., 2007. Cashing in palm oil for conservation. *Nature*, 448, 993–4.

Kudoh, H., 2015. Molecular phenology in plants: in natura systems biology for the comprehensive understanding of seasonal responses under natural environments. *New Phytol.*, 210, 399–412. doi:10.1111/nph.13733

Lamade, E. and Setiyo, E., 1996. Variation in maximum photosynthesis of oil palm in Indonesia: comparison of three morphologically constrasting clones. *Plant. Rech. Dév.*, 3, 429–38.

Lamade, E., Tcherkez, G., Darlan, N. H., Rodrigues, R. L., Fresneau, C., Mauve, C., Lamothe-Sibold, M., Sketriené, D. and Ghashghaie, J., 2016. Natural 13C distribution in oil palm (Elaeis guineensis Jacq.) and consequences for allocation pattern. *Plant Cell Environ.*, 39, 199–212. doi:10.1111/pce.12606

Legros, S., Mialet-Serra, I., Caliman, J.-P., Siregar, F. A., Clément-Vidal, A. and Dingkuhn, M., 2009a. Phenology and growth adjustments of oil palm (Elaeis guineensis) to photoperiod and climate variability. *Ann. Bot.*, 104, 1171–82. doi:10.1093/aob/mcp214

Legros, S., Mialet-Serra, I., Caliman, J.-P., Siregar, F. A., Clement-Vidal, A., Fabre, D. and Dingkuhn, M., 2009b. Phenology, growth and physiological adjustments of oil palm (Elaeis guineensis) to sink limitation induced by fruit pruning. *Ann. Bot.*, 104, 1183–94. doi:10.1093/aob/mcp216

Legros, S., Mialet-Serra, I., Clement-Vidal, A., Caliman, J.-P., Siregar, F. A., Fabre, D. and Dingkuhn, M., 2009c. Role of transitory carbon reserves during adjustment to climate variability and source–sink imbalances in oil palm (Elaeis guineensis). *Tree Physiol.*, 29, 1199–211. doi:10.1093/treephys/tpp057

Lei, X., Xiao, Y., Xia, W., Mason, A. S., Yang, Y., Ma, Z. and Peng, M., 2014. RNA-Seq Analysis of Oil Palm under Cold Stress Reveals a Different C-Repeat Binding Factor (CBF) Mediated Gene Expression Pattern in Elaeis guineensis Compared to Other Species. *PLoS ONE*, 9, e114482. doi:10.1371/journal.pone.0114482

Li, Y., Ye, W., Wang, M. and Yan, X., 2009. Climate change and drought: a risk assessment of crop-yield impacts. *Clim. Res.*, 39, 31–46.

Liu, Z. and Mara, C., 2010. Regulatory mechanisms for floral homeotic gene expression. Semin. *Cell Dev. Biol.*, 21, 80–6. doi:10.1016/j.semcdb.2009.11.012

Luquet, D., Clément-Vidal, A., Fabre, D., This, D., Sonderegger, N. and Dingkuhn, M., 2008. Orchestration of transpiration, growth and carbohydrate dynamics in rice during a dry-down cycle. *Funct. Plant Biol.*, 35, 689. doi:10.1071/FP08027

Masmoudi-Allouche, F., Châari-Rkhis, A., Kriaâ, W., Gargouri-Bouzid, R., Jain, S. M. and Drira, N., 2009. In vitro hermaphrodism induction in date palm female flower. *Plant Cell Rep.*, 28, 1–10. doi:10.1007/s00299-008-0611-0

Matthes, M., Singh, R., Cheah, S. C. and Karp, A., 2001. Variation in oil palm (Elaeis guineensis Jacq.) tissue culture-derived regenerants revealed by AFLPs with methylation-sensitive enzymes. *Theor. Appl. Genet.*, 102, 971–9.

Miguel, C. and Marum, L., 2011. An epigenetic view of plant cells cultured in vitro: somaclonal variation and beyond. *J. Exp. Bot.*, 62, 3713–25. doi:10.1093/jxb/err155

Mirouze, M. and Paszkowski, J., 2011. Epigenetic contribution to stress adaptation in plants. *Curr. Opin. Plant Biol.*, 14, 267–74. doi:16/j.pbi.2011.03.004

Morcillo, F., Cros, D., Billotte, N., Ngando-Ebongue, G.-F., Domonhédo, H., Pizot, M., Cuéllar, T., Espéout, S., Dhouib, R., Bourgis, F., Claverol, S., Tranbarger, T. J., Nouy, B. and Arondel, V., 2013. Improving palm oil quality through identification and mapping of the lipase gene causing oil deterioration. *Nat. Commun.*, 4, 2160. doi:10.1038/ncomms3160

Murphy, D. J., 2014. The future of oil palm as a major global crop: opportunities and challenges. *J. Oil Palm Res.*, 26, 1–24.

Ó'Maoiléidigh, D. S., Graciet, E. and Wellmer, F., 2014. Gene networks controlling Arabidopsis thaliana flower development. *New Phytol.*, 201, 16–30. doi:10.1111/nph.12444

Ong-Abdullah, M., Ordway, J. M., Jiang, N., Ooi, S.-E., Kok, S.-Y., Sarpan, N., Azimi, N., Hashim, A. T., Ishak, Z., Rosli, S. K., Malike, F. A., Bakar, N. A. A., Marjuni, M., Abdullah, N., Yaakub, Z., Amiruddin, M. D., Nookiah, R., Singh, R., Low, E.-T. L., Chan, K.-L., Azizi, N., Smith, S. W., Bacher, B., Budiman, M. A., Van Brunt, A., Wischmeyer, C., Beil, M., Hogan, M., Lakey, N., Lim,

C.-C., Arulandoo, X., Wong, C.-K., Choo, C.-N., Wong, W.-C., Kwan, Y.-Y., Alwee, S. S. R. S., Sambanthamurthi, R. and Martienssen, R. A., 2015. Loss of Karma transposon methylation underlies the mantled somaclonal variant of oil palm. *Nature*, 525, 533–7. doi:10.1038/nature15365

Pallas, B., Mialet-Serra, I., Rouan, L., Clément-Vidal, A., Caliman, J.-P. and Dingkuhn, M., 2013. Effect of source/sink ratios on yield components, growth dynamics and structural characteristics of oil palm (Elaeis guineensis) bunches. *Tree Physiol.*, 33, 409–24. doi:10.1093/treephys/tpt015

Parveez, G. K. A., Rasid, O. A., Masani, M. Y. A. and Sambanthamurthi, R., 2014. Biotechnology of oil palm: strategies towards manipulation of lipid content and composition. *Plant Cell Rep.*, 34, 533–43. doi:10.1007/s00299-014-1722-4

Pinheiro, C. and Chaves, M. M., 2011. Photosynthesis and drought: can we make metabolic connections from available data? *J. Exp. Bot.*, 62, 869–82. doi:10.1093/jxb/erq340

Rai, M. K., Kalia, R. K., Singh, R., Gangola, M. P. and Dhawan, A. K., 2011. Developing stress tolerant plants through in vitro selection—An overview of the recent progress. *Environ. Exp. Bot.*, 71, 89–98. doi:10.1016/j.envexpbot.2010.10.021

Richards, E. J., 2011. Natural epigenetic variation in plant species: a view from the field. *Curr. Opin. Plant Biol.*, 14, 204–9. doi:10.1016/j.pbi.2011.03.009

Rival, A., et al., 2000. Somatic embryogenesis in oil palm, in, Jain, S. M., Gupta, P. K. and Newton, R. J. (Eds.), *Somatic Embryogenesis in Woody Plants*, 249–90. Kluwer Academic Publishers, Dordrecht, The Netherlands.

Rival, A. and Levang, P., 2014. *Palms of Controversies: Oil Palm and Development Challenges*, CIFOR.

Roongsattham, P., Morcillo, F., Fooyontphanich, K., Jantasuriyarat, C., Tragoonrung, S., Amblard, P., Collin, M., Mouille, G., Verdeil, J.-L. and Tranbarger, T. J., 2016. Cellular and Pectin Dynamics during Abscission Zone Development and Ripe Fruit Abscission of the Monocot Oil Palm. *Front. Plant Sci.*, 7, 540. doi:10.3389/fpls.2016.00540

Roongsattham, P., Morcillo, F., Jantasuriyarat, C., Pizot, M., Moussu, S., Jayaweera, D., Collin, M., Gonzalez-Carranza, Z., Amblard, P., Tregear, J. W., Tragoonrung, S., Verdeil, J.-L. and Tranbarger, T. J., 2012. Temporal and spatial expression of polygalacturonase gene family members reveals divergent regulation during fleshy fruit ripening and abscission in the monocot species oil palm. *BMC Plant Biol.*, 12, 150. doi:10.1186/1471-2229-12-150

Santoso, A., McGregor, S., Jin, F.-F., Cai, W., England, M. H., An, S.-I., McPhaden, M. J. and Guilyardi, E., 2013. Late-twentieth-century emergence of the El Nino propagation asymmetry and future projections. *Nature*, 504, 126–30. doi:10.1038/nature12683

Shrestha, R., Gómez-Ariza, J., Brambilla, V. and Fornara, F., 2014. Molecular control of seasonal flowering in rice, arabidopsis and temperate cereals. *Ann. Bot.*, 114, 1445–58. doi:10.1093/aob/mcu032

Silva, P. A., Oliveira, I. V., Rodrigues, K. C. B., Cosme, V. S., Bastos, A. J. R., Detmann, K. S. C., Cunha, R. L., Festucci-Buselli, R. A., DaMatta, F. M. and Pinheiro, H. A., 2015. Leaf gas exchange and multiple enzymatic and non-enzymatic antioxidant strategies related to drought tolerance in two oil palm hybrids. *Trees*, 30, 203–14. doi:10.1007/s00468-015-1289-x

Smith, B. G., 1989. The Effects of Soil Water and Atmospheric Vapour Pressure Deficit on Stomatal Behaviour and Photosynthesis in the Oil Palm. *J. Exp. Bot.*, 40, 647–51. doi:10.1093/jxb/40.6.647

Somyong, S., Poopear, S., Jomchai, N., Uthaipaisanwong, P., Ruang-areerate, P., Sangsrakru, D., Sonthirod, C., Ukoskit, K., Tragoonrung, S. and Tangphatsornruang, S., 2014. The AKR gene family and modifying sex ratios in palms through abiotic stress responsiveness. *Funct. Integr. Genomics*, 15, 349–62. doi:10.1007/s10142-014-0423-y

Somyong, S., Poopear, S., Sunner, S. K., Wanlayaporn, K., Jomchai, N., Yoocha, T., Ukoskit, K., Tangphatsornruang, S. and Tragoonrung, S., 2016. ACC oxidase and miRNA 159a, and their involvement in fresh fruit bunch yield (FFB) via sex ratio determination in oil palm. *Mol. Genet. Genomics*, 291, 1243–57. doi:10.1007/s00438-016-1181-4

Spinoni, J., Naumann, G., Carrao, H., Barbosa, P. and Vogt, J., 2014. World drought frequency, duration, and severity for 1951–2010. *Int. J. Climatol.*, 34, 2792–804.

Teh, H. F., Neoh, B. K., Hong, M. P. L., Low, J. Y. S., Ng, T. L. M., Ithnin, N., Thang, Y. M., Mohamed, M., Chew, F. T., Yusof, H. M., Kulaveerasingam, H. and Appleton, D. R., 2013. Differential Metabolite Profiles during Fruit Development in High-Yielding Oil Palm Mesocarp. *PLoS ONE*, 8, e61344. doi:10.1371/journal.pone.0061344

Tranbarger, T. J., Dussert, S., Joët, T., Argout, X., Summo, M., Champion, A., Cros, D., Omore, A., Nouy, B. and Morcillo, F., 2011. Regulatory Mechanisms Underlying Oil Palm Fruit Mesocarp Maturation, Ripening, and Functional Specialization in Lipid and Carotenoid Metabolism. *Plant Physiol.*, 156, 564–84. doi:10.1104/pp.111.175141

Us-Camas, R., Rivera-Solís, G., Duarte-Aké, F. and De-la-Peña, C., 2014. In vitro culture: an epigenetic challenge for plants. *Plant Cell Tissue Organ Cult. PCTOC*, 118, 187–201. doi:10.1007/s11240-014-0482-8

Wang, Q.-M. and Wang, L., 2012. An evolutionary view of plant tissue culture: somaclonal variation and selection. *Plant Cell Rep.*, 31, 1535–47. doi:10.1007/s00299-012-1281-5

Wellmer, F. and Riechmann, J. L., 2010. Gene networks controlling the initiation of flower development. *Trends Genet.*, 26, 519–27. doi:10.1016/j.tig.2010.09.001

Wich, S. A., Garcia-Ulloa, J., Kühl, H. S., Humle, T., Lee, J. S. H. and Koh, L. P., 2014. Will Oil Palm's Homecoming Spell Doom for Africa's Great Apes?. *Curr. Biol.*, 24, 1659–63. doi:10.1016/j.cub.2014.05.077

Xiao, Y., Zhou, L., Xia, W., Mason, A. S., Yang, Y., Ma, Z. and Peng, M., 2014. Exploiting transcriptome data for the development and characterization of gene-based SSR markers related to cold tolerance in oil palm (Elaeis guineensis). *BMC Plant Biol.*, 14, 384. doi:10.1186/s12870-014-0384-2

Chapter 6

Diversity in the genetic resources of oil palm

N. Rajanaidu, A. Mohd Din, M. Marhalil, A. Norziha, O. A. Meilina, A. M. Fadila, A. B. Nor Azwani, L. Adelina, H. Zulkifli, S. Wan Salmiah and A. Kushairi, Malaysian Palm Oil Board, Malaysia

1 Introduction

The oil palm is a monocotyledonous plant classified in the Arecaceae family, formerly known as *Palmae*. There are two major species of oil palm, namely *Elaeis guineensis* and *Elaeis oleifera*. *Elaeis guineensis and Elaeis oleifera* are native to West Africa and South America, respectively (Corley and Tinker, 2003). Extensive natural or semi-wild varieties of *Elaeis guineensis* (Fig. 1) are found along the west coast of Africa from Senegal to Angola (Zeven, 1967), whereas *Elaeis oleifera* (Fig. 2) is distributed in Honduras, Nicaragua, Costa Rica, Panama, Colombia, Venezuela, Suriname, Ecuador, Brazil and Peru (Meunier, 1975; Ooi et al., 1981; Escobar, 1982; Rajanaidu, 1983; Julian Barba, 2012). These two species are able to hybridize to produce fertile fruits. Hybrids are short with desirable oil quality and fresh fruit bunch (FFB) yield.

This chapter discusses the rather narrow genetic base of current breeding materials and the work undertaken by the Malaysian Palm Oil Board (MPOB) with the aim of broadening

http://dx.doi.org/10.19103/AS.2017.0018.05

Figure 1 African *Elaeis guineensis*.

Figure 2 American *Elaeis oleifera*.

this genetic base. It looks at ways of assessing genetic diversity in oil palm, through the analysis of fruit forms and types, morphological traits and genetic markers. It also reviews methods for the conservation of oil palm collections and ways of utilizing germ plasm in order to develop improved varieties.

2 The genetic base of current breeding materials

Oil palm planting materials are based on an extremely narrow genetic base. The discovery of the genetic inheritance of shell thickness by Beirnaert and Vanderweyen (1941) formed the foundation for the production of *dura* × *pisifera* (DxP) planting material. It has been well documented that female sources are restricted to four Deli *dura* palms planted in Bogor Botanical Gardens, Indonesia, in 1848 (Hartley, 1988). The Deli *dura* population is the backbone of every oil palm breeding programme developed in the world. The genetic base of *pisiferas* is also limited. Both Deli *dura* and AVROS *pisifera* were imported to Malaysia via Indonesia. To date, the DxP hybrid known as *tenera* is the main planting material used and has resulted in a significant increase in oil palm yields. The selection of the thin-shelled *teneras* is due to the high oil-bearing content of the mesocarp (75–85%) compared to *duras* (20–65%) (Kushairi and Rajanaidu, 2000). Other important sources of *pisiferas* are the Djongo palm, Yangambi, Calabar, Ekona, La Me, URT and Pure SP540 (Fig. 3).

It has been generally recognized that the narrowness of the gene pool had been a major obstacle towards increasing yields in many crops including oil palm. The need to broaden the genetic base for continuous yield improvement has been recognized by the oil palm industry worldwide. This concern provided the initial impetus for the MPOB to carry out several expeditions to the various centres of diversity for oil palm in West Africa and South America (Rajanaidu et al., 2000). From 1973 to 2010, the MPOB team collected *Elaeis guineensis* germ plasm from the centres of origin in 11 countries, viz. Nigeria, Cameroon,

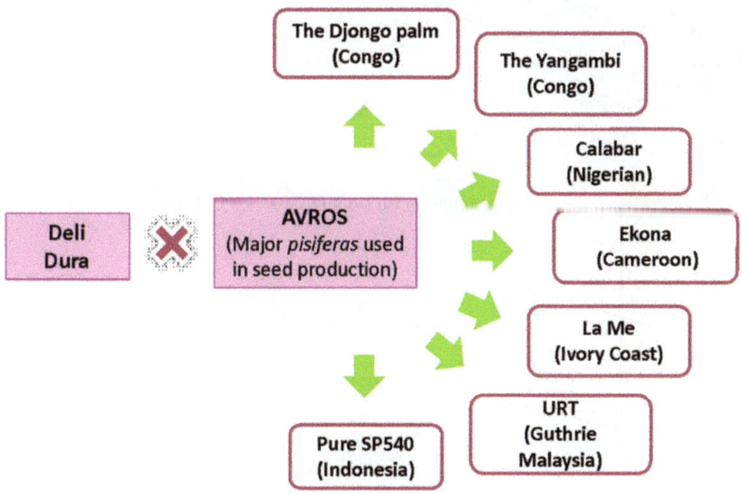

Figure 3 Major breeding materials in the seed production.

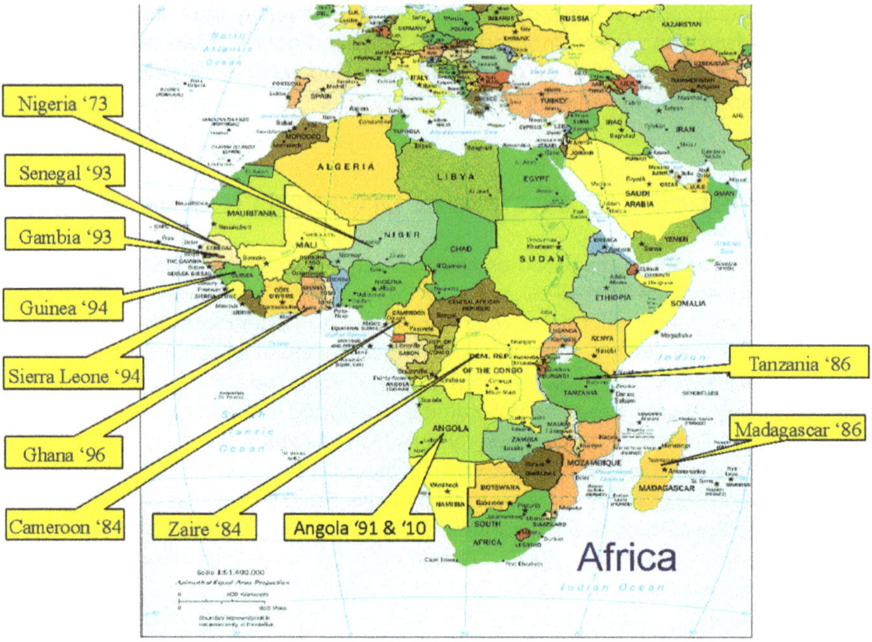

Figure 4 Germ plasm collection (*Elaeis guineensis*) sites in Africa.

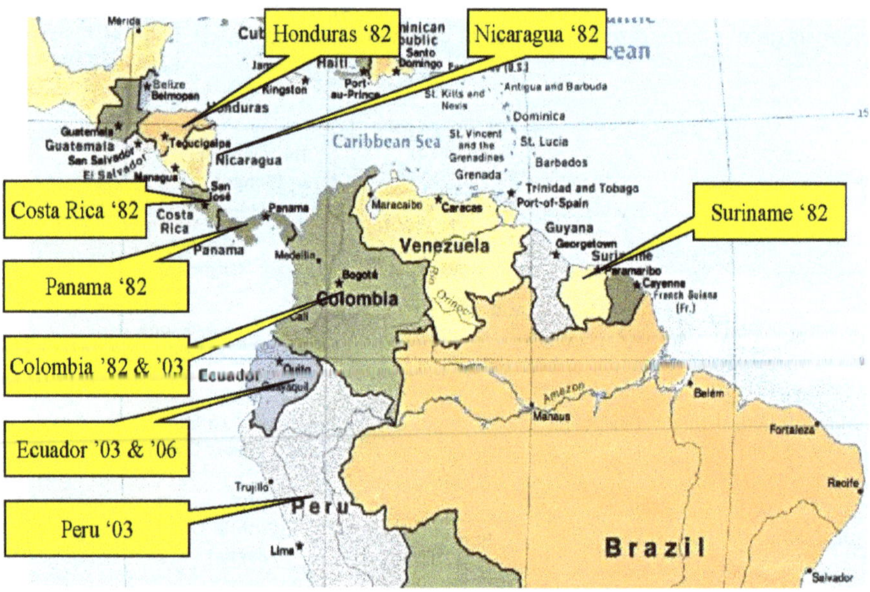

Figure 5 Germ plasm collection (*Elaeis oleifera*) sites in South America.

Zaire, Tanzania, Madagascar, Angola, Senegal, Gambia, Sierra Leone, Guinea and Ghana (Fig. 4). In the case of *Elaeis oleifera*, the collection effort covered eight countries, namely Honduras, Panama, Costa Rica, Nicaragua, Suriname, Ecuador, Peru and Colombia (Fig. 5). In addition, palms of economic interest such as *Bactris*, *Jessenia*, *Oneocarpus*, *Euterpe* and *Babassu* were also sampled. Introducing these genetic materials will allow new traits of economic interest to be discovered and subsequently introgressed into current breeding materials in order to broaden the genetic base. Table 1 summarizes the germ plasm collection activities developed at MPOB.

Table 1 List of germ plasm collections (Rajanaidu, 1994)

Country	Year	No. of accessions	Collectors
Elaeis guineensis			
Nigeria	1973	919	Rajanaidu, Arasu and Obasola
Cameroon	1984	95	Rajanaidu and Unilever
Zaire	1984	369	Rajanaidu and Unilever
Tanzania	1986	60	Rajanaidu and Ministry of Agriculture Officials, Tanzania
Madagascar	1986	17	Rajanaidu and Ministry of Agriculture Officials, Madagascar
Angola	1991, 2010	54, 127	Rajanaidu, Jalani Sukaimi, Mohd Din Amiruddin, Marhalil Marjuni and Ministry of Agriculture Officials, Angola
Senegal	1993	104	Rajanaidu, Jalani Sukaimi and Ministry of Agriculture Officials, Senegal
Gambia	1993	45	Rajanaidu, Jalaini Sukaimi and Ministry of Agriculture Officials, Gambia
Sierra Leone	1994	56	Rajanaidu, Ahmad Kushairi Din and Ministry of Agriculture Officials, Sierra Leone
Guinea	1994	61	Rajanaidu, Ahmad Kushairi Din and Ministry of Agriculture Offcials, Guinea
Ghana	1996	58	Rajanaidu, Mohd Rafii Yusof and Ministry of Agriculture Offcials, Ghana
Elaeis oleifera			
Honduras	1982	14	Rajanaidu
Nicaragua	1982	18	Rajanaidu
Costa Rica	1982	61	Rajanaidu
Panama	1982	27	Rajanaidu
Colombia	1982	41	Rajanaidu
Suriname	1982	6	Rajanaidu
Peru	2004	4	Rajanaidu and Ahmad Kushairi Din
Ecuador	2004/06	10	Rajanaidu, Ahmad Kushairi Din and Noh Ahmad
Jessenia, Oenocarpus, Bactris, Euterpe			
Peru	1981	50	Rajanaidu
Colombia	1989	182	Rajanaidu

3 Genetic diversity in oil palm: fruit forms

Genetic diversity in oil palm will be considered with reference to its distribution, fruit forms and fruit types, agronomic and morphological characters as well as using molecular markers. The phenotypic expression of traits in oil palm depends on around 32 000 genes, as estimated when sequencing the oil palm genome (Singh et al., 2013).

The oil palm fruit is a drupe, which consists of mesocarp, shell and kernel. A cross section of the fruit reveals the three fruit forms of the oil palm based on the presence or absence of shell: *dura*, *pisifera* and *tenera* (Fig. 6). *Dura* is homozygous-dominant ($sh^+ sh^+$) with a thick shell. *Pisifera* ($sh^- sh^-$) is a homozygous-recessive allele with no shell. As mentioned earlier, the cross between *dura* and *pisifera* produces a heterozygous hybrid, *tenera* ($sh^+ sh^-$), with a thin shell surrounded by a fibre ring (Beirnaert and Vanderweyen, 1941).

It is of paramount interest to explore genetic diversity in the natural oil palm groves of Africa in order to identify new economically important traits for further breeding. Recent oil palm collections provide information on the actual diversity of fruit forms that exist in such natural oil palm groves (Bakoumé, 2016). Based on MPOB oil palm germ plasm collection, the greatest proportion of fruit types in natural oil palm groves in Africa was *duras* (73.32%) followed by *teneras* (26.63%). Only one *pisifera* (0.05%) was found during collection expeditions in Sierra Leone (Table 2). Indeed, *pisifera* is considered rare by local communities and its scarcity has been attributed to the preference by local people for the other two varieties. Farmers tend to keep the *tenera* bunches for home consumption while selling the heavier *dura bunches* at markets because growers are paid by bunch weight (Rajanaidu, 1986). The frequency of dominant allele 'D' is higher (86.63%) in all the collections in Africa.

The data in Table 2 are further illustrated in Fig. 7 to show the frequency and distribution of dominant allele 'D' and recessive allele 'd' in various countries.

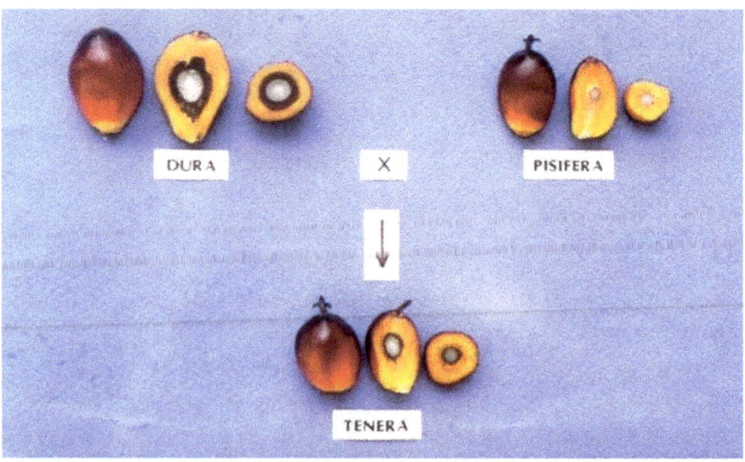

Figure 6 Characteristics of fruit forms (Source: MPOB).

Table 2 Fruit form frequency in natural oil palm populations

Year	Country	No of accessions	Genotype (%)			Frequency of alleles	
			Dura	Tenera	Pisifera	Allele 'D'	Allele 'd'
1973	Nigeria	919	595 (64.74)	324 (35.26)	0 (0.00)	1514 (0.82)	324 (0.18)
1984	Cameroon	95	58 (61.05)	37 (38.95)	0 (0.00)	153 (0.81)	37 (0.19)
1984	Zaire	369	283 (76.69)	86 (23.31)	0 (0.00)	652 (0.88)	86 (0.12)
1986	Tanzania	59	42 (71.19)	17 (28.81)	0 (0.00)	101 (0.86)	17 (0.14)
1986	Madagascar	17	17 (100.00)	0 (0.00)	0 (0.00)	34 (1.00)	0 (0.00)
1991, 2000	Angola	181	137 (75.69)	44 (24.31)	0 (0.00)	318 (0.88)	44 (0.12)
1993	Senegal	104	104 (100.00)	0 (0.00)	0 (0.00)	208 (1.00)	0 (0.00)
1993	Gambia	45	45 (100.00)	0 (0.00)	0 (0.00)	90 (1.00)	0 (0.00)
1994	Sierra Leone	56	52 (92.86)	3 (5.36)	1 (1.79)	107 (0.96)	5 (0.04)
1994	Guinea	61	58 (95.08)	3 (4.92)	0 (0.00)	119 (0.98)	3 (0.02)
1996	Ghana	58	49 (84.48)	9 (15.52)	0 (0.00)	107 (0.92)	9 (0.08)
Total		**1964**	**1440**	**523**	**1**	**3403**	**525**
(%)		**100**	**73.32**	**26.63**	**0.05**	**86.63**	**13.37**

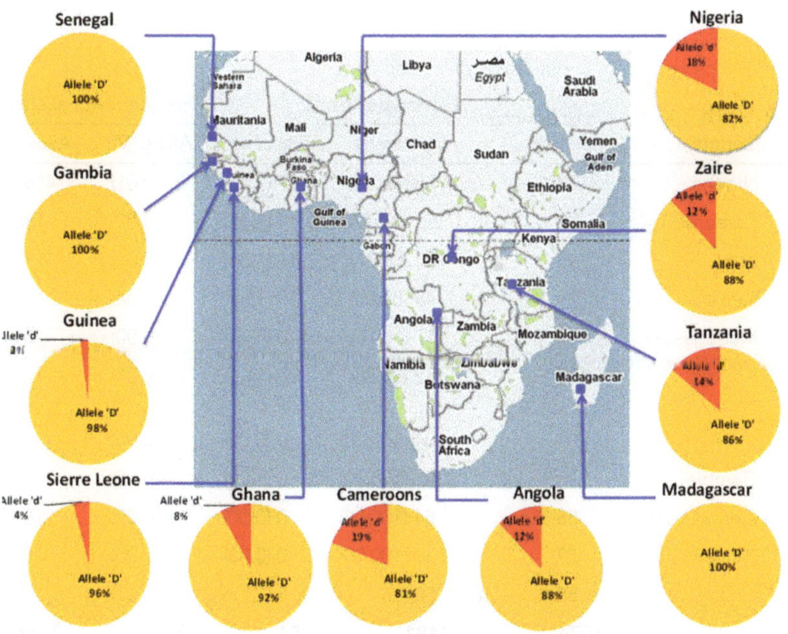

Figure 7 The frequency distribution of 'D' and 'd' alleles in various germ plasm countries.

4 Genetic diversity in oil palm: fruit types

There are two types of fruits according to external exocarp colour, namely *nigrescens* and *virescens*. It appears that *virescens* is controlled by a single dominant gene. The homozygous-dominant and heterozygous hybrids express the *virescens* phenotype. Based on the MPOB germ plasm collection, the frequency of *virescens* in natural oil palm groves is extremely low even though the gene is dominant (Table 3). The percentage of *virescens* was found to be 2.63% in natural oil palm groves, whereas the *nigrescens* proportion reached 97.34%. It is hypothesized that *virescens* is a recent mutation from NIGRESCENS (recessive) to VIRESCENS (dominant). The highest frequency of *virescens* has been found in Tanzania (16.95%) followed by Angola (14.12%) (Table 3). Hartley (1988) found a 0.7% frequency for *virescens* in Angola. Rajanaidu (1986) found a high incidence of *virescens* palms at certain sites in Cameroon (36%) and Democratic Republic of Congo (50%). Hartley (1988) found an open-pollinated *virescens* bunch in Nigeria with 46% *virescens* and 54% *nigrescens.*

Studies of phylogenetic characteristics and the transcriptome of *virescens* and *nigrescens* suggest that the *virescens* gene controls exocarp colour by regulating the expression of genes in the anthocyanin biosynthetic pathway (Singh et al., 2014). It is perhaps not surprising that the gene might be found under the heterozygous condition since the frequency of the recessive allele in natural oil palm germ plasm is significantly higher (98.69%) compared to the dominant allele (1.31%). Based on the MPOB germ plasm collections, 3 out of 5 independent mutant alleles of the VIR gene in oil palm are present in material from Angola and Democratic Republic of the Congo (Singh et al., 2014). The identification of alleles

Table 3 Fruit type frequency in natural oil palm population

Year	Country	No of bunches observed	Genotype (%)		Frequency of alleles	
			Nigrescens	*Virescens*	Allele 'V'	Allele 'v'
1973	Nigeria	2112	2075 (98.25)	37 (1.75)	37 (0.01)	4187 (0.99)
1984	Cameroon	95	91 (95.79)	4 (4.21)	4 (0.02)	186 (0.98)
1984	Zaire	795	785 (98.74)	10 (1.26)	10 (0.01)	1580 (0.99)
1986	Tanzania	59	49 (83.05)	10 (16.95)	10 (0.08)	108 (0.92)
1986	Madagascar	17	17 (100.00)	0 (0.00)	0 (0.00)	34 (1.00)
1991, 2000	Angola	177	152 (85.88)	25 (14.12)	25 (0.07)	329 (0.93)
1993	Senegal	104	104 (100.00)	0 (0.00)	0 (0.00)	208 (1.00)
1993	Gambia	45	45 (100.00)	0 (0.00)	0 (0.00)	90 (1.00)
1994	Sierra Leone	55	54 (100.00)	0 (0.00)	0 (0.00)	108 (1.00)
1994	Guinea	59	59 (100.00)	0 (0.00)	0 (0.00)	118 (1.00)
1996	Ghana	58	50 (86.21)	8 (13.79)	8 (0.07)	108 (0.93)
Total		**3576**	**3481**	**94**	**94**	**7056**
Mean (%)		**100**	**97.37**	**2.63**	**1.31**	**98.69**

responsible for the *virescens* genotype allows breeders to develop paternal (*pisifera*) lines that are homozygous *virescens* when crossed with homozygous Deli *duras* to produce a 100% *virescens* phenotype for future commercial planting. Planters prefer to plant *virescens* because they find it hard to identify when *nigrescens* bunches move from being unripe (black) to ripe (dark red). In contrast, *virescens* are green when unripe but turn to orange when ripe, making it easy to identify ripe bunches, even in tall palms.

The data shown in Table 3 can be compared to Fig. 8 which shows the frequency and distribution of the dominant allele 'V' and recessive allele 'v' in various countries. We further analysed the frequency of *virescens* in *dura* and *tenera* populations where most of the *dura* and *tenera* palms are *nigrescens* (93.88% and 89.92%, respectively) (Table 4). The mean frequency of *virescens* in *dura* and *tenera* palms is extremely low, at 6.12% and 10.08%, respectively.

4.1 *Albescens* (mescocarp colour)

The *albescens* fruit, which is characterized by a very low carotene level in the mesocarp, has been found in Sierra Leone (1.79%). Corley and Tinker (2016) highlighted the presence of *albescens* palms in Ghana, Democratic Republic of Congo, Angola, Nigeria and Cote d'Ivoire, but it is very rare. Rajanaidu (1986) encountered one *albescens* palm at the Yangambi Research Station, Democratic Republic of Congo (Fig. 9).

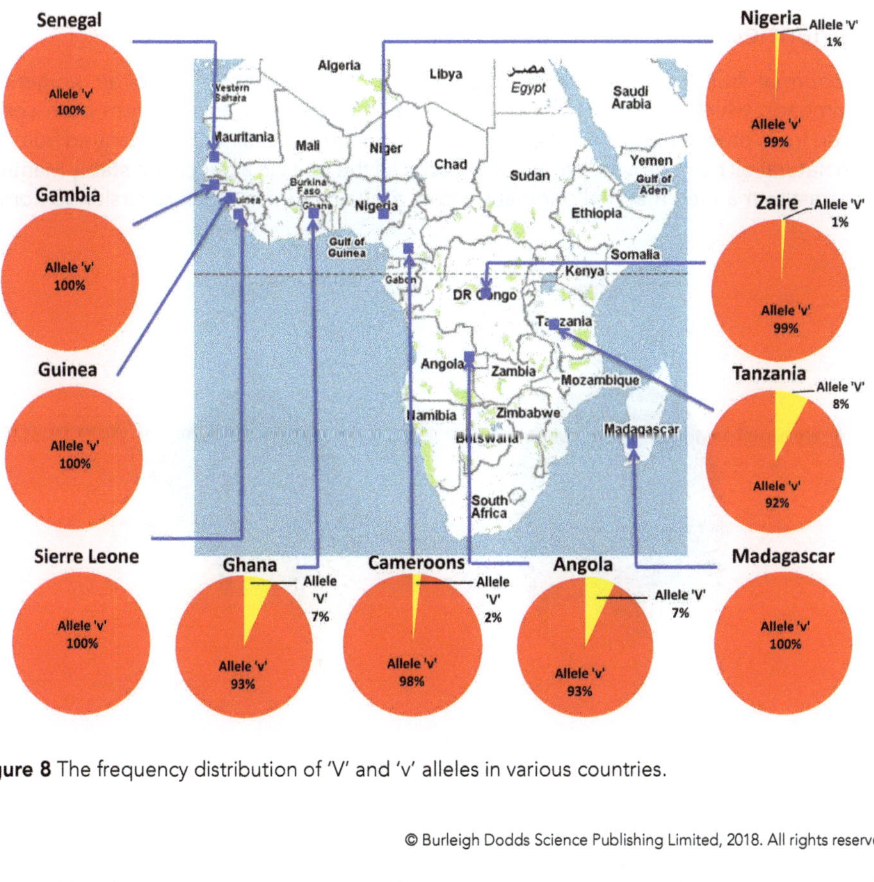

Figure 8 The frequency distribution of 'V' and 'v' alleles in various countries.

Table 4 Fruit type frequency in *dura* and *tenera*

	Fruit type			
	Dura		Tenera	
Country	*Nigrescens*	*Virescens*	*Nigrescens*	*Virescens*
Cameroon	50	3	41	1
Tanzania	36	6	13	4
Madagascar	17	0	0	0
Angola	119	18	38	6
Senegal	104	0	0	0
Gambia	45	0	0	0
Sierra Leone	52	0	3	0
Guinea	58	0	3	0
Ghana	41	7	9	1
TOTAL	522	34	107	12
MEAN (%)	**93.88**	**6.12**	**89.92**	**10.08**

4.2 Mantled palm

An abnormal fruit known as mantled is a somaclonal variant due to an epigenetic phenomenon which causes a malformation of floral organs in flowers from both sexes. Mantled fruits produce little or no oil in the bunch (Shearman et al., 2013). Mantled fruits are rare in natural oil palm groves. Where they occur, all the fruits in a bunch are stably mantled, in contrast to mantled material originating from tissue culture in which reversion to normal

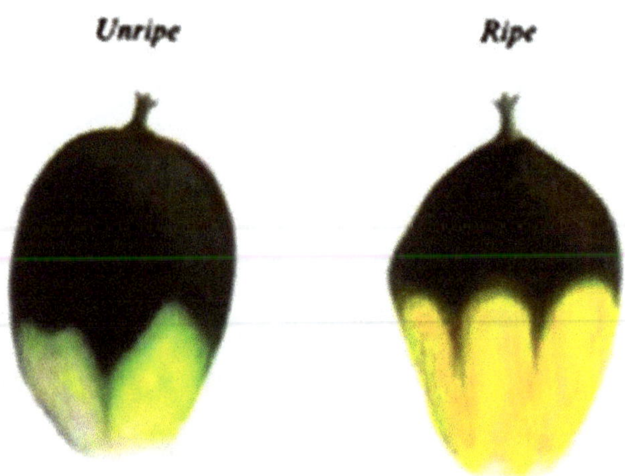

Figure 9 *Albescens* fruit (Reproduced from Hartley (1988)).

Figure 10 (a) Normal fruit. (b) Cross section of normal fruit. (c) Mantled fruit. (d) Cross section of mantled fruit.

phenotype can occur and phenotypes of fruits can be diverse on the same bunch (Jaligot et al., 2000). Rajanaidu (1986) observed a natural mantled palm in Zaire and Nigeria (at site 10), while Zeven (1973) found mantled fruit constituting about 0.1% of bunches harvested in Nigeria and 0.16% in Angola. Hartley (1988) found some spontaneous mantled palms in Congo. When selfed, mantled palm resulted in the 75–100% mantled progeny. When crosses were made between mantled and normal palms, 50% of the progeny was mantled (Hartley, 1988). DNA hypomethylation was found to be related to the determinism of mantled flower than fruit formation by several authors (Jaligot et al., 2000, 2014; Ong-Abdullah, 2015) (Fig. 10).

4.3 Leaflet variation (*Idolatrica* palm)

The leaflets in the *idolatrica* palm do not separate normally and they are fused. The centre of distribution of the *idolatrica* palm lies is Ghana, Benin and western Nigeria (Corley and Tinker, 2016). Rajanaidu (1986) found a number of *idolatrica* palms in the Boteka area, Democratic Republic of the Congo, and also noticed *idolatrica* palms along the Edea/ Douala Road in Cameroon (Fig. 11).

Figure 11 The *idolatrica* palm with fused leaflets (Adopted from Corley and Tinker (2016)).

5 Genetic diversity in oil palm: morphological traits

Although genetic diversity can be measured by using protein- and DNA-based markers, morphological characteristics are still successfully used as a method to assess genetic diversity in oil palm. Indeed, natural oil palm populations collected from Nigeria (1973) by MPOB showed some considerable genetic variations in morphological traits, allowing breeders to select individual palms with economically important traits (Table 5).

Genetic diversity within and among oil palm populations is measured using variance component statistics. The genetic structure of oil palm populations shows that the variation within population (σ^2w) is much higher than between population σ^2p. σ^2w varies from 56.39 to 90.82% and σ^2p from 9.18 to 43.61%. The magnitude of σ^2p and σ^2w depends on the traits. The highly heritable traits exhibit a higher level of σ^2p. This information is useful to oil palm breeders in helping them to allocate resources in the sampling of oil palm germ plasm in various countries.

Principal component analysis (PCA) and cluster analysis have been used to understand the oil palm populations collected at various sites and the influence of the environment. Data involving 41 morphological traits of *dura* palms from 11 countries (Nigeria, Cameroon, Zaire, Tanzania, Madagascar, Angola, Senegal, Gambia, Sierra Leone, Guinea and Ghana) were retrieved from the MPOB Breeding Information System and analysed using PCA and cluster analysis. In the PCA, the correlation matrix based on all traits was calculated with respective eigenvalues.

The patterns of components 1 and 2 are shown in Fig. 12. A PCA plot of PC1 and PC2 showed that oil palm populations were grouped into four distinct clusters indicating the differentiation among germ plasm material. The cluster analysis of the oil palm germ plasm

Table 5 Variation for morphological characteristics in Nigerian collection (Trial 0.151) –*Duras* (Rajanaidu and Rao, 1988)

Bunch and vegetative traits	Variance components (%)	
	Between populations (σ^2p)	Within populations (σ^2w)
Bunch characteristics		
Fruit-to-bunch ratio (F/B) (%)	9.18	90.82
Oil-to-bunch ratio (O/B) (%)	27.96	72.04
Kernel-to-fruit ratio (K/F) (%)	36.00	64.00
Mean fruit weight (MFW) (g)	30.10	69.90
Mesocarp-to-fruit ratio (M/F) (%)	32.57	67.43
Vegetative characteristics		
Rachis length (RL) (m)	43.61	56.39
Leaflet number (LN)	23.56	76.44
Height/yr (H/yr) (m)	34.78	65.22
Harvest Index (HI)	17.67	82.33
Leaf area (LA)	37.91	62.09
Fresh fruit bunch (FFB) (kg/palm/yr)	20.00	80.00
Bunch no.	14.59	85.41
A bunch wt	31.81	68.19
Frond production (FP)	33.72	66.28

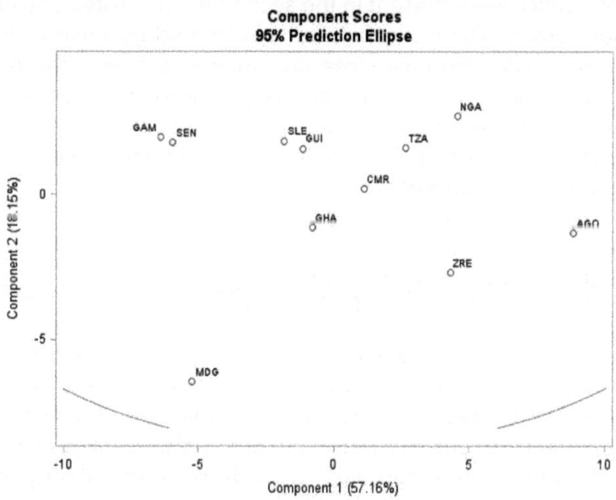

Figure 12 PCA scores extracted from PC1 and PC2 based on 95% prediction ellipse.

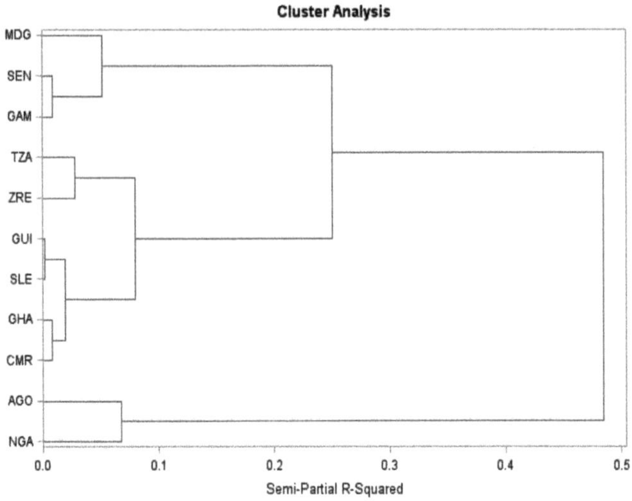

Figure 13 Dendogram of oil palm (cluster analysis tree chart) showing the relationship among 11 countries of germ plasm collections using 41 diverse characteristics.

material revealed that the populations from Madagascar were distinct as measured by the characteristics used in the study. Surprisingly, the grouping of the populations reflects the original geographical collection sites.

Ward's method was used for the hierarchical clustering analysis since it is considered very efficient (Siracli et al., 2013). The samples were grouped according to their nearness and similarity (Hossain et al., 2011). Based on the PCA, Ward's hierarchy clustered analysis resulted in two major clusters (Fig. 13). The first cluster includes all the countries except Angola and Nigeria which were present in the second major cluster. The first major cluster consists of two subclusters. The first subcluster was formed by a minor cluster of Senegal and Gambia materials reflecting their close proximity as shown in the African map, with Madagascar being singled out. The oil palm populations from Tanzania and Zaire are closely related since they are grouped together in the second subcluster. Oil palm from Ghana and Cameroon as well as Guinea and Sierra Leone shared some similarities, reflecting their geographical closeness.

6 Genetic diversity in oil palm: genetic markers

Morphological and agronomic characteristics are not always informative enough to characterize diversity and they are often influenced by the environment. It is therefore essential to use neutral markers such as molecular markers to investigate the genetic structure of these materials to ensure a better utilization of present germ plasm in breeding programmes. This information is also vital to guide breeders when planning breeding programmes focused on crosses in order to develop new populations with sufficiently broad genetic base and degree of adaptability. A genetic marker is a DNA sequence that is commonly recognizable by a restriction enzyme indicating a given chromosome

(Hoelzel and Dover, 1991). A molecular marker or DNA-based marker is a DNA fragment associated with a certain location within the genome.

Molecular markers used to analyse oil palm populations are as follows:

- Microsatellite or Simple Sequence Repeat (SSR) (Billotte et al., 2000; Zulkifli et al., 2008).
- Amplified Fragment Length Polymorphism (AFLP) (Kularatne et al., 2001).
- Isozymes (Hayati et al., 2004).
- Restriction Fragment Length Polymorphism (RFLP) (Barcelos et al., 2000; Rajanaidu et al., 2000).
- Randomly Amplified Polymorphic DNA (RAPD) (Rajanaidu et al., 2000).
- Single Nucleotide Polymorphisms (SNPs) (Ong et al., 2015).

These markers have proven to be useful in the investigation of oil palm genetic variability. Each marker allows the reporting of allelic frequencies and other genetic variability parameters such as the mean number of alleles per locus (*A*o), mean effective number of alleles per locus (Ae), the percentage of polymorphic loci (*P*), observed (*H*o) and expected (*H*e) heterozygosity. Table 6 summarizes the genetic diversity for oil palm populations using SSR, isozyme, AFLP and RFLP.

Significant intra-population genetic diversity has been found in natural oil palm populations, irrespective of both the country of origin and the genetic markers used (Table 7). Kularatne et al. (2000) employed AFLP markers when studying oil palm germ plasm belonging to 11 African countries and Malaysia Deli *dura* and they revealed that the higher proportion of total diversity was observed within populations (55%) than between populations (45%). Similarly, the RAPD analysis of natural oil palms revealed a higher proportion of diversity which could be attributable to within-population differences, regardless of species (Rajanaidu et al., 2000).

Table 6 Genetic diversity of oil palm breeding and germ plasm material revealed by genetic markers

Oil palm materials	References	Genetic markers	Genetic diversity				
			A_O	A_e	P	H_o	H_e
49 Natural oil palm populations from 10 African countries and three breeding materials of *Elaeis guineensis*	Bakoume et al. (2015)	SSR	5.0	2.8	–	0.460	0.644
26 populations of *Elaeis guineensis* from 100 African countries and Deli *dura*	Hayati et al. (2004)	Isozyme	1.80	1.35	54.5	0.186	0.184
Four populations of breeding materials maintained at IOPRI (Indonesia)	Purba et al. (2000)	AFLP	19.2	–	61.0	–	–
		Isozyme	1.81	–	36.4	0.332	0.300
Natural oil palm collections from 11 African countries and Deli *dura*	Maizura et al. (2006)	RFLP	1.66	–	53.0	–	0.199

Table 7 Genetic diversity of oil palm breeding and germ plasm materials revealed by AFLP and RAPD

Oil palm materials		Genetic marker	Diversity within population (H_{POP}/H_{SP})	Diversity between population [$(H_{SP} - H_{POP})/H_{SP}$]	References
Natural oil palm collections from 11 African countries and Malaysia Deli *dura*		AFLP	0.55	0.45	Kularatne et al. (2000)
Natural oil palm populations and Deli *dura*	*Elaeis guineensis* populations	RAPD	0.53	0.47	Rajanaidu et al. (2000)
	Elaeis oleifera populations		0.52	0.48	

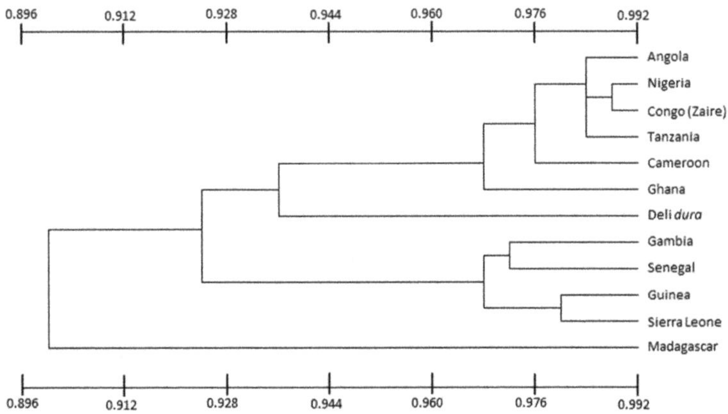

Figure 14 Dendogram of oil palm from different countries and Deli *dura* based on UPGMA analysis and AFLP of simple matching similarities (Kularatne, 2001).

The cluster analysis by Kularatne et al. (2001) resulted in oil palm populations divided into three major groups (Fig. 14). The first major cluster belongs to oil palm from Ghana, Nigeria, Cameroon, Democratic Republic of the Congo, Angola and Tanzania. Similar results were observed when the analysis was carried out using isozymes by Hayati et al. (2004) and RFLP by Rajanaidu et al. (2000). The palms originating from West Africa (Senegal, Gambia, Guinea and Sierra Leone) were clustered in the second group indicating their geographical closeness. Results based on the SSR analysis (Zulkifli et al., 2008) indicated that Gambia and Senegal, as well as Guinea and Sierra Leone, were closely related to each other. The uniqueness of Madagascar oil palm formed the third cluster, exhibiting low genetic similarity with oil palms originating from other countries. The distinct relationship of Madagascar oil palm from other countries was somewhat comparable to the results based on other molecular analyses.

Phenotypic clustering analysis is slightly different compared to molecular analysis. Indeed, Madagascar oil palm in the phenotypic analysis is related to Senegal and Gambia, while Madagascar is totally distinct from other countries (Bakoume et al., 2015). Madagascar is an island isolated from mainland Africa. The movement of pollen and dispersal of seed through geographical barriers was thus limited, reducing gene flow. This result shows the uniqueness of Madagascar oil palm as compared to other countries. Palms from Madagascar are generally short with a high iodine value and linoleic acid (C18:2) in oils (Rajanaidu et al., 2000).

Both phenotypic and molecular analyses showed similar results for oil palm from Gambia and Senegal, as well as from Guinea and Sierra Leone. The grouping of Gambia and Senegal has a strong geographical association since Gambia is actually a political enclave of Senegal. The similarities between Guinea and Sierra Leone suggest that the main oil palm belt begins in Guinea before spreading south through Sierra Leone, Ghana and other central African countries (Zeven, 1967). Significant genetic similarities of oil palm between Tanzania and Zaire, and between Ghana and Cameroon, using both phenotypic and molecular analyses, suggest a high migration rate between the regions. However, the closer relationship between oil palm from Angola and Nigeria is quite difficult to explain.

Even though there are some differences in the clustering of populations using phenotypic and molecular analyses, the information obtained from both analyses is important for formulating a sampling strategy in order to establish a core collection of oil palm germ plasm. Such analyses allow the conservation of oil palm germ plasm to be improved to a manageable level to ensure the diversity of collections. Populations with sufficient diversity can be used successfully for the development of new breeding populations.

7 Conservation of oil palm collections

Oil palm populations collected in various parts of the world can be preserved in *in situ* and *ex situ* collections (Engelman, 1991). *In situ* conservation can be defined as the conservation of species in their natural habitat, which is considered the most appropriate way of preserving and managing biodiversity. In contrast, *ex situ* conservation relies on the preservation of components of biological diversity outside their natural habitats.

7.1 *Ex situ* conservation

Ex situ conservation techniques include the following:

- Field genebanks
- Cryopreservation
- DNA preservation
- *In vitro* methods

The field genebank method is used for species which produce recalcitrant seed. Field genebanks have the benefit that material is readily available for use. Field genebanks require high maintenance costs, a large land area and they face possible exposure to diseases and extreme weather conditions. In the case of oil palm, only 148 palms are planted per hectare. To date, MPOB has successfully conserved approximately 100 000 palms collected from West Africa and Latin America in an area covering nearly 500 ha (Fig. 15).

Cryopreservation is an alternative method to preserve oil palm tissues under extremely low temperatures (−196°C) in liquid nitrogen. It was found possible to revive and germinate oil palm zygotic embryo after freezing in liquid nitrogen (Grout et al., 1983). The plants

Figure 15 Large-scale field planting of oil palm germ plasm in Malaysia.

regenerated from frozen embryoids are predominantly normal as shown by Konan et al. (2005). A simple desiccation stage should be conducted to reduce the moisture content in the embryoids in order to prevent the loss of viability when stored in liquid nitrogen. In comparison to *ex situ* preservation, this technique requires less space; lowers maintenance costs; and protects the genetic materials from pests, diseases and extreme weather. For long-term conservation, MPOB routinely stores zygotic embryos in liquid nitrogen. More than 34000 zygotic embryos of accessions have been stored in the MPOB's cryotank (Rajanaidu and Ainul, 2013).

DNA banking is an emerging technique which is aimed at ensuring the availability of genetic resources for oil palm improvement. Currently, MPOB runs a programme to preserve DNA samples of oil palm germ plasm in order to support molecular-based oil palm breeding. Readily available DNA samples are used for the verification of specific genotypes/crosses. DNA materials must be maintained under a stable condition at −4°C for long-term preservation (Mohd Din et al., 2014). Conservation of DNA from older palm and valuable genitors from seed production schemes is essential for future pedigree verification.

In order to preserve germ plasm, *in vitro* methods have also been used via embryo rescue, cloning and *in vitro* storage (seeds, embryo, polyembroid cultures and excised roots) (Rohani et al., 2000). The long-term preservation of oil palm collections via seed storage is not practical as the seeds can only be stored for two years (Rajanaidu, 1980).

7.2 *In situ* conservation

A sample of natural oil palm groves can be conserved on a long-term basis to represent various ecological niches. Natural palm groves in Africa harbour genetic diversity which is essential to plant breeders to develop high-yielding planting material. Such resources are rapidly disappearing in Africa because of the development and population pressures. Under the guidance of FAO, MPOB has sought to preserve these resources for posterity and safeguard the long-term interests of the oil palm industry in Malaysia and elsewhere (Rajanaidu et al., 2000). Local population has supported the conservation of natural oil palms since income can be generated and a variety of products be produced. Locals have taken measures to protect and use natural oil palm groves, even though no or few plantings have been carried out.

The concept of a core collection has been extensively studied by Brown (1989) and Odong et al. (2013). As mentioned earlier, MPOB is conserving approximately 100 000 palms from West Africa and Latin America in a germ plasm collection occupying nearly 500 ha. The development of a core collection is cost-effective in increasing the efficiency of conservation, characterization, use and regeneration of germ plasm. Various approaches are used to preserve genetic diversity with a minimum number of accessions. As recommended by Odong et al. (2013), MPOB studied the genetic distance of collections using both molecular and phenotypic data. The study of genetic distances highlighted gross differences between populations for core collections. In addition, breeders have also included some elite germ plasms in the core collection.

8 Utilization of germ plasm

MPOB runs the largest oil palm germ plasm collection in the world. Samples from all across the oil palm belt have been sampled in order to capture the full range of genetic variation. It is essential that germ plasm collections are maintained for the future. However, since this strategy is costly, it is also essential that these collections are used effectively (Corley and Tinker, 2016; Spraque, 1986). At MPOB, oil palm germ plasm has been evaluated and is being utilized in a number of ways. These are discussed below.

8.1 Direct selection of individual elite palms

The breeding and selection of oil palm are time-consuming and difficult. Normally, newly introduced genetic material has no potential for direct commercial use. It has to be introgressed into another advanced material. The new gene pool will only be considered for immediate use if it possesses interesting traits and acceptable yields. Elite palms which are high-yielding and compact, such as P126, can be cloned.

8.2 Broadening the genetic base of Deli *duras* and *teneras*/ *pisiferas*

Crossing Deli *duras* with Nigerian *duras*/*teneras* broadens overall genetic variability. Increasing the amount of heritable variation for desired traits of these introgressed populations is the basis for further selection and breeding. Selected palms need to possess one or a combination of desirable traits such as high bunch yield, superior oil and kernel content, reduced height and superior oil quality.

8.3 Progeny testing of *duras* and *pisiferas*

Outstanding Nigerian *tenera* varieties were progeny-tested with a range of commercial Deli *duras* varieties. The T × D or D × T hybrids and their *dura* and *tenera* parents are selfed simultaneously. The purpose of the crossing programme is to progeny-test the Nigerian *teneras* to study their combining ability. By using reciprocal recurrent selection, the selfs can be used in the seed production (Jacquemard et al., 1981). The *pisiferas* varieties are also progeny-tested to evaluate their combining ability with Deli *duras*.

8.4 New foundation of breeding programmes

Germ plasm is used to initiate entirely new breeding programmes aimed at producing superior alternatives to Deli *duras* and modern breeding *teneras*. This is because of the presence of *teneras* comparable to the best current material, produced from Nigerian *dura* × Nigerian *pisifera*.

8.5 Development of elite DxP planting materials

The main objective of oil palm germ plasm collections is to make use of outstanding palms with useful attributes and to introgress such genotypes into current breeding material. The emphasis is on the attributes such as short stems, more liquid oil composition, high carotene content, long stalks and high kernel yield. Since 1992, 14 distinct genotypes have been distributed successfully to the oil palm industry in Malaysia by MPOB (Table 8). Recently, MPOB has released the PS14 genotype related to high protein kernel as an alternative to imported ingredients for animal feed (Noh et al., 2015).

After undergoing extensive progeny-testing, the MPOB has been able to breed shorter planting materials (PS 1.1) based on the MPOB–Nigerian population No. 12. Some of such crosses were found to produce high oil yield and to be 30% shorter than the DxP control (Rajanaidu et al., 2013). This characteristic makes bunch harvesting easier and more amenable to mechanization, thus reducing production costs.

Table 8 Commercial oil palm planting materials and breeding populations transferred by MPOB to the industry

No.	*PORIM series no.*	*Trait*	*Year of transfer*	*Intended use of material*
1	PS1	Dwarf	1992	Planting material
2	PS2	High iodine value	1992	Planting material
3	PS3	Large kernel	1996	Breeding population
4	PS4	High carotene (*E. oleifera*)	2002	Breeding population
5	PS5	Thin-shelled *tenera*	2003	Breeding population
6	PS6	Large-fruit *dura*	2003	Breeding population
7	PS7	High bunch index	2004	Breeding population
8	PS8	High vitamin E	2004	Breeding population
9	PS9	*Bactris gasipaes*	2004	Economic palm
10	PS10	Long stalk	2006	Breeding population
11	PS11	High carotene (*Elaeis guineensis*)	2006	Breeding population
12	PS12	High oleic	2006	Breeding population
13	PS13	Low lipase	2008	Breeding population

Source: Mohd Din et al. (2014).

9 Conclusion

International cooperation and collaboration are vital to collect and conserve oil palm germ plasm which is the lifeline of the industry. International treaties such as the Convention on Biodiversity and the International Treaty for Plant Genetic Resources, as well as the work of the FAO Access and Benefit Sharing Working Group, govern the regulations on the collection of oil palm germ plasm in foreign countries. Preservation of natural oil palm germ plasm *in situ* needs international cooperation between developed and developing countries in order to finance long-term conservation projects which are largely located in farmers' fields.

10 Where to look for further information

Further information on the contents of this chapter can be obtained on the websites of:

- Malaysian Palm Oil Board
- Indonesian Oil Palm Research Institute
- CIRAD
- CINEPALMA, Colombia
- ASD Costa Rica
- International Treaty on Plant Genetic Resources for Food and Agriculture and Access and Benefit Sharing
- Convention on Biological Diversity (CBD)
- International Society for Oil Palm Breeders (ISOBP)

11 Acknowledgement

The authors wish to thank MPOB for the permission to publish this paper.

12 References

Bakoumé, C., (2016). Genetic diversity, erosion and conservation in oil palm (*Elaeis guineensis* Jacq.). In: *Genetic Diversity and Erosion in Plants* (Eds.: Ahuja, M. R. and Mohan Jain, S.). Springer, Switzerland, pp. 1–33.

Bakoumé, C., Wickneswari, R., Siju, S., Rajanaidu, R., Kushairi, A. and Billotte, N. (2015). Genetic diversity of the world's largest oil palm (*Elaeis guineensis* Jacq.) field genebank accessions using microsatellite markers. *Genet. Res. Crop Evol.* 62(3), 349–60.

Barcelos, E., Amblard, P., Berthaud, J. and Seguin, M. (2000). The genetic diversity of the American oil palm, *Elaeis oleifera* (Kunth), Cortes, revealed by nuclear RFLP markers. In: *Int. Symp. Oil palm genetic ressources and utilization*, 8–10 June 2000 (Eds: Rajanaidu, N. and Ariffin, D.). Malaysian Palm Oil Board, Kuala Lumpur, pp. 544–55.

Beirnaert, A. and Vanderweyen, R. (1941). Influence de l'origine variétale sur les rendements. In: *Publ. Inst. Nat. Etude agron. Congo Belge*, Comm. 3 sur le Palmier à huile, pp. 1–26.

Billotte, N., Risterucci, A. M., Barcelos, E., Noyer, J. L., Amblard, P. and Baurens, F. C. (2000). Development and characterisation of oil palm (*Elaeis guineensis* Jacq.) microsatellite markers. In: *Int. Symp. Oil Palm Genetic Resources and Utilization,* 8–10 June 2000 (Eds: Rajanaidu, N. and Ariffin, D.). Malaysian Palm Oil Board, Kuala Lumpur, pp. 1–17.

Brown, A. D. H. (1989). Core collection: A practical approach to genetic resource management. *Genome* 31:818–24.

Corley, R. H. V. and Tinker, P. B. (2003). *The Oil Palm* (4th ed.). Blackwell, Oxford, UK.

Corley, R. H. V. and Tinker, P. B. (2016). *The Oil Palm* (5th ed.). Blackwell, Oxford, UK.

Engleman, F. 1991. Current development of cryopreservation for oil palm somatic embryos. In: *Proceedings of the XV111th International Refrigeration Congress*, vol. 1V, 10–17 August 1991. International Refrigeration Congress, Montreal, pp. 1676–80.

Escobar, R. (1982). Preliminary results of the collection and evaluation of the American oil palm *Elaeis oleifera* (HBK, Cortes) in Costa Rica. In: *The Oil Palm in Agriculture in the Eighties, Vol.* 1. Incorp. Soc. Planters, Kuala Lumpur, pp. 79–93.

Grout, B. W. W., Shelton, K. and Pritchard, H. W. (1983). Orthodox behavior of oil palm seed and cryopreservation of the excised embryo for genetic conservation. *Ann. Bot.*, 52, 381–4.

Hartley, C. W. S. (1988). *The Oil Palm.* Third ed. Longman, London, p. 71.

Hayati, A., Wickneswari, R., Rajanaidu, N. and Maizura, I. (2004). Genetic diversity of oil palm (*Elaeis guineensis Jacq.*) germplasm collections from Africa: Implications for improvement and conservation of genetic resources. *Theor. Appl. Genet.* 108 (7): 1274–84.

Hoelzel, A. R. and Dover, G. A. (1991). *Molecular Genetic Ecology*. Oxford University Press, Oxford.

Hossain, M., Patras, A., Barry-Ryan, C., Martin-Diana, A. B. and Brunton, P. (2011). Application of principal analysis and hierarchical cluster analysis to classify different spices based on in vitro antioxidant activity and individual polyphenolic antioxidant compounds. *J. Funct. Food* 3, 179–89.

Jacquemard, J. C., Meunier, J. and Bonnot, F. (1981). Etude Genetique de la production du croisement chex le palmier a huile *Elaeis guineensis. Oleagineux* 36(7): 543–52.

Jaligot, E., Rival, A., Beulé, T., Dussert, S. and Verdeil, J.-L. (2000) Somaclonal variation in Oil Palm (*Elaeis guineensis* Jacq.): The DNA methylation hypothesis. *Plant Cell Reports* 19 (7): 684–90.

Jaligot, E., Hooi, W. Y., Debladis, E., Richaud, F., Beulé, T., Collin, M., Agbessi M. D. T., Sabot F., Garsmeur O., D'Hont A., Syed Alwee S. S. R. and Rival A. (2014). DNA methylation and expression of the EgDEF1 gene and neighboring retrotransposons in mantled somaclonal variants of oil palm. *PloS ONE* 9(3), e91896.

Julian Barba, R. (2012). Ecuadorian oleiferas minimize losses caused by the bud-rot in Latin America. In: *ISOPB Proceedings International Seminar Proceedings on 'Breeding for Oil Palm Disease Resistance and Field Visits'*, 21–24 September 2012. International Society for Oil Palm Breeders, Bogota, Colombia.

Konan, K. E., Rival, A., Kouadio, Y. J., Duval, Y., Adon, B., Pene, C. and Durand-Gasselin, T. (2005). Assessment of various strategies for the preservation of clonal oil palm genetic resources. In: *Proc. Int. Palm Oil Cong. Technological Breakthroughs and Commercialization- the Way Forward*, 25–29 September. Malaysia Palm Oil Board, Kuala Lumpur, pp. 1168–88.

Kularatne, R. S., Shah, F. H. and Rajanaidu, N. (2000). Investigation of genetic diversity in African natural oil palm populations and Deli *dura* using AFLP markers. In: *Int. Symp. Oil Palm Genetic Resources and Utilization*, 8–10 June 2000 (Eds: Rajanaidu, N. and Ariffin, D.). Malaysian Palm Oil Board, Kuala Lumpur, pp. 497–531.

Kularatne, R. S., Shah, F. H. and Rajanaidu, N. (2001). The evaluation of genetic diversity of Deli dura and African oil palm germplasm collection by AFLP technique. *Trop. Agric. Res.* 13: 1–12.

Kushairi, A. and Rajanaidu, N. (2000). Breeding populations, seed production and nursery management. In: *Advances in Oil Palm Research Vol.* 1 (Eds: Basiron, Y., Jalani, B. S. and Chan, K. W.. Malaysian Palm Oil Board, Kuala Lumpur, pp. 39–96.

Maizura, I., Rajanaidu, N., Zakri, A. H. and Cheah, S. C. (2006). Assessment of genetic diversity in oil palm (Elaeis guinennsis Jacq.) using Restriction Fragment Length Polymorphism (RFLP). *Genet. Resour. Crop. Evol.* 53, 187–95.

Meunier, J. (1975). The American (oil palm) *Elaeis melanococca. Oleagineux* 30(2), 51–61.

Mohd Din, A., Nookiah, R., Ong Abdullah, M., Singh, R., Yaakub, Z., Abdullah, N. and Tarmizi, A. H. (2014) Planting materials and yield potential. Presented at the 11th NATSEM 2014 – ISP National Seminar in Kuantan, 23–25 June 2014.

Noh, A., Rajanaidu, N., Kushairi, A., Mohd Din, A. and Wan Nooraida, W. M. (2015). Oil palm breeding population selected for high protein kernel. MPOB Information Series No. 566, Malaysian Palm Oil Board, Kuala Lumpur.

Odong, T. L. Jansen, J., van Eauwijk, F. A. and van Hintum, T. J. L. (2013). Quality of core collections for effective utilization of genetic resources: Review, discussion and interpretation. *Theor. Appl. Genet.* 126, 289–305.

Ong-Abdullah, M., Ordway, J. M., Jiang, N., Ooi, S. E., Kok, S. Y., Sarpan, N., Azimi, N., Hashim, A. T., Ishak, Z., Rosli, S. K., Malike, F. A., Bakar, N. A., Marjuni, M., Abdullah, N., Zulkifli, Y., Mohd Din, A., Rajanaidu, N., Singh, R., Low, E. T., Chan, K. L., Azizi, N., Smith, S. W., Bacher, B., Budiman, M. A., Van Brunt, A., Wischmeyer, C., Beil, M., Hogan, M., Lakey, N., Lim, C. C., Arulandoo, X., Wong, C. K., Choo, C. N., Wong, W. C., Kwan, Y. Y., Alwee, S. S., Sambanthamurthi, R. and Martienssen, R. A. (2015). Loss of *Karma* transposon methylation underlies the mantled somaclonal variant oil palm. *Nature* 525, 533–7.

Ong, P. W., Maizura, I., Abdullah, N. A. P., Rafii, M. W., Ooi, L. C. L., Low, E. T. L. and Singh, R. (2015). Development of SNP markers and their application for genetic diversity analysis in the oil palm (*Elaeis guineensis*). *Genet. Mol. Res.* 14(4), 12205–16.

Ooi, S. C., Silva, E. B. da, Muller, A. A. and Nascimento, J. C. (1981). Oil palm genetic resources. Native *Elaeis oleifera* populations in Brazil offer promising sources. *Pesquis Agropecuaria Brasisileira* 16(No. 4).

Purba, A. R., Noyer, J. L., Baudouin, L., Perrier, X., Hamon, S. and Lagoda, P. J. L. (2000). A new aspect of genetic diversity of Indonesian oil palm (*Elaeis guineensis* Jacq.) revealed by isoenzyme and AFLP markers and its consequences for breeding. *Theor. Appl. Genet.* 101, 956–61.

Rajanaidu, N. (1980). Oil palm genetic resources: Current methods of conservation. Paper presented at International Symposium on Conservation Inputs from Life Sciences. MPOB, Bangi, pp. 25–30.

Rajanaidu, N. (1983). *Elaeis oleifera* collection in South and Central America. Plant Genetic Resources-Newsletter. No 56. FAO, Rome, pp. 52–1.

Rajanaidu, N. (1986). The oil palm (*Elaeis guineensis*) collections in Africa. In: *Proc. Int. Workshop Oil Palm Germplasm & Utilisation.* Palm Oil Res. Inst. Malaysia, Kuala Lumpur, pp. 59–83.

Rajanaidu, N. and Rao, V. (1988). Oil palm genetic collections: Their performance and use to the industry. In: *Proc. 1987 Int. Oil Palm Conf. Progress and Prospects.* Palm Oil Res. Inst. Malaysia, Kuala Lumpur, pp. 59–85

Rajanaidu, N. (1994). *PORIM Oil Palm Genebank – Collection, Evaluation, Utilisation and Conservation of Oil Palm Genetic Resources.* Palm Oil Res. Inst. Malaysia, Kuala Lumpur, pp. 1–19.

Rajanaidu, N., Maizura, I. and Cheah, S. C. (2000). Screening of oil palm natural populations using RAPD and RFLP molecular markers. In: *Int. Symp. Oil Palm Genetic Resources and Utilization*, 8–10 June 2000 (Eds: Rajanaidu, N. and Ariffin, D.). Malaysian Palm Oil Board, Kuala Lumpur, pp. 1–28.

Rajanaidu, N., Kushairi, A., Rafii, M. Y., Din, M. A., Maizura, I., Isa, Z. A. and Jalani, B. S. (2000). Oil palm genetic resources and their utilisation – a review. In: *Int. Symp. Oil Palm Genetic Resources and Utilization*, 8–10 June 2000 (Eds: Rajanaidu, N. and Ariffin, D.). Malaysian Palm Oil Board, Kuala Lumpur, pp. 1–55.

Rajanaidu, N., Kushairi, A., Marhalil, M., Mohd Din, A., Fadila, A. M., Noh, A., Meilina, A., Raviga, S. and Isa, Z. A. (2013). Breeding for oil palm for strategic requirement of the industry. In: *MPOB International Palm Oil Congress (PIPOC 2013)*, 19–21 November 2013. Kuala Lumpur.

Rajanaidu, N. and Ainul, M. M. (2013). Conservation of oil palm and coconut genetic resources. In: *Conservation of Tropical Plant Species* (Eds: Normah, M. N., Chin, H. F. and Barbara, M. R.). Springer, New York, pp. 189–212.

Rohani, O., Tarmizi, A. H. and Zamzuri, I. (2000). Application of in *vitro* methods for oil palm germplasm conservation. In: *Proceedings of the International Symposium on Oil Palm Genetic Resources and Their Utilization*. 8–10 June 2000. Malaysian Palm Oil Board, Kuala Lumpur, Malaysia.

Singh, R., Ong-Abdullah, M., Low, E. T., Manaf, M. A., Rosli, R., Rajanaidu, N., Ooi, L. C., Ooi, S. E., Chan, K. L., Halim, M. A., Azizi, N., Nagappan, J., Bacher, B., Lakey, N., Smith, S. W., He, D., Hogan, M., Budiman, M. A., Lee, E. K., DeSalle, R., Kudrna, D., Goicoechea, J. L., Wing, R. A., Wilson, R. K., Fulton, R. S., Ordway, J. M., Martienssen, R. A. and Sambanthamurthi, R. (2013). Oil palm genome sequence reveals divergence of interfertile species in Old and New world. *Nature* 500(7462), 335–9.

Singh, R., Low, E. T., Ooi, L. C., Ong-Abdullah, M., Rajanaidu, N., Ting, N. C., Marjuni, M., Chan, P. L., Ithnin, M., Manaf, M. A., Nagappan, J., Chan, K. L., Rosli, R., Halim, M. A., Azizi, N., Budiman, M. A., Lakey, N., Bacher, B., Van Brunt, A., Wang, C., Hogan, M., He, D., MacDonald, J. D., Smith, S. W., Ordway, J. M., Martienssen, R. A., Sambanthamurthi, R. (2014). The oil palm VIRESCENS gene controls fruit colour and encodes a R2R3-MYB. *Nat. Comm.* 5(4106), 1–8.

Siracli, S., Dogan, N. and Dogan, I. (2013). Comparison of hierarchical cluster analysis methods by cophenetic correlation. *J. Inequal. Appl.* 203, 1–8.

Shearman, J. R., Jantasuriyarat, C., Sangsrakru, D., Yoocha, T., Vannavichit, A., Tragoonrung, S. and Tangphatsornruang, S. (2013). Transcriptome analysis of normal and mantled developing oil palm flower and fruit. *Genomics* 101, 306–12.

Sprague, E. W. (1986). Maize germplasm utilization in Asia. In: *New Frontiers in Breeding Researches* (Eds: Napompeth, B. and Suhadrabandhu, S.. Fac. Of Agric., Kasetsart University, Bangkok, Thailand.

Zeven, A. C. (1967). *The Semi-wild Oil Palm and Its Industry in Africa*. PUDOC, Wageningen.

Zeven, A. C. (1973). The 'mantled' oil palm (*Elaeis guineensis* Jacq.). *J. West. Afr. Inst. Oil Palm Res.* 5(18), 31–3.

Zulkifli, Y., Maizura, I. and Singh, R. (2008). Molecular screening pf oil palm germplasm collections (*Elaeis guineensis*) using EST-SSR markers. In: *Proceedings of the 3rd Seminar on Performance of MPOB PS1 and PS2 Materials and Elite Germplasm*. Malaysian Palm Oil Board, Selangor, pp. 168–85.

Chapter 7

Advances in conventional breeding techniques for oil palm

Benoît Cochard and Tristan Durand-Gasselin, PalmElit SAS, France

1 Introduction

The oil palm has been grown in Africa for thousands of years. The oldest traces of oil palm, in the form of pollen, date back to the Eocene period (Zaklinskaya and Prokofyev, 1971). Pollen has been found in all the current areas of oil palm cultivation in Africa from the Miocene period (23.03 Ma to 7.246 Ma) (Zeven, 1964). Oil palm has been cultivated in Egypt since 3000 BC (Friedel 1897 in Zeven, 1967).

Oil palm has played a fundamental role in the nutrition, technological, economic and social development of many African forest peoples. For these populations, the oil palm is central. It provides wine, oil and sauce from seeds. It is used to build houses, provide flooring and weave hunting nets and traps as well as build dams to catch fish. In short, 'the oil palm has so many uses that cannot be counted' (Haxaire, 1992). The oil palm is also part of founding myths. For the Gouro people from Côte d'Ivoire, 'That Bali, the creator God has given to man before putting on earth is the palm. He gave it to eat... The oil palm and man are the same, the palm is on earth, and humans are on earth. God has given us oil palm to watch men' (Haxaire, 1992). Despite such an ancient and strong relationship between the oil palm and the populations of African forests, the oil palm is mostly only semi-domesticated. No attempts were made to

http://dx.doi.org/10.19103/AS.2017.0018.06

propagate the palm by sowing seeds and/or transplanting seedlings (Zeven, 1972). Hence, African populations almost never planted oil palms. This was the case in Guinea (Madelaine, 2005), Côte d'Ivoire (Blanc-Pamard, 1980; Haxaire 1992), Togo (Tchamie, 1995), Cameroon (Baeke, 1996) and the Democratic Republic of Congo (Eggert, 1993). When old oil palm stands were present, seedlings spontaneously sprouted and were protected by clearing regularly around them so that they could receive enough light (Haxaire, 1996). This clearing and elimination of some oil palm seedlings allowed a natural expansion of the oil palm. Maley (1999) described two types of palm groves used by the African populations: (i) forest regrowth where oil palm is a leading pioneer species and (ii) sub-spontaneous oil palm groves near the villages, in fallows and other type of vegetation disturbed by human activity.

At the beginning of the twentieth century, various authors reported that Ébriés in Côte d'Ivoire (Bret, 1911 in Zeven 1967), and Togo (Grüner, 1904 in Zeven 1967), were able to sow unselected seeds. Transplanting unselected palm was also done in Dahomey (now Benin) (Zeven 1967) and in Ghana. At this time, selecting and planting seeds was a quite rare practice, although it was done in Nigeria, Angola and Congo. In the middle of the twentieth century, Tovey (1947) observed that the ratio of Tenera[1] palms in Ufuma (Nigeria) was higher than that in natural oil palm groves. Sparnaaij and Blaak (1963) had observed the same situation in Nigeria. These abnormal ratios have also been described in Angola and Congo (Zeven, 1967).

Therefore, oil palm could be considered as having not been subjected to breeding historically, meaning that at the beginning of the twentieth century it was a semi-wild crop. In the middle of the nineteenth century, oil palm was planted in different botanical gardens and breeding started at the beginning of the twentieth century both in Africa and Asia. In the present chapter, we will present the story of oil palm breeding which occurred between the beginning of the twentieth century and the end of the 1950s, and then the current state of oil palm breeding. Present breeding strategies will be described via the various breeding schemes at work, traits observed, mating designs, and the experimental designs used. The organization of seed production for oil palm will also be reviewed. This chapter will mainly focus on oil palm breeding carried out by Institut de Recherche pour les Huiles et Oléagineux (IRHO), CIRAD and now PalmElit, although oil palm breeding activities by other companies will also be discussed. Finally, the future trends in oil palm breeding will be explored.

2 Early history of oil palm breeding

Oil palm breeding is based on populations developed in different countries or in different estates from the 1920s, and generally originated from a limited number of founders. These populations have been named 'breeding population of restricted origin' (BPRO) by Rosenquist (1986). Such populations are still now at the base of the distinction between the principal seed suppliers. The main BPROs are presented in the subsequent sections.

1. Tenera palms: oil palms with a thin shell. The shell thickness genetic determinism is important for oil palm breeding. It is explained in Section 3.1.

2.1 Populations from Africa

Côte d'Ivoire

The botanical garden at Bingerville was established in 1900. Besides various food crops, forestry species and introduction of accessions from several countries, the botanical garden included a natural oil palm grove. In 1915, under the direction of Paul Teissonnier, the garden focused its works on cocoa, oil palm, coffee, kola, rubber plants, plantain, forest, fruit and ornamentals. In 1915 and 1917, selected oil palm were planted on 3.5 ha (Tourte, 2005; Houard et al., 1926). The first survey undertaken in the Bingerville area (Bingerville botanical garden, Blachon concession, Granjean concession, Drevet concession, Necker concession in M'Bato) resulted in the collection of 133 oil palms. In 1925, the La Mé Experimental Station assessed 43 palms from which 20 palms were selected (Houard, 1926; Gascon et de Berchoux, 1964). Finally, 19 palms were 'selfed' and planted in La Mé Experimental Station between 1924 and 1930 (Cochard et al., 2006). In 2006, of these 19 palms, 6 selfed palms, namely M12, B205, B212, B247, TEIS3 and BRT10, survived. BRT10 is the base of the current breeding programme developed by CIRAD and PalmElit.

Cameroon

In 1904, trials and breeding of the Lisombé[2] were undertaken in the botanical garden of Victoria (Limbé) by Preuss. In some reports, it is said that 100 000 palms of Lisombé were received in the Kribi area and planted along the Kribi-Longji road. However, oil palm was rare in the Kribi area (Annet, 1921).

Before 1910, only a few concessions existed. Among these, the Deutsche Kautschuk Aktien Gesellschaft had planted oil palms in 1914, in Mpondo on the Mongo River near Ekona (Annet, 1921). The Ikassa Estate belonging to the Gesellschaft Süd Kamerun (Courade, 1978) was probably planted in 1913, using seeds from the Ekona[3] district (Corley and Tinker, 2003).

The history of Ekona BPRO has been described in detail by Rosenquist (1986). The United Africa Company planted oil palms from 1928 to 1933. Seeds were obtained from Calabar (Nigeria) and Ikassa Estate and selection started in 1933 in N'Dian. Seeds from controlled pollinations were planted in N'Dian and the Cowan estates in Nigeria. In addition, 800 palms were used to produce open-pollinated seeds for Cowan estate. Selection on the two estates led to the selection of 19 teneras, 6 duras and 2 fertile pisiferas. From this programme, 18 crosses and 5 selfs were planted in 1954/1955 in the Lobé Estate in Cameroon which was newly designed for oil palm plantation. This was the basis of the so-called Ekona population.

Democratic Republic of Congo

Here, breeding began with the creation of the Botanical Garden of Eala in 1900. The Yangambi Research Station was created after the First World War (Drachoussof, 1989) and, in 1921, Ringoet, the director of Yangambi, developed an interest in palms from N'Gazi. Mr Tihon, a chemist from Eala, identified the famous 'Djongo' palm in the

2. Lisombé: Disombé or Lisombé was the name given by the Bakondo people to *Elaeis guineensis* var. Tenera, Beccari (Annet, 1921).
3. Possibly, it is the Ekona mentioned by Annet (1921).

'clonal' gardens. Ringoet planted these palms in a 4 ha field called the small 'plantation de la Rive', now known as 'Palmeraie de la Rive'. Of the 1880 palms planted, 583 were kept after analysis (Ergo, 2013). This plantation would be complemented by four plantings:

- The first one was done in 1924 with palms from N'Gazi, Yawenda and Isangi (Demol et al., 2002) and excess from the nursery dedicated to 'Plantation de la Rive'.
- The second one was carried out in 1927 using seeds from N'Gazi.
- The third one, in 1929, constituted of illegitimate seeds originating from oil palms of the 'Plantation de la Rive'.
- Finally, the fourth one was planted in 1930 with one illegitimate progeny from the 'Plantation de la Rive' (Ergo, 2013).

These five selected oil palm groves constituted the basis of the Yangambi breeding programme.

Nigeria

From 1911 to 1915, 800 oil palms were planted by E. Smith from the Nigerian Department of Agriculture. Data for yield and bunch compositions were collected from 1922 to 1928 and nine dura and ten tenera palms were selected. Twelve palms were self-pollinated to form the Calabar F1 generation. The progenies were planted between 1930 and 1935 on four stations (Hartley, 1967).

In the 1920s, an oil palm selection programme was initiated in a 11 ha grove located near Aba. Several thousand out-pollinated palms were established at the oil palm research station near Benin City in 1939/1941 (Hartley, 1967).

Benin

Between 1922 and 1927, 38 oil palms were chosen in the sub-spontaneous oil palm groves located near Porto Novo and Pobè (Gascon et de Berchoux 1964), and gathered in the Pobè Research Station (Institut National de Recherche Agronomique du Bénin - INRAB). Thirty-five palms were 'selfed' and planted in Pobè. Fourteen palms are still represented in Pobè in 2016 by their 'self' (AG1, AS2, ED2, FAG1, J1, L10, Lo10, M9, MK5, OD1, OD2, OL4, S2, S4). However, this population was never used in the current breeding programme.

2.2 Other populations

Indonesia and Malaysia

Open pollination from the Djongo palm from the Eala botanical garden in Congo was distributed to Sungei Panchur Research Station in Indonesia. Only 13 seedlings survived (8 duras and 5 teneras). One of the tenera was the SP540T. This palm is one of the pillars of the Marihat (Indonesian Oil Palm Research Institute – IOPRI) breeding programme. SP540T is also the basis of the so-called AVROS (Algemeene Vereniging van Rubber Planters ter Oostkust van Sumatra) population; this population is also crossed by the name BM119, planted in 1957 in Banting (Malaysia) (Corley and Tinker, 2003).

Deli populations

In 1848, four 1-year-old oil palm seedlings were introduced in Indonesia, at the Bogor (Buitenzorg) Botanical Gardens. These four oil palms came from Bourbon Island (now La Réunion Island, a French overseas region). Two seedlings came directly and two others arrived via the Hortus Botanicus of Amsterdam (Rosenquist (1986) quoting Toovey and Broekmans (1955), Hunger (1917) and Rutgers, (1922)). Hunger (1924) explained in detail why there was confusion between Réunion Island and Mauritius, and therefore why he was almost certain that the palms came from Réunion Island. Seeds from the Bogor Botanical Gardens were sent to Sumatra in 1874 or 1875 to the Deli Maatschappij. Hunger (1924) and Jagoe (1952) described in detail the early history of the development of the Deli populations.

3 Oil palm breeding from the beginning of the 20th Century

3.1 Shell thickness

Chevalier and Beccari in the beginning of the twentieth century identified a number of oil palm varieties which were classified by Professor Jumelle (Anet, 1921) according to shell thickness:

- Thick kernel (2–4 mm) from black fruit before ripening and red after, normal mesocarp of 2–3 mm, non-angular fruits, 35–40 mm: var. *dura* Becc.
- Thin kernel (<2 mm) from black fruit before ripening and red after, fruit not beaked on the top, unwelded leaf segment, fruit length from 22 to 28 mm and fruit width from 18 to 22 mm, kernel throughout the length of the fruit: var. *tenera* Becc.
- Thin kernel (<2 mm) from black fruit before ripening and red after, fruit not beaked on the top, unwelded leaf segment, fruit length from 23 to 28 mm and fruit width from 16 to 19 mm, reduced kernel across the top of the fruit: var. *pisifera* Chev.

African people also knew these different fruit forms, for instance:

- Dura was called Dipobe (near Douala, Balong, Bakondo) in Cameroon (Annet 1921), possibly Dé-Yaya in Benin and Adé-Quoi for Ébriés in Côte d'Ivoire (Houard et al., 1927).
- Tenera was called Sombé (Douala, Balong, Abo, Nyombé, Disombé or Lisombé (Bakondo) in Cameroon (Annet, 1921), Dé-Kla or Dé-Gbakoun in Benin, and Aquoi-Sran or Adé-Sran for Ébriés in Côte d'Ivoire (Houard, 1927).
- Pisifera was called Djoungoumea (Douala) in Cameroon (Annet, 1921), Votchi in Benin, and nothing for the Ébriés in Côte d'Ivoire (Houard, 1927).

It is clear from the above facts that botanists and the African people were able to distinguish between different oil palm fruit forms, but did not understand the relationship between them. It was only in 1941 that Beirnaert and Vanderweyen described how a single gene controls the shell thickness of oil palm. Hence, the dura form was defined by sh^+/sh^+, tenera by Sh^+/Sh^- and pisifera by sh^-/sh^-. However, these authors observed in tenera x tenera crosses or selfings some divergences from the 25% dura: 50% tenera: 25% pisifera ratio. In addition, it is known that within such classifications, there is a polygenic variation of the fruit form.

3.2 The 'Experience Internationale'

Until the Second World War, breeding programmes implemented in Asia and Africa were mainly carried out on local populations. Oil palm being a cross-pollinated plant, it was considered interesting to assess crosses made between different populations. From 1946, the 'Expérience Internationale' was initiated with the aim of first testing different combinations between African populations, and between the Deli population and African populations. They were also interested in researching why the Deli dura generated better yields (Ferrand, 1946 – internal notes):

- Breeding dura is better than breeding tenera,
- The weather conditions are better in Asia than in Africa for oil palm.

Genetic trials undertaken between 1950 and 1954 revealed that African populations produced twice as many bunches than the Deli population. The average bunch weight (ABW) in the Deli populations were found to be higher than that of the African populations. Deli populations also produced the highest mesocarp to fruit (%M/F) ratio.

The fresh fruit bunch (FFB) measured in inter-origin crosses was significantly higher than that of the parental origin. Number of bunches (NB) and ABW of inter-origin crosses were intermediate to that of parental origins. The percentage of M/F and the percentages of palm oil and of total oil in inter-origin crosses were intermediate to that of parental origins.

Finally, inter-origin crosses displayed a better annual oil yield than that of the parental origins (Table 1) (Gascon and de Berchoux, 1964).

The improved performance of inter-origin crosses has been recorded, and breeding has been directed towards improving these crosses. Hence, the reciprocal recurrent selection (RRS) strategy presented by Comstock et al. (1949) was adapted with full-sib progenies

Table 1 Results from the 'Experience Internationale' – oil yield (kg/palm/year) for inter-origin crosses and their parental origin. Average 5–6 years (Gascon and de Berchoux, 1964)

Crosses	Number of progenies	Oil/palm/year (kg)	
La Mé x Deli	22	Dura	18.9
		Tenera	27.0
Yangambi/Sibiti x Deli	5	Dura	22.0
		Tenera	30.2
Deli x Deli	12	Dura	16.6
La Mé x La Mé	15	Dura	12.7
		Tenera	17.8
Yangambi/Sibiti x Yangambi/Sibiti	26	Dura	15.8
		Tenera	22.0

instead of half-sib progenies and with the addition of selfs (Gascon and de Berchoux, 1964).

3.3 Current breeding

Two major breeding methods were followed:

- The RRS was adopted in 1957 by IRHO (Meunier and Gascon, 1972). The first cycle has been planted in the La Mé Station between 1959 and 1968 (Gascon et al., 1988). Between 1975 and 1986, the first set of the second cycle was planted in Aek Kwasan (Indonesia) and in La Mé (Côte d'Ivoire). It was followed by several sets of the second cycle of selection from 1995 until now.
- The FIS was defined by Hardon (1970). This method was mainly adopted by South Asian companies and research institutes.

Some minor breeding methods also exist for oil palm breeding:

- Pedigree selection has also been practised in Nigeria, Indonesia and Malaysia (Rosenquist, 1990). In several breeding programmes, a F3 generation (two generations of self) was obtained. This strategy is designed for the long term, with the aim of obtaining pure lines after 10 generations (80–100 years). As oil palm is an outcrossing species, the strategy takes into account the inbreeding depression effect.
- Backcross breeding is also used for specific programmes. Dumpy Deli was introgressed into AVROS, which had been developed by backcrossing SP540T x Bangun with SP540T (Soh et al., 2006). This method is also used for specific traits such as dwarfness (Adon et al., 2001) and it is also commonly used with *Elaeis oleifera* by introgressing *E. oleifera* into *E. guineensis*.

4 Breeding objectives and methods

4.1 Breeding objectives

CIRAD and PalmElit have described major traits that need to be taken into account for breeding programmes.

The first aim concerns increasing the oil yield. This is the first priority as it is the main profitability factor for plantation companies. To do so, two principal components of the oil yields are taken into account, namely FFB which is the weight of bunches per hectare for a given period, and oil extraction rate (OER) which governs the quantity of oil in the bunch.

- FFB is also the result of two components: the bunch number (BN) and the ABW,
- OER is the result of three components: percentage of fruit to bunch (%F/B), percentage of mesocarp to fruit (% M/F) and percentage of oil to mesocarp (%O/M).

The second aim is to select for resistance to diseases. The first works were carried out on *Fusarium* wilt which is mainly located in Africa. Early screening tests have been developed from the initial work of Prendergast (1963) at the nursery stage. Renard et al. (1972) adapted screening tests to the pre-nursery stage and this method is still used. In South East Asia, and increasingly in Africa, basal stem rot affects plantations and an early screening test has been developed for *Ganoderma* (Breton et al., 2006). Finally, in Latin America, a complex of lethal bud rot damages plantations. It is possible that the causal agent is *Phytophthora* at least in Colombia (Martinez et al., 2009; Torres et al., 2010), but this hypothesis still needs to be confirmed. No early screening tests are routinely available and screenings are done through planting in bud rot-infested zones. These resistance breeding objectives are presented in detail in this book.

Tolerance to marginal cultivation conditions, especially to dry conditions, is becoming increasingly significant due to climate change, and the necessity to develop this crop under marginal conditions.

Other breeding objectives are to reduce height growth in order to increase the economic life of a plantation, to reduce tree size in order to increase the density and therefore the oil yield, and to reduce lipase activity in order to have more time between harvesting and processing in the oil mill.

Soh et al. (2009) presented the priority list of desirable traits for the genetic improvement of oil palm in Malaysia. It includes characteristics such as high palm oil yield, dwarf stature, *Ganoderma* resistance and high oleic acid.

4.2 Breeding methods

Reciprocal recurrent selection (RRS)

RRS is based on the approach developed by maize breeders (Comstock et al., 1949), and relies on the development of two complementary populations. This method concentrates the desirable traits by increasing the frequency of these traits in the populations. It is well adapted to produce improved hybrids, and particularly to enhance low-heritable traits. This method is composed of two cycles: a cycle of recombination or self, and a cycle of progeny testing (Fig. 1). These two cycles can be repeated over time.

Initially, populations have been divided into two groups (Gascon et de Berchoux, 1964):

- Group A includes dura palms of the Deli origin, which have a small number of large bunches,
- Group B includes pisifera of La Mé and Yangambi/Sibiti origins, which have a large number of smaller bunches.

Some other populations have been added later in the process (Meunier et Gascon, 1972)

- Angola and Cameroon have been added to group A
- Pobè, Yocoboué, Nigeria and also Cameroon have been added to group B

The test phase consists of assessing crosses between palms of group A and palms of group B. The DxT or DxP hybrids obtained are planted in comparative trials. This enables the ranking of crosses (progenies) and corresponding parents, and highlights particular combining abilities.

The recombination phase consists of crossing or selfing the best parents of the test cycle. In group A, crosses are realized between the best duras (DxD) and selfs. In group B, crosses are realized between the best tenera and pisifera, (TxT and TxP), and selfs of tenera.

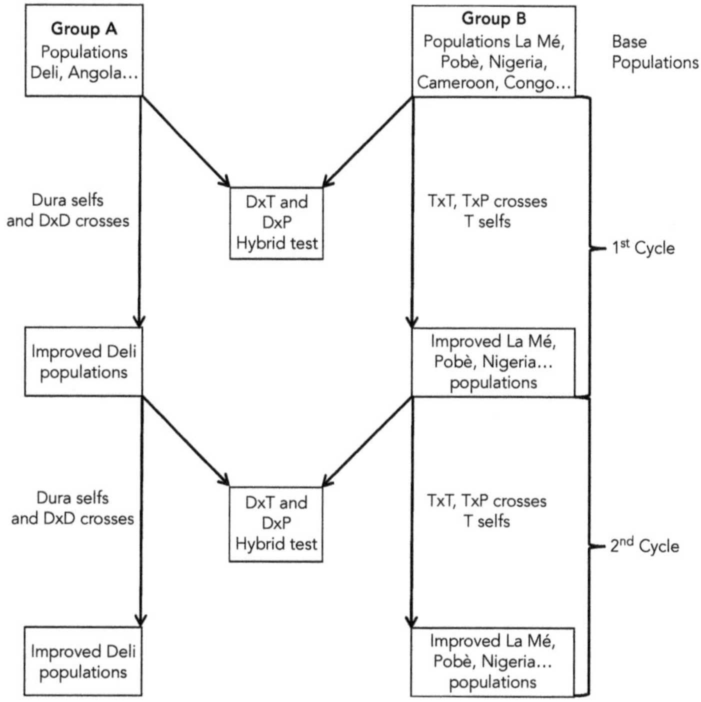

Figure 1 Recurrent reciprocal selection programme in PalmElit/Cirad.

This method, with very minor modifications, was used by INEAC (Institut National pour l'Etude Agronomique du Congo Belge) (Picel, 1956; Sparnaaij, 1958). The WAIFOR[4] (West African Institute for Oil Palm Research) adopted the same method, but the studied population was more diverse and less inbred (Sparnaaij et al., 1963). Finally, Dami Oil Palm Research Station also used this breeding method in Papua New Guinea (Dumortier, 2003).

Backcross: dwarfness

The introgression of slow vertical growth in improved oil palm populations is an example of a backcross breeding programme for *E. guineensis*. In group A, the Dumpy Deli population is known for its dwarfness. In group B, there is a progeny of the palm 1–2T (or PO452T) originating from the Akpadanou region in Benin (Adon et al., 2001).

Amongst different crosses, Deli x Dumpy Deli and PO452T x La Mé crosses were found to display slow vertical growth. In trials planted in 1983–1985 at La Mé Station, Dumpy Deli x Deli showed a vertical growth which was 28% lower than the control cross. For PO425T self x La Mé self, the vertical growth rate was even lower, showing less than 42.7% to the control cross. However, these crosses had a low oil yield (Adon et al., 2001), thus it appeared necessary to develop a programme of backcross in order to combine the high yield of the Deli and La Mé material with the dwarfness of Dumpy Deli and PO452T (Fig. 2).

4. WAIFOR is becoming the NIFOR, Nigerian Institute For Oil palm Research.

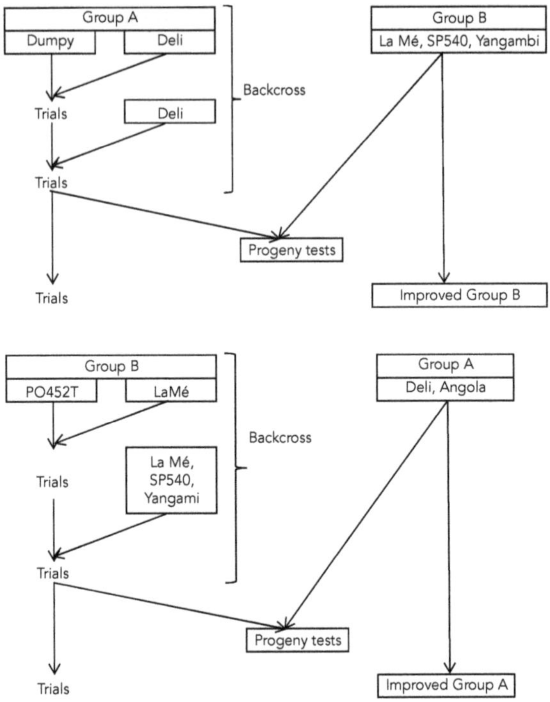

Figure 2 Introgression of Dwarfness of Dumpy Deli and PO452T in improved populations.

Backcross: *Elaeis oleifera*

E. oleifera has been the focus of attention of oil palm breeders for a long time. *E. oleifera* was introduced at Eala (DR Congo) in 1927 and its progenies at Yangambi in 1940/41 (Vanderweyen and Roels, 1949). Elsewhere this happened in 1952 and 1956 in Indonesia (Lubis, 1988) and then in La Mé Station in 1960 (Meunier, 1975). This palm presents interesting properties such as tolerance to lethal bud rod in Latin America, oil fluidity, and slow vertical growth. In addition, some ecotypes seemed to display some good resistance to vascular wilt (Renard et al., 1980).

First, an interspecific hybrid improvement programme was set up: this has been presented in details by Le Guen et al. (1991) and was initially designed by Meunier et al. (1976). It consisted of:

- Tests measuring combining ability between ecotypes, to determine which populations combined well with each other
- Test of individual combining ability, to identify the best parents from both *E. guineensis* and each *E. oleifera*.

In La Mé (Côte d'Ivoire), Aek Kwasan (Indonesia), Pobè (Benin) and Rio Urubu (Brazil), 11 common *E. oleifera* and 7 *E. guineensis* have been tested. Quickly, it appeared that the partial sterility of the interspecific hybrid was important (Arnaud, 1980). Bunch production can be comparable to that of *E. guineensis*, although OER remained low in the interspecific hybrid. However, interspecific hybrids are still the only solution for plantations affected by lethal bud rot, through the introgression of *E. oleifera*'s desirable qualities into the African oil palm. A breeding strategy for interspecific hybrids and for backcrossing, with the aim of introgressing *E. oleifera*'s desirable qualities into *E. guineensis*, has been described by Le Guen et al. (1991).

Many programmes of backcrossing have been undertaken, for example, in Nigeria (Obasola et al., 1977), Costa Rica (Sterling et al., 1988; Escobar and Alvarado, 2003) and Malaysia (Tam et al., 1977; Rajanaidu et al., 1983).

The major problem posed by interspecific backcrosses is the assessment of the yield. In first generations, the value of the family does not have real significance (Durand-Gasselin et al., 2009) and the problem is the same at the individual level (Soh, 2003).

Family and individual selection (FIS)

This method, called family and individual selection (or FIS; Hardon, 1970) or modified recurrent selection (Soh et al., 2003), is widely used in South East Asia. The first step is a mass selection: the selection of Deli dura parents is based on family and individual palm performances. These parents can be used for both seed production and breeding programmes. The tenera parent selection scheme follows the same method. However, for the seed production, pisifera are needed, so they are selected on the basis of the tenera sibling performance. Then a DxP progeny trial is implemented. This confirms the choice of pisifera selected for seed production (Soh et al., 2009).

The choice of this method was based on several hypotheses. The number of palms constituting the base population favours less genetic drift and inbreeding. The latter can be considered as a reverse selection and it is also important to broaden the genetic base. Under such conditions, the eigenvalue and the breeding value were found to show a good correlation. It is for this reason that the method uses crosses between different BPROs, and why relatedness is systematically avoided. Therefore, this resulted in an exacerbated GCA (general combining ability[5]), and forced the introduction of new germplasm (Soh et al., 2009; Corley and Tinker, 2016).

4.3 Mating design

The test phase

The biparental mating design was used during the first cycle of the RRS and at the beginning of the second cycle. At the end of the 1970s, NCM1 (North Carolina Model 1) was used for a limited number of progeny testing trials. Since then, only incomplete factorial design was used, and this was later complemented by adding testers. This model allowed estimations of both the GCA and the SCA, although the GCA was largely predominant.

The recombination phase

In RRS, the focus is on the recombination phase, where most importantly the hybrid crosses between group A and group B are assessed. This is because commercial seeds are a representation of the best hybrids. For this step, the biparental mating design was mainly used. There were a few attempts at using diallel design, but this did not last long.

5. General combining ability: it is the average genetic effect transmitted by an individual to its offspring.

The biparental mating design is mainly used as it allows the assessment of many more crosses compared to the diallel mating design.

Soh et al. (2009) explained that for the FIS method, the biparental mating design is used in the parental crosses. All the known genetic mating designs are used for dura x dura, tenera x tenera and tenera x pisifera within population crosses and also for the dura x tenera and the dura x pisifera inter-population progeny test crosses (biparental crosses, NCM1, NCM2, diallels).

4.4 Experimental design

The test phase

The first experimental design was based on row planting: progenies were planted in rows, generally with 3 replicates of 3 rows. Nowadays, the main experimental design is the lattice, which is a balanced incomplete block design. Indeed, 16 or 25 crosses are tested, using 5 or 6 replicates of 12 or 16 palms per replicate. When the number of crosses differs from 16 or 25, the randomized complete block design is used with 6 replicates. Finally, the D-optimal design has been tested which is an unbalanced incomplete block design. Instead of having 5 trials with connections or control crosses between trials, all the crosses can be tested in the same trial. However, this design has proved difficult to manage in practice.

The recombination phase

All the crosses are planted by rows. In group A recombinations, progenies are often represented by 13 palms. However, progenies of the seed garden can also be considered as recombinations. In this case, the number of palm per progeny can reach 300. In group B recombinations, progenies are often represented by ca. 26 palms. As for group A, progenies from the seed garden can also be considered as recombinations. Here, the number of palms can reach 78. The maximum number was for the self of LM2T with 644 palms.

5 Data collection methods

5.1 Oil yield

The oil yield is measured through its two major components: the FFB yield and the oil to bunch (O/B) yield.

FFB yield is collected for each palm, from the third year after planting to the end of the ninth year. Every 7–15 days of the harvesting cycle, the number and weight of the ripe bunches are recorded. This allows to record the NB, the total FFB and the ABW for a palm each month and so for each year.

The O/B is estimated differently. For the test phase, O/B is determined at the progeny level. For dura x tenera, all the teneras are analysed once at 5 years of age and once when they reach 6 years old. For the dura x pisifera, 40 teneras are analysed when 5 and 6 years old. In the recombination phase, for the dura x dura, all the palms are analysed at 5 and 6 years of age. For the tenera x tenera and tenera x pisifera, only the teneras are analysed when 5 and 6 years old. Depending on the progeny value, the number of bunch analyses can be increased from 2 to 4, and even to 8 in some cases.

The bunch analysis method was derived from the method developed by Blaak et al. (1963) at Waifor. For bunches under 14 kg, the ratio of the fresh fruit weight to the bunch (fresh fruit + spikelet + stalk) weight (F/B) is determined for the whole bunch. For bunches over 14 kg, F/B is estimated from half of the bunch. A sample of 30 random normal fruits is taken and weighed. The mesocarp is scraped off. Nuts are then weighed. The ratio of the mesocarp to the fruit (M/F) is then estimated. Finally, the oil to mesocarp (O/M) ratio is measured by extracting oil with a solvent (hexane) in a giant Soxhlet apparatus. F/B x M/F x O/M give the O/B. Instead of using the O/B ratio, we prefer to use the OER which is the O/B x 0.855. This correction factor corresponds to an estimation of losses during extraction under mill conditions.

Palm oil yield is the result of NB x ABW x F/B x M/F x O/M x 0.855 or FFB x OER.

In addition, the kernel oil yield is determined. Kernel weight of the 30 fruits is recorded. The kernel oil content is estimated to 50%. Kernel OER is the result of F/B x K/F x 0.5.

5.2 Oil quality

Iodine value is measured on oil samples after the bunch analysis procedure. This method relies on a titration with a Hanus reagent and thiosulphate, which allows the unsaturated fatty acid content to be estimated.

5.3 Lipase activity

After harvesting, a rapid acidification of bunches occurs. This is due to a high endogenous lipase activity that releases free fatty acids (FFA) in the mesocarp of bruised ripe fruits (Sambanthamurthi et al, 1995). Low FFA is a requirement for good quality palm oil. A FFA content of more than 5% is considered to be inadequate for human consumption (Ngando-Ebongue et al., 2008). Identifying the genetic variability of this trait, is important in order to facilitate harvesting and limit oil acidity. The oil acidity is estimated by FFA titration. For this, the pressed mesocarp is placed in solution in a solvent mixture. FFAs are titrated with an ethanol solution of potassium hydroxide containing phenolphthalein.

5.4 Vegetative traits

Vegetative traits are measured for each individual palm tree. Height is measured according to Jacquemard (1979), mainly at 6 and 9 years old, although measurements at 12 and 15 years old can also be added. Canopy and leaf area index are measured at 10 years old. Canopy is the average of the projection on the ground of the longest three leaves at three directions measured at 1.5 m height. Leaf area index was adapted from the method described by Tailliez and Ballo Koffi (1992).

6 Impact of reciprocal recurrent selection (RRS) on oil yield

6.1 The Expérience Internationale

In La Mé, Côte d'Ivoire, at the end of the 1940s, the oil yield per hectare reached 1.9 tons. The main result of the Expérience Internationale was to highlight the interest of Deli x Africa crosses with an oil yield increase of 58%, thus reaching 3 tons/ha (Gascon and Wuidart, 1975).

6.2 The first RRS cycle

The first cycle was implemented from 1959 to 1968 in La Mé and it was called 'Bloc 500'. On 500 hectares, 392 crosses of Deli x Africa were assessed. Almost 62 crosses were selected, with an average oil yield per hectare and per year of 3.9 tons. Compared to the Expérience Internationale, the gain was about 30%. Compared to the best crosses from the Expérience Internationale, LM2T x DA10D, the gain was 18% (Gascon et al., 1981).

On these 62 crosses, a selection was carried out based on height growth. The average height growth in the studied population ranged from 36 to 76 cm. It was decided to select only crosses displaying a height growth lower or equal to 50 cm/year. Finally, the height growth of the selected crosses was around 45 cm/year. However, this selection ended up in the rejection of a large number of Deli x Yangambi crosses (Gascon et al., 1981).

The fatty acid composition of fruits sampled from 169 crosses originating from the first cycle was studied. Palms from La Mé origin displayed the best unsaturated fatty acid rate reaching 54.1%. This rate was 52.5% for Deli and only 49.2% for Yangambi origins. These results were consistent with those from Eckey (1954), who described a gradient between Sierra Leone (57.1%) and Zaire (52%) material (Gascon and Wuidart, 1975). These differences were also shown in the Deli x Africa crosses. Indeed, the unsaturated fatty acid rate was the highest in Deli x La Mé crosses (52.8%) and it was 50.6% for Deli x Sibiti, 50.1% for Deli x Nigeria and 48.9% for Deli x Yangambi (Gascon and Wuidart, 1975).

6.3 The second RRS cycle

The second RRS cycle was implemented in eight locations between 1975 and 2012 (Table 2). Unfortunately, due to a severe drought, all the trials planted in Benin at Pobè Research Station could not be fully monitored. For La Dibamba and Rio Urubu, only a very few number of trials were completed.

The first results obtained for oil yield presented an average progress of 15% between the second cycle and first cycle of RRS (Gascon et al., 1981). This progress was higher when the tested Deli parents came from recombinations (Table 3). Crosses from selfed progenies reflected more improvement of the cross than from the second cycle. However, these results showed that improving a cross is possible, because there was enough variability within a first cycle cross. These second cycle trials highlighted the importance of recombinations between Deli parents. Such recombinations within group A are also based on the complementarity between the number and the average weight bunch and these allowed the best progress. Recombinations between parents of group B did not generate the same progress than that obtained for recombinations between Deli parents. Indeed, in this case, recombinations were not based on the complementarity between bunch yields components (Cochard et al., 1993). On average, and compared to the standard cross LM2T x DA10D:

- crosses between selfed progenies generated a progress of 3.9%
- crosses between self of group A progenies and recombination between group B progenies displayed a progress of 1.4%
- crosses between recombinations of group A progenies and self of group B progenies have enabled the highest progress, which was 13.6%
- crosses between recombination of group A and group B progenies did not generate any progress (Fig. 3). However, there were only two examples of recombinations.

Table 2 Experimental designs of the second RRS cycle conducted since 1975

Estate	Company	Country	Years	No. of trials	No. of crosses	No. of deli dura	No. of tenera/ pisifera	Origins of tenera/pisifera
Aek Kwasan	PT Socfin Indonesia	Indonesia	1975–1979	16	327	294	110	La Mé, Yangambi, SP540
La Mé	IRHO, now CNRA	Côte d'Ivoire	1975–1991	21	714	389	242	La Mé, Yangambi
La Dibamba	IRHO, now IRAD	Cameroon	1975–2000	12	318	257	121	La Mé, Yangambi
Bangun Bandar	PT Socfin Indonesia	Indonesia	1981–1991	3	152	122	46	La Mé, Yangambi
Rio Urubu	Embrapa	Brazil	1984–1988	12	256	123	89	La Mé, Yangambi
Aek Loba Timur	PT Socfin Indonesia	Indonesia	1995–2000	25	571	142	154	La Mé, Yangambi
Pobè	CRA-PP de l'INRAB	Bénin	1995–2002	9	167	94	96	La Mé, Yangambi
Aek Kwasan II	PT Socfin Indonesia	Indonesia	2005–2012	25	660	175	147	La Mé, Yangambi, Ekona, Ekona x AVROS

Table 3 Progress between second cycle and first cycle for crosses from selfed progenies and recombined progenies (from Gascon et al., 1981)

Selfed progenies	Progress	Recombined progenies	Progress
DA115D self x LM2T self	5–15%	(LM404DxDA10D) x LM2T self	8–17%
DA10D self x LM2T self	9–12%	(DA5DxDA3D) x LM2T self	13–21%
LM404D self x LM2T self	5–11%	(DA10DxDA3D) x LM2T self	25–32%
Average	6–13%	Average	15–23%

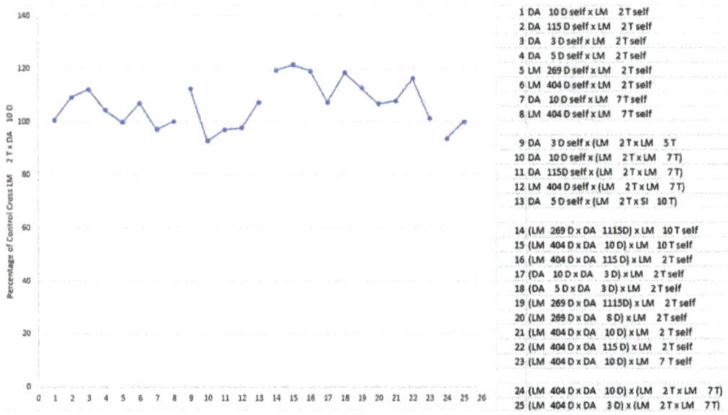

Figure 3 Oil yield as a percentage of control cross: LM2T x DA10D. Results of the second cycle of RRS mainly planted in Aek Kwasan (Indonesia) and La Mé (Côte d'Ivoire). From Cochard et al. (1993).

Table 4 Average and general combining abilities (GCA) of oil yield, FFB, OER and growth for the 4 independent groups tested in the Aek Loba Timur experimental design. GCA estimated for the x % best crosses for oil yield. Oil yield: Tons/ha/year – FFB: kg/palm/year – OER: industrial percentage of oil per bunch – i: intensity of selection

Group		nb parents	i	Oil yield	FFB	OER	Growth
Angola x La Mé	Average			7.56	241.8	23.2	51.8
	Angola	11	10%	+8.7 %	+4.0 %	+4.8 %	+3.1 %
	La Mé	11	10%	+6.9 %	−3.2 %	+10.9 %	+9.1 %
BBSP crosses	Average			7.42	208.8	26.3	63.8
	Deli	16	12.5%	+5.3 %	+5.0 %	+0.4 %	+4.7 %
	Group B	24	12.5%	+8.5 %	+10.1 %	−1.2 %	+6.9 %
Deli x La Mé	Average			7.53	213.5	26.2	52.5
	Deli	83	10%	+7.7 %	+2.7 %	+4.5 %	+4.8 %
	La Mé	102	10%	+11.3 %	+6.0 %	+5.1 %	+6.5 %
Deli x Yangambi	Average			7.45	210.1	26.3	59
	Deli	19	10.5%	+8.1 %	+8.4 %	−0.1 %	−3.7 %
	Yangambi	8	12.5%	+8.3 %	+3.0 %	+4.9 %	−8.0 %

Table 5 Experimental designs of the third RRS cycle established since 2010

Estate	Company	Country	Years	No. of trials	No. of crosses	No. of Deli Dura	No. of Tenera/Pisifera
Cole	Murrin	Ecuador	2010–2015	23	413	125	96
Presco	Siat	Nigeria	2010–2016	29	690	182	116

The last second cycle RRS experimental design for which results were obtained was the Aek Loba Timur's trials. For 571 crosses tested, the average oil palm yield was 7.4 tons/ha/year. This production level was equivalent to the standard cross LM2TxDA10D, showing all the genetic improvement achieved (Jacquemard et al., 2010). The best 10% crosses for palm oil yield displayed a production of 8.5 tons/ha/year, which still represent a gain of 15%. The best cross reached an oil yield of 9.2 tons/ha/year, which means a 24% improvement. For these best 10% crosses, FFB was 229 kg/tree/year (around 32.7 tons of bunches/ha/year) and the OER was 27.8. For the best cross, FFB was 231 kg/tree/year (around 33.0 tons of bunches/ha/year) and OER was 28.1. This result reflected the difficulties to combine both traits in order to obtain the best cross. Indeed, in this genetic experimental design, the best FFB was 256 kg/tree/year (around 36.6 tons of bunches/ha/year) and the best OER was 30.5.

Estimations of GCA were carried out in this experimental design (Jacquemard et al., 2010). They allowed the value of each tested parent to be assessed. When the additive value of each parent was compared to the average of all crosses, the value of non-tested crosses by adding the value of each parent, or approximately 10% of the best crosses, the oil yield gain was at least + 5.3%. Therefore, by combining the best crosses, the minimum gain was around +10%. This total gain can be largely the result of an improvement in FFB, or rather a gain on the OER, but also an almost equivalent gain for these two components (Table 4). However, this oil yield increase was accompanied by a height growth increase. Progenies that have a better FFB, a better OER, or a smaller growth than presented in Table 4 exist, but with a lower oil yield improvement. Accordingly, the commercial supply was segmented based on defined typologies.

6.4 The third SRR cycle

The third SRR cycle is presently being planted. The two first experimental designs have been planted as rom 2010, in both Ecuador and Nigeria (Table 5). In addition, in order to assess improved second cycle parents of Deli, La Mé and Yangambi origins, this third cycle will have incorporated new Nigeria and La Mé x SP540 recombinations.

7 Impact of oil palm breeding on other traits and using other methods

7.1 Disease resistance

The second major objective of the oil palm breeding programme is to select for resistance to major disease: *Ganoderma* for South East Asia, *Ganoderma* and *Fusarium* wilt for Africa, and the complex of bud rot for Latin America.

7.2 Iodine value

The iodine value of palm oil usually ranges between 50 and 58 (Meunier and Boutin, 1975). Breeding has increased the average iodine value. Final results from the Aek Loba Timure experimental design (Jacquemard et al., 2010) reported an average iodine value of 53.8. The iodine value of the Deli material was approximately 52.6, but reached 56 for progenies with LM404D as founder. The iodine value in palms of Yangambi origin was lower than that in palms of La Mé origin (50–54 compared to 57–60). In each origin, it was possible to obtain a 5% improvement for the iodine value.

7.3 Lipase

FFA and oil acidity were measured for several palms from various crosses by Ngando Ebongue et al. (2008). The two *E. oleifera* palms from Monteria (Brazil) provided oil with low acidity and a low % of FFA. For *E. guineensis*, few palms display a low % of FFA and low acidity, on different origins, Deli x La Mé, La Mé and Deli. The differences were not at the origin or at the progeny level but only at the palm level for some progenies. Morcillo et al. (2013) explained that segregation of the low acidity trait was monogenic and recessive. These studies were performed on a small number of progenies and origins and it is now necessary to study the genetic variation of this trait throughout the available genetic diversity.

Wong et al. (2016) have screened the Malaysian Palm Oil Board's wild oil palm germplasm for lipase activity. At the population level, low FFA levels were measured in palms originating from Nigeria (2.55%), Guinea (7.99%) and Senegal (8.38%). Angola, Tanzania and Zaire displayed a high mean FFA value, which was higher than 20%.

7.4 Results from FIS

Dura x pisifera progeny tests are based on family and individual assessments of the dura x dura and tenera x tenera tests. Pisifera choice for the progeny test is based on the sib tenera performance. As the Deli dura is assumed to show little variability, work was focused on pisifera parent breeding value. Several performances of pisifera have been described and analysed (Hardon et al., 1987, Breure and Konimor, 1992; Lim et al., 2003; Teo et al., 2003; Noh et al., 2010; Junaidah et al., 2011; Noh et al., 2012).

Recently, results were published by Arolu et al. (2016) for Deli dura x Nigeria pisifera progeny tests. With a NCM I breeding design using 24 Deli dura nested into 10 Nigerian pisifera, and a randomized completed block design with two replicates of 16 palms per progeny, the authors described the performance of dura x pisifera crosses and GCA of the pisifera parents. Amongst the 10 pisifera tested, palms with the highest GCA for the FFB, displayed the highest GCA for the BN. This result suggests once more that the easiest way to increase FFB is to increase the BN, as Beirnaert already explained in 1935. When compared to the trials average results, the best pisiferas show an increase of 12% for oil yield.

Teo et al. (2003) described performance results of dura and pisifera tested under the design explained by Lim et al. (2003). They showed that AVROS pisiferas had higher oil yield but also displayed a rapid height increment and strong vegetative growth. Ekona pisiferas were equivalent or superior to those from AVROS origins, and Binga pisiferas were not as good as Ekona, on average. Pisiferas from Ulu Remis were found to display the poorest oil yield. Yields of duras, African duras, Deli duras or mixed duras have also been analyzed. The African and mixed duras produced more vigorous and higher yielding dura x pisifera progenies. Therefore, the authors concluded that there is no justification for persisting exclusively with pure Deli duras as seed parents.

8 Seed production

8.1 Methods

The first method for seed production established by IRHO, was based on the reproduction of the most successful crosses from the test phase of the first cycle of the RRS through the selfing of parents (Jacquemard et al. 1982). The most successful tested pisiferas were first crossed with the most successful teneras, and then pisiferas obtained from these most successful tenera x pisifera crosses were used for seed production. A minimum of twelve duras and twelve pisiferas were necessary to adequately reproduce the original cross. Between 12 and 20 crosses were enough to reproduce the initial cross (Jacquemard et al. 1981). This method has the advantage of producing a large number of seeds representing the initial cross.

For the second cycle, it was difficult to implement this method because it required a larger surface and a high number of crosses. In addition, Soh et al. (1989, 1990) found that a trial of 6 replicates of 16 palms did not allow differentiation between crosses with less than a 15% difference under Malaysian conditions. Under conditions in Indonesia or Cote d'Ivoire, trials can differentiate between crosses with at lease a 10% difference. Therefore, it is not possible to build the seed production only from the most successful crosses, but from a set of the best crosses from each trial, thus increasing the number of crosses involved in the seed production. The seed production based on the second cycle of SRR has evolved: indeed, when possible, crosses are reproduced as the first cycle, otherwise it is balanced at the grandparent level (Fig. 4).

With the FIS breeding method, the seed production is based on the duras selected on their phenotypic performances, and on pisiferas selected on the phenotypic performances of the duras and teneras sibs, or after a generation of progeny testing. Thus, seed production is simply the combination of most successful parents identified on their phenotypes or direct reproduction of the most successful crosses identified in the progeny tests.

8.2 Pollen availability

The CIRAD/PalmElit seed production process has the advantage of relying on many pisiferas. However, they are not always enough to guarantee sufficient pollen production. To overcome the problem of pollen deficiency, a drastic pruning method has been developed which ensures a pollen production 20–24 months after pruning, depending on the oil palm age (Durand-Gasselin, et al. 1999).

8.3 Seed preparation

Unlike in most plants, oil palm seeds cannot be delivered as such, seeds must be delivered as germinated seeds. In order to obtain seed germination simultaneously, a long and complex method has been developed.

Bunches originating from controlled pollinations are harvested as soon as one fruit becomes naturally detached. At the seed production facility, bunches are chopped in order to separate

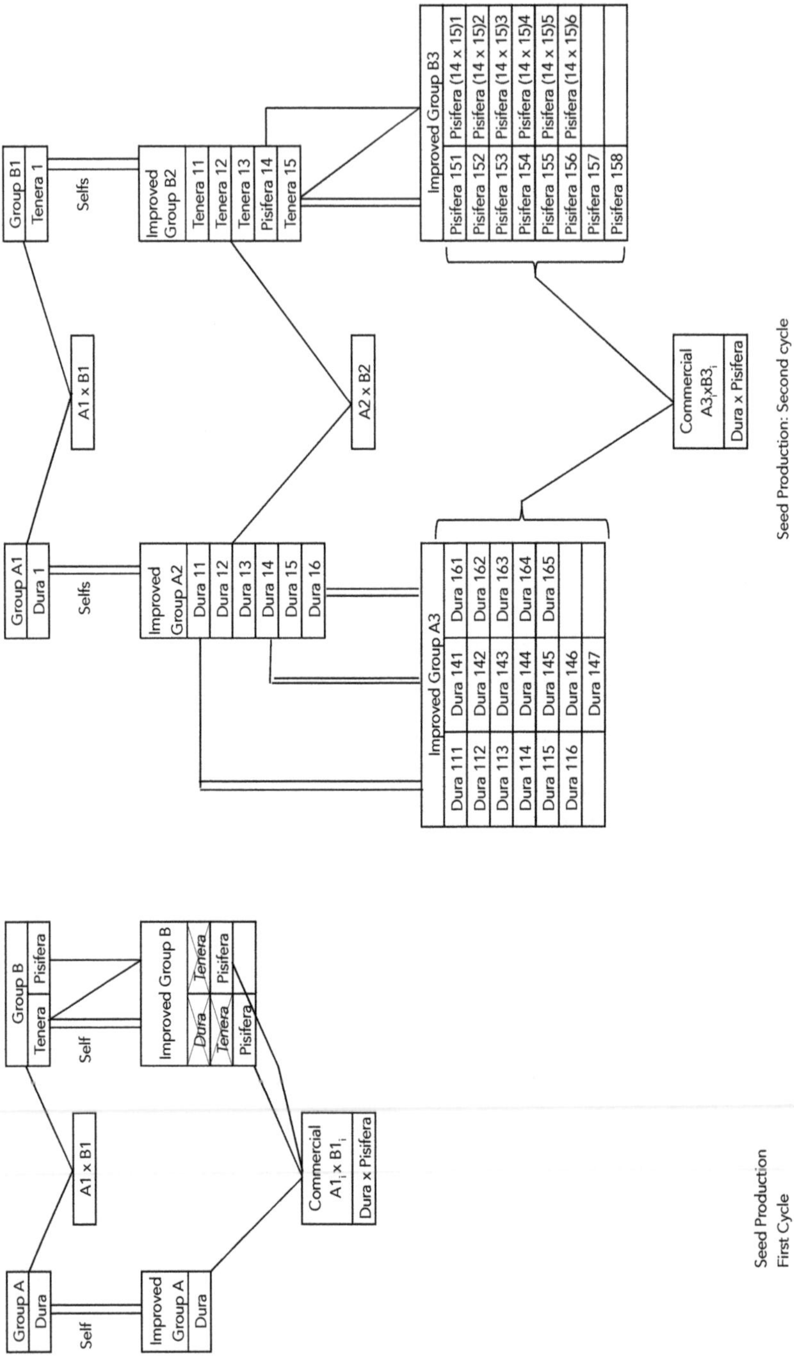

Figure 4 Organization of the seed production in the first and second cycle of SRR. For the second cycle, it may be possible to reproduce a cross, for instance self of Dura 14 with the self of tenera 15; it may also be possible to balance all the crosses between the self of dura 11, dura A4, dura 15 and tenera 15, and tenera 15 x pisifera 15. This is a reproduction of an improved second cycle of the grandparents dura 1 and tenera 1.

spikelets and rachis. Spikelets are put in boxes for 3–4 days and humidified during this period in order to facilitate their destemming. Alternatively, chopping is continued in order to separate fruit spikelets. Fruits are then placed into pools for 7 days to facilitate pulp removal. The mesocarp of the fruits is removed using a depulping machine. Fresh seeds are then cleaned of the adhering fibres, and white seeds, small seeds or floating seeds are eliminated. Seeds of one-fifth of the crosses are observed for embryo conformity. If a cross shows less than 90% of normal embryos, it is eliminated. Seeds are then counted and treated with a fungicide.

Seeds are stored in an air-conditioned room. Hygrometry is maintained at 60–65% and the temperature kept at 22°C. Seeds can be stored for a maximum of 18–24 months otherwise their germination capacity becomes reduced.

The germination method includes two steps (Corrado and Wuidart, 1990):

- the first step is to break dormancy by reducing moisture, down to 17 and 19% moisture on kernel fresh weight or 18 ± 0.5% moisture of the whole seed dry weight through heating
- the second step is to ensure the development of embryos by placing the seeds at room temperature with a moisture rate on kernel fresh weight between 23 and 25% of a whole seed fresh 22 ± 0.5%

Seeds are removed from the storage room and soaked for 7 days. They are then wiped until they reach a dark grey colour, indicating a moisture content of 17–19% of fresh kernel weight. Seeds are then placed in transparent polyethylene bags which are sealed and placed in a germinator at a temperature of 39°C ± 0.5. They remain here for 80 days for the Deli x La Mé crosses or 60 days for Deli x Yangambi crosses.

After heating, seeds are rehydrated by soaking for 5–7 days. Seeds are wiped a second time until they reach a matte black colour, indicating a moisture content of 23–25% of fresh kernel weight. Seeds are once again put in transparent polyethylene bags at room temperature (25–27°C). Ten days later, the first set of seeds are collected followed by a weekly collection during the next 4–5 weeks. Finally, collected seeds are packed in batches of 100–200 germinated seeds inside plastic bags.

8.4 Seed delivery

CIRAD seeds are delivered as various categories based on a mix of crosses with the same grandparents (Durand-Gasselin et al., 1999, 2000). Standard material is Deli x La Mé, characterized by a high FFB and CPO production, and higher oil content of unsaturated fatty acids compared to other planting materials. It also has a significant drought resistance capacity and a low height increment which enables an economically profitable operation over more than 25 years. During the past decades, CIRAD and PalmElit breeders have improved the Deli x La Mé material, in the aim of generating less bulky palms for high-density plantation, resistance against diseases and low lipase material. The other elite materials on display are Deli x Yangambi and the interspecific *E. oleifera* x *E. guineensis* hybrid (Turnbull et al., 2016).

In the future, the labelling of planting material will advertise specific characteristics:

- resistant to bud rot disease, *Ganoderma*, *Fusarium* wilt, or drought
- those with a very low height increment, with less bulkiness
- those with a low lipase activity.

9 Future trends

For the future genetic improvement of the oil palm, conventional breeding will profit from various different research area themes.

9.1 Breeding for resistance to drought

Countries and planters are increasingly interested in oil palm cultivation. Except in some countries from Africa and Latin America, most of the remaining areas display unfavourable environmental conditions. In addition, climatic change leads to an increase in uncertainty about environmental parameters. It is therefore increasingly important to breed for the resistance to drought.

Experiments have been implemented in Pobè (Benin), a place with a water deficit of 700 mm per year. In these conditions, the most relevant criteria is the percentage of mortality. The Deli x Yangambi showed less mortality, but FFB was lower (43.5 kg/palm/year) than in Deli x La Mé (65.8 kg/palm/year) (Houssou, 1985). A comparison between crosses planted in Benin and Indonesia (North Sumatra) showed that OER components stayed relatively unaffected (Nouy et al., 1999). For Deli x Yangambi, and Deli x La Mé which display the same FFB yield in Indonesia, oil and FFB yields were found to be 50% higher for Deli x La Mé compared to Deli x Yangambi in Benin (Nouy et al., 1999).

New trials are now implemented in different locations (Nigeria, Benin, Indonesia), with different water deficits (200, 700 and 400 mm). Periodic observations are made on this experimental design such as phenological monitoring of the appearance and development of organs, morphological and biomass monitoring of all organs (including roots). Also, bunch analysis, leaf photosynthetic capacity, and levels of non-structural sugars in vegetative organs are measured. The intra-annual variations in phenology, growth, functioning and production are thus analysed in the light of changes to climatic factors such as water, temperature and radiation.

9.2 Improving the yield distribution over the year

Oil palms, when planted in locations where water deficit occurs, display a peak of production. With no water deficit, the bunch production is spread over the year. To widen this production peak, there is a means of estimating and representing the distribution of production using the Gini coefficient and illustrating it by Lorenz curves. Hence, Cros et al. (2013) showed differences between crosses as Gini coefficient varied between 0.382 and 0.699 for the bunch harvest. This coefficient summarizes the distribution of production and also highlights existing variability parameter to be taken into account for breeding, provided that crosses are at equivalent production levels.

9.3 Improving fertilizer's efficiency

The cost of fertilizer constitutes up to 60% of the plantation upkeep cost. Different methods are followed to reduce this cost, by recycling palm oil mill effluents, empty fruits bunches or compost. Foliar diagnosis is also used to assess nutrient status and fertilizer requirements (Foster 2003). The future of oil palm breeding can be based on improving the ability of palms to make better use of fertilizers. Genetic variability has been shown for nitrogen uptake at the nursery stage (Law et al. 2012). Ollivier et al. (2013) described contrasted leaflet concentrations for elements such as N, P, K and Mg between different genetic origins at measured plantations. Hence, there are optimum levels of nutrients for each genetic origin. For the same production level, the amount of mobilized nutrients may vary by up to 30–40% between genetic origins. Efforts must concentrate in the future on the fine-tuning of the critical levels using contrasted oil palm varieties to specifically adapt fertilization (Ollivier et al. 2017).

9.4 Improving pollination at a young age

In 1989, Meunier et al. showed that increases in oil yields mainly comes from the NB. Corley and Tinker (2016) showed that the modern populations, after four generations of selection, can display 16 bunches per palm per year. At 5–10 years of age the best pisiferas can now produce more than 18 bunches per year (Arolu et al., 2016). In the Aek Loba Timur project (PT Socfindo, Indonesia), on average, the NB per year in palms aged 6–9 years was around 17. At the young age (3–5 years), this NB accounted for one-third, with around 28 bunches per year. In these conditions, the number of male inflorescences was rare at the young age. This lack of pollen and consequently a bad fruit set of the female bunch, is being seen more frequently in the commercial planting material, regardless of breeding strategy. In order to limit this drawback, two possibilities exist:

- The first option is to limit the NB at the young age while increasing the ABW. However, this solution also limits the NB at the adult age and therefore limits the yields.
- The second option is to develop dioecious palms. Two kinds of palm will be developed: some with only female bunches and some with only male inflorescences. The latter, called supermale palms, will ensure adequate pollination of the other palms. The targets are to develop a supermale breeding programme and an appropriate design, with a defined and limited number of supermale palms per hectare.

9.5 Improving data analyses

Experimental and genetic designs have evolved considerably. As a consequence, data analyses have also changed. At the beginning, the aim was to find the best crosses using ANOVA. With genetic designs, the aim was to find the best parents by estimating mainly GCAs and breeding values. Breure and Verdooren (1995) described different genetic designs adapted for oil palm to estimate the GCAs and the breeding values of the tested parents. In 1994, Soh estimated breeding value using BLUP (best linear unbiased prediction). This method can integrate unbalanced data from experimental and mating designs and it was also used to predict hybrid performance (Purba et al., 2001). With the BLUP method, estimation of relatedness is facilitated by the knowledge of the oil palm pedigree. Oil palm breeding methods are increasingly similar to animal breeding, hence in breeding programmes BLUP approaches will be used more frequently. In addition, such an approach in a well-studied pedigree could prove useful for the detection of QTLS (Cochard et al., 2015; Tisné et al., 2015). This enables the implementation of marker-assisted selection or genomic selection (Cros el al., 2015a,b), which may help to shape the future of oil palm breeding.

10 References

Adon, B., Cochard, B., Flori, A., Potier, F., Quencez, P. and Durand-Gasselin, T. (2001). Introgression of slow vertical growth in improved oil palm (E. guineensis Jacq) populations. In *Cutting-edge technologies for sustained competitiveness: Proceedings of the 2001 PIPOC International Palm Oil Congress,* Agriculture Conference, Kuala Lumpur, Malaysia, 20–22 August 2001. (pp. 210–17). Malaysian Palm Oil Board (MPOB).

Annet, M. E. (1921). Le palmier à huile au Cameroun et en Afrique tropical. Variétés – Culture – Exploitation. Emile Larose, Libraire, Editeur, Paris.

Arolu, I. W., Rafii, M. Y., Marjuni, M., Hanafi, M. M., Zulkifli, S., Harun, A. R., Olalekan, K. K., Abidin, M. I. Z., Amiruddin, M. D., Kushairi Din, A. and Rajanaidu, N. (2016). Genetic variability analysis and selection of pisifera palms for commercial production of high yielding and dwarf oil palm planting materials. *Ind. Crops Prod.*, 90, 135–41.

Baeke, V. (1996). *Le temps des Rites. L'univers magico-religieux des Wuli (Mfumte du Camerounoccidental).* Thèse Sc. Sociale, Univ. Libre Bruxelles.

Beirnaert, A. and Vanderweyen, R. (1941). Contribution à l'étude génétique et biométrique des variétés d'*Elaeis guineensis*Jacquin. *Publ. Inst. Nat. Etude agron. Congo Belge, Ser. Sci.*, 27, 1–101.

Blaak, G., Sparnaaij, L. D. and Menendez, T. (1963). Breeding and inheritance in the oil palm (*Elaeis guineensis* Jacq.) Part II. Methods of bunch quality analysis. *J. W. Afr. Inst. Oil Palm Res.*, 4, 146–55.

Blanc-Pamard, C. (1980). De l'utilisation de trois espèces de palmiers dans le sud du 'V' Baoulé (Côte d'Ivoire). In L'Arbre en Afrique Tropicale, la Fonction et le Signe, *Cahier ORSTOM, Sc. Humaines*, 17, 247–55.

Breton, F., Hasan, Y., Hariadi, Lubis, Z. and de Franqueville, H. (2006). Characterization of parameters for the development of an early screening test for basal stem rot tolerance in oil palm progenies. *J. Oil Palm Res.*, Special Issue, 24–36.

Breure, C. J. and Konimor, J. (1992). Parent selection for oil palm clonal seed gardens. In: *Proceedings Workshop 'Yield potential in the oil palm'* (Ed. by V. Rao, I. E. Henson and N. Rajanaidu), pp. 122–44, Palm Oil Res. Inst. Malaysia, Kuala Lumpur.

Breure, C. J. and Verdooren, L. R. (1995). Guidelines for testing and selecting parent palms in oil palm. Practical aspects and statistical methods. *ASD Oil Palm Papers* 9, 1–101.

Cochard, B., Durand-Gasselin, T. and Adon, B. (2006). Oil palm genetic resources in the Côte d'Ivoire – composition, assessment and use. Paper presented at Int. Symp. Oil palm genetic resources and utilization, Malaysian Palm Oil Board, Kuala Lumpur, 8–10 June 2000.

Cochard, B., Noiret, J. M., Baudouin, L., Flori, A. and Amblard, P. (1993). Second cycle de sélection récurrente réciproque chez le palmier à huile Elaeis guineensis Jacq.: résultats des tests d'hybrides Deli x La Mé. *Oléagineux*, 48, 441–51.

Cochard, B., Carasco-Lacombe, C., Pomiès, V., Dufayard, J. F., Suryana, E., Omoré, A., Durand-Gasselin, T. and Tisné, S. (2015). Pedigree-based linkage map in two genetic groups of oil palm. *Tree Genet. Genomes*, 11, 68. doi:10.1007/s11295-015-0893-7.

Comstock, R. E., Robinson, H. F. and Harvey, P. H. (1949). A breeding procedure designed to make maximum use of both general and specific combining ability. *Agron. J.*, 41, 360.

Corley, R. H. V. and Tinker, P. B. (2016). *The Oil Palm*. Wiley Blackwell: Oxford.

Corley, R. H. V. and Tinker, P. B. (2003). *The Oil Palm*. Fourth Edition. Blackwell Sceince Ltd.: Oxford.

Corrado, F. and Wuidart, W. (1990). Germination des graines de palmier à huile (*Elaeis guineensis*) en sac de polyéthylène. Méthode par 'chaleur sèche'. *Oléagineux*, 45, 511–18.

Courade, G. (1978). Les plantations industrielles d'Unilever en Cameroun. *Cahiers d'ONAREST*, 1(2), 91–159.

Cros, D., Denis, M., Bouvet, J. M. and Sánchez, L. (2015b). Long-term genomic selection for heterosis without dominance in multiplicative traits: case study of bunch production in oil palm. *BMC Genomics*, 16, 651. doi:10.1186/s12864-015-1866-9.

Cros, D., Denis, M., Sánchez, L., Cochard, B., Flori, A., Durand-Gasselin, T., Nouy, B., Omoré, A., Pomiès, V., Riou, V., Suryana, E. and Bouvet, J. M. (2015a). Genomic selection prediction accuracy in a perennial crop: case study of oil palm (*Elaeis guineensis* Jacq.). *Theor. Appl. Genet.*, 128, 397–410.

Cros, D., Flori, A., Leifi, N., Omoré, A. and Nouy, B. (2003). Differential response to water balance and bunch load generates diversity of bunch production profiles among oil palm crosses (Elaeis guineensis). *Trop. Plant Biol.*, 6, 26–36.

Demol, J., Baudoin, J. P., Louant, B. B., Marechal, R., Mergeai, G. and Otoul, E. (2002). *L'amélioration des plantes. Application aux principales espèces cultivées en régions tropicales.* Les presses Agronomiques de Gembloux, Gembloux, Belgique. ISBN 2-87016-067-4, p. 584.

Drachoussof, M. V. (1989). Historique des recherches en agronomie tropicale africaine. In AUPELF-UREF (Ed.), *Plantes vivrières tropicales*. John Libbey Eurotext: Paris, pp. 5–12.

Dumortier, F. (2003). Breeding for high yielding progenies at Dami OPRS. In: *Proc. Agric. Conf. Palm oil: the power-house for the global oils and fats economy*, pp. 51–74, Malaysian Palm Oil Board, Kuala Lumpur, 24–28 August.

Durand-Gasselin, T., Corredor, J., Sanz, J. I., de Franqueville, H. and Amblard, P. (2009). Future vision of oil palm genetic improvement in Latin America: several resolutions on cooperation in Colombia for planting material improvement. Paper presented at XVI Conf. Int. sobre Palma de Aceite, Cartagena, Colombia, 22–25 September.

Durand-Gasselin, T., Kouamé Kouamé, R., Cochard, B., Adon, B. and Amblard, P. (2000). Diffusion variétale du palmier à huile (Elaeis guineensis Jacq.). *Oléagineux Corps gras Lipide*, 7, 207–14.

Durand-Gasselin, T., Noiret, J. M., Kouamé Kouamé, R., Cochard, B. and Adon, B. (1999). Disponibilité de pollen performant pour la production de semences améliorées de palmier à huile (Elaeis guineensis Jacq.). *Plantation, Recherche, Développement*, 6, 264–76.

Eckey, E. W. (1954). Vegetable Fats and Oils. Edition New York Rheinhold. Publ. Corp. Am. Chem. Soc. Scientifique n°27 Efficiency of Different Oil Palm Genotypes Using 15N Isotope Labelling Method. *Pertanika J. Trop. Agric. Sci.*, 35(4), 743–54.

Eggert, M. K. (1993). Central Africa and the archaeeology of the equatorial rainforest: reflectionson some major topics. In Shaw, Th., Sinclair, P., Andah, B. and Okpoko, A. (Eds), *The Archaeology of Africa: Food, Metals and Towns.* Routledge publ.: London et New-York, pp. 289–329.

Ergo, A. B. (2013). Les recherches relatives au Palmier à Huile (Elaeis guineensis Jacq.) à l'INEAC de 1933 à 1960, p. 31. abergo1.e-monsite.com/medias/files/palmineac.doc.

Escobar, R. and Alvarado, A. (2003). Strategies in production of oil palm compact clones and seeds. In: *Proc. Agric. Conf. Palm oil: the power-house for the global oils & fats economy*, pp. 75–90, Malaysian Palm Oil Board, Kuala Lumpur, 24–28 August.

Foster, H. (2003). Assessment of oil palm fertilizer requirements. In Fairhurst, T. and Hardter, R. (Eds), *Oil Palm: Management for Large and Sustainable Yields.* Potash & Phosphate Institute (PPI)/ Potash & Phosphate Institute of Canada (PPIC), Singapore, ISBN 9810484852, p. 382.

Gascon, J. P. and de Berchoux, C. (1964). Caractéristiques de la production d'*Elaeis guineensis* (Jacq.) de diverses origines et leurs croisements. Application à la sélection du palmier à huile. *Oléagineux*, 19, 75–84.

Gascon, J. P. and Wuidart, W. (1975). Amélioration de la production et de la qualité de l'huile d'*Elaeis guineensis* Jacq. *Oléagineux*, 30, 1–4.

Gascon, J. P., jacquemard, J. C., Houssou, M., Boutin, D., Chaillard, H. and Kanga Fondjo, F. (1981). La production de semences sélectionnées de palmier à huile. *Oléagineux*, 36, 475–86.

Gascon, J. P., Le Guen, V., Nouy, B., Asmady, H. and Kamga, F. (1988). Résultats d'essais de second cycle de sélection récurrente réciproque chez le palmier à huile *Elaeis guineensis* Jacq. *Oléagineux*, 43, 1–7.

Hardon, J. J. (1970). Inbreeding in populations of the oil palm (*Elaeis guineensis* Jacq.) and its effects on selection. *Oléagineux*, 25, 449–56.

Hardon, J. J., Corley, R. H. V. and Lee, C. H. (1987). Breeding and selecting the oil palm. In Abbott, A. J. and Atkin, R. K. (Eds), *Improving Vegetatively Propagated Crops*. Academic Press: London, pp. 63–81.

Hartley, C. W. S. 1967. *The Oil Palm (Elaeis guineensis Jacq.)*. Longmans: London, pp. 71–136.

Haxaire, C., 1992. Le palmier à huile chez les Gouro de Côte d'Ivoire. *Journal des africanistes*, 62 (1), 55–77.

Houard, A., Castelli, L. and Lavergne, J. (1927). Contribution à l'étude du Palmier à huile en Afique Occidentale Française. Notes sur quelques travaux effectués en 1922 et au début de 19223 dans les stations expérimentales du Palmier à huile à la Côte d'Ivoire et au Dahomey. Bibliothèque de l'institut national d'agronomie coloniale. Libraire Emile Larose, Paris, p. 104.

Houard, A., Lavergne, J. and Blondeleau. (1926). *La sélection du Palmier à Huile à la Station Expérimentale de la Mé*. Rapport technique de 1925. Librairie Larose: Paris, p. 232.

Houssou, M. 1985. Amélioration du palmier à huile (Elaeis guineensis J.) en zone peu humide. Résultats récents obtenus au Bénin. Thèse, Université Paris-Sud, Orsay, France.

Hunger, F. W. T. (1924), *De Oliepalm (Elaeis guineensis). Historisch onderzoek over den oliepalm in Nederlandsch-indië*. Edited by Boekhandel, N. V. and Drullerij, E. N. Leiden, Netherland.

Jacquemard, J. C. (1979). Contribution à l'étude de la croissance en hauteur du stipe d'Elaeis guineensis Jacq. Etude du croisement LM2T x DA10D. *Oléagineux*, 34, 492–7.

Jacquemard, J. C., Meunier, J. and Bonnot F. (1981). Etude génétique de la reproduction d'un croisement chez le palmier à huile *Elaeis guineensis*. *Oléagineux*, 36, 343–52.

Jacquemard, J. C., Suryana, E., Cochard, B., de Franqueville, H., Breton, F., Syaputra, I., Derwaman, E. and Permadi, P. (2010). Intensification of oil palm (*Elaeis guineensis* Jacq.) plantation efficiency through planting material: New results and developments. In: International Oil Palm Conference (IOPC 2010): transforming oil palm industry, Yogyakarta, Indonesia, 1–3 June. IOPRI, p. 36.

Jagoe, R. B. (1952). Deli oil palms and early introductions of *Elaeis guineensis* to Malaysia. *Malay. agric. J.*, 35, 3–11.

Junaidah, J., Rafii, M. Y., Chin, C. and Saleh, G. (2011). Performance of Tenera oil palm population derived from crosses between Deli Dura and Pisifera from different sources on inland soils. *J. Oil Palm Res.*, 23, 1210–21.

Le Guen, V., Amblard, P., Omore, A., Koutou, A. and Meunier, J. (1991) Le programme hybride interspécifique *Elaeis oleifera x Elaeis* de l'IRHO. *Oléagineux*, 46, 12, 479–87.

Lim, C. C., Teo, K. W., Rao, V. and Chia, A. C. C. (2003). Performance of *pisiferas* of Binga, Ekona, URT and Angolan origin: Part 1 – Breeding background and fruit bunch traits. *J. Oil Palm Res.*, 15, 21–31.

Lubis, A. A. (1988). History and origin of oil palm breeding material used in Marihat research station. ISOPB Workshop on dura and tenera population, Marihat.

Madelaine, C. (2005). Analyse du fonctionnement et de la dynamique de la palmeraie sub-spontannée en Guinée forestière. Cas du village de Nienh. Mémoire de Stage. Ecole Nationale du Génie Rural, des Eaux et Forêts. Centre de Montpellier.

Maley, J. (1999). L'expansion du palmier à huile (*Elaeis guineensis*) en Afrique Centrale au cours des trois derniers millénaires: nouvelles données et interprétations. In Bahuchet, S., Bley, D., Pagezy, H. and et al. (Eds), *L'homme et la forêt tropicale. Trav. Soc. Ecologie Humaine*, Publ. du Bergier: Paris, pp. 237–54.

Martinez, G., Sarria, G. A., Torres, G. A., Varon, F., Romero, H. M. and Sanz, J. I. (2009). Advances in the research of *Phytophthora palmivora*, the causal agent of bud rot in oil palm in Colombia. In: *Proc. Int. Palm Oil Conf. Agriculture, Biotechnology and Sustainability*, Vol. II, pp. 875–890, Malaysian Palm Oil Board, Kuala Lumpur.

Meunier, J. (1975). Le palmier à huile américain *Elaeis melanococca*. *Oléagineux*, 30, 51–62.

Meunier, J. and Gascon, J. P. (1972). Le schéma général d'amélioration du palmier à huile à l'IRHO. *Oléagineux*, 40, 365–72.

Meunier, J. and Boutin, D. (1975). L'*Elaeis melanoccoca* et l'hybride l'hybride *Elaeis melanoccoca* x *Elaeis guineensis*. Premières données. *Oléagineux*, 30, 5–8.

Meunier, J., Potier, F., Amblard, P. and Tailliez, B. (1989). Relations entre la production d'huile et le nombre de régimes chez le palmier à huile (Elaeis guineensis Jacq.). Conséquences pour la pollinisation dans les jeunes plantations. *Olégineux*, 11, 25–34.

Meunier, J., Vallejo, G. and Boutin, D. (1976). L'hybride *E. melanococca* x *E. guineensis* et son amélioration. Un nouvel avenir pour le palmier à huile. *Oléagineux*, 31, 519–28.

Ngando-Ebongue, G. F., Koona, P., Nouy, B., Zok, S., Carriere, F., Zollo, P. H. A. and Arondel, V. (2008). Identification of oil palm breeding lines producing oils with low acid values. *Eur. J. Lipid Sci. Technol.*, 110, 505–9.

Noh, A., Rafii, M. Y., Saleh, G. and Kushairi, A. (2010). Genetic performance of 40 Deli Dura x Avros *Pisifera* Full-sib families. *J. Oil Palm Res.*, 22, 781–95.

Noh, A., Rafii, M. Y., Saleh, G., Kushairi, A. and Latif, M. A. (2012). Genetic Performance and General Combining Ability of Oil Palm dura x AVROS pisifera Tested on Inland Soils. *Sci. World J.*, 212, Article ID 792601, 8.

Nouy, B., Baudouin, L., Djégui, N. and Omoré, A. (1999). Oil palm under limiting water supply conditions. *Plantation Recherche, Developpement*, 6, 31–45.

Obasola, C. O., Obisesan, I. O. and Opute, F. I. (1977). Breeding of short-stemmed oil palm in Nigeria. In Earp, D. A. and Newall, W. (Eds), *International developments in oil palm*. Incorp. Soc. Planters: Kuala Lumpur, pp. 68–94.

Ollivier, J., Lamade, E., Dubos, B., Erwanda S., Permadi, P., Suryana, E., Flori, A., Cochard, B. and Jacquemard, J. C. (2013). Towards an accurante nutritional diagnosis for oil palm taking into account the origin of the Planting Material. *Palmas*, 34(1), 203–20.

Pichel, R. (1956). L'amélioration du palmier à huile au Congo Belge. *Bull. Agron.*, 14, 59–66.

Purba, A. R., Flori, A., Baudouin, L. and Hamond, S. (2001). Prediction of oil palm (*Elaeis guineensis* Jacq.) agronomic perfomances using best linear unbiased predictor (BLUP). *Theor. Appl. Genet.* 102, 787–92.

Rajanaidu, N., Rao, V. and Tan, B. K. (1983). Analysis of fatty acid composition (FAC) in *Elaeis guineensis, Elaeis oleifera*, their hybrids and its implications in breeding. *Palm Oil Res. Inst. Malaysia, Bull.*, 7, 9–20.

Renard, J. L., Gascon, J. P. and Bachy, A. (1972). Recherches sur la fusariose du palmier à huile. *Oléagineux*, 27, 581–91.

Renard, J. L., Noiret, J. M. and Meunier, J. (1980). Sources et gammes de résistance à la fusariose chez les palmiers à huile *Elaeis guineensis* et *Elaeis melanococca. Oléagineux*, 35, 387–93.

Rosenquist, E. A. (1986). The genetic base of oil palm breeding populations. In: *Proc. Int. Workshop Oil palm germplasm & utilisation*, pp. 27–56, Palm Oil Research Institute of Malaysia, Kuala Lumpur.

Rosenquist, E. A. (1990). An overview of breeding technology and selection in *Elaeis guineensis*. In: *Proceedings 1989 Int. Palm Oil Dev.Conf. – Agriculture* (Ed. by B. S. Jalani et al.), pp. 5–25, Palm Oil Res. Inst. Malaysia, Kuala Lumpur.

Sambanthamurthi, R., Oo, K. C. and Parman, S. H. (1995). Factors affecting lipase activity in *Elaeis guineensis* mesocarp. *Pl. Physiol. Biochem.*, 33, 353–9.

Soh, A. C. (1994). Ranking parents by best linear unbiased prediction (BLUP) breeding values in oil palm. *Euphytica*, 76, 13–21.

Soh, A. C., Kee, K. K. and Goh, K. J. (2006). Research and innovation towards sustainable palm oil production. *J. Sci. Technol. Trop.*, 2, 77–95.

Soh, A. C., Wong, C. K., Ho, Y. W. and Choong, C. W. (2009). Oil Palm. In Vollmann J. and Rajcan, I. (Eds), *Oil Crops*. Springer: New York, pp. 333–67.

Soh, A. C., Wong, G., Hor, T. Y., Tan, C. C. and Chew, P. S. (2003). Oil palm genetic improvement. In Janick, J. *Plant Breeding Reviews*, Volume 22. John Wiley & Sons Inc.: New York, pp. 165–219.

Sparnaaij, L. D. (1958). The palmeries of the Kwango district. In Hartley, C. W. S. (Ed.), *Notes on a visit to research station, oil palm plantations and palmeries in the Belgian Congo*, April 1958, Mimeographed report. W. Afr. Inst. Oil Palm Res.: Benin, Nigeria.

Sparnaaij, L. D., Menendez, T. and Blaak, G. (1963). Breeding and inheritance in the oil palm (*Elaeis guineensis* Jacq.) Part I The design of a breeding programme. *J. West Afr. Inst. Oil Palm Res.*, 4, 126–155.

Sterling, F., Richardson, D. L. and Chavez, C. (1988). Some phenotypic characteristics of the descendants of 'QB49:238', an exceptional hybrid of oil palm. In *Proceedings 1987 Int. Oil PalmConf. 'Progress and Prospects'* (Ed. by A. Halim Hassan et al.), pp. 94–102, Palm Oil Res. Inst. Malaysia, Kuala Lumpur.

Tailliez, B. and Ballo Koffi, C. (1992). Une méthode de mesure de la surface foliaire du palmier à huile. *Oléagineux*, 47, 537–45

Tam, T. K., Lim, C. S., Yeoh, G. H. and Ooi, S. C. (1977). The oil and other characteristics of interspecific (*Elaeis guineensis x Elaeis oleifera*) hybrids planted in Malaysia and their significance for future breeding programmes. In Earp, D. A. and Newall, W. (Eds), *International Developments in Oil Palm*. Incorp. Soc. Planters: Kuala Lumpur, pp. 27–38.

Tchamie, T. T. K. (1995). Contribution à l'étude de la palmeraie du pays Kabye-Losso-Lamba (Nord-Togo). *Cahiers d'Outre-Mer*, 48(192), 501–18.

Teo, K. W., Rao, V., Chia, C. C. and Lim, C. C. (2003). Performance of some *pisiferas* of Binga, Ekona, URT and Angolan origins: Part 2 – Fruit bunch yields, vegetative growth and physiological traits. *J. Oil Palm Res.*, 16(1), 22–38.

Tisné, S., Denis, M., Cros, D., Pomiès, V., Riou, V., Syahputra, I., Omoré, A., Durand-Gasselin, T., Bouvet, J. M. and Cochard, B. (2015). Mixed model approach for IBD-based QTL mapping in a complex oi palm pedigree. *BMC Genomics*, 16: 798. doi:10.1186/s12864-015-1985-3.

Torres, G. A., Sarria, G. A., Varon, F., Coffey, M. D., Elliott, M. L. and Martinez, G. (2010). First report of bud rot caused by *Phytophthora palmivora* on African oil palm in Colombia. *Plant Dis.*, 94, 1163.

Tourte R., (2005). Histoire de la recherche agricole. La période coloniale et les grands moments des jardins d'essais: 1885/1890–1914/1918. Ed FAO. ISBN: 92-5-205407-3.

Tovey F. W. (1947). 7th Annual Report of the oil palm research station, Benin, 1946, 1947, p. 111.

Turnbull N., Cazemajor M., Guerin C., Louise C., Amplard P., Cochard B. and Durand-Gasselin T. (2016). Oil Palm breeding and seed production in PalmElit. International Seminar on Oil Palm Breeding & Seed Production and Field Visits, 29th-30th September 2016 – Kisaran – Indonesia.

Vanderweyen, R. and Roels, O. (1949). Les varietes d'*Elaeis guineensis* Jacquin du type *albescens* et *Elaeis melanococca* Gaertner (em. Bailey). Note preliminaire. *Publ. Inst. Nat. Etude agron. Congo Belge, Ser. Sci.*, 42, 1–24.

Wong, Y. T., Kushairi, A., Rajanaidu, N., Osman, M., Wickneswari, R. and Sambanthamurthi, R. (2016). Screening of wild oil palm (Elaeis guineensis) germplasm for lipase activity. *J. Agric. Sci.*, 154, 1241–52.

Zaklinkaya Y. D. and Prokofyev S. S. (1971), New data on the Cenozoic flora of the southeast coast of the Republic of Guinea (translation). *Doklady Akad. Nauk SSSR*, 201, 1171–3.

Zeven A. C. (1964). On the origin of the oil palm. *Grana palynol.*, 5, 50.

Zeven A. C. (1967). *The Semi-Wild Oil Palm and Its Industry in Africa*. Wageningen University: Wageningen.

Zeven A. C. (1972). The partial and complete domestication of the oil palm (*Elaeis guineensis*). *Econ. Bot.*, 26, 274.

Chapter 8

Advances in marker-assisted breeding of oil palm

Rajinder Singh, Chan Pek Lan, Maizura Ithnin and Umi Salamah Ramli, Malaysian Palm Oil Board, Malaysia

1 Introduction

Oil palm is mostly cultivated in the tropical belt that lies between 10 degrees North and South of the Equator (Godswill et al., 2016). Oil palm is unique, as it produces two types of oil: palm oil from the mesocarp and palm kernel oil (PKO) from the kernel. Both oils have different fatty acid profiles and hence separate applications. Palm oil from the mesocarp is a major source of vegetable oils, while PKO has wide industrial applications, such as to make soap, detergents and personal care products among others.

The oil palm industry has shown spectacular improvement in productivity over the past 50 years, with increases in production from 1.26 million tonnes in the early 1960s to about 45 million tonnes in the late 2000s (Abdullah and Wahid, 2010). Such a rapid improvement was basically the result of (i) improved planting materials obtained via breeding, (ii) increase in acreage cultivated with oil palm and (iii) more extensive and efficient use of fertilizers. Among the three reasons, only the first is sustainable in the long term and is an area of R & D across many oil palm–producing countries.

http://dx.doi.org/10.19103/AS.2017.0018.08

Thus far, conventional plant breeding has been in the forefront of producing elite planting materials, helping the industry to achieve its status as the main source of vegetable oils worldwide. Based mostly on phenotypic selection, commercial oil palm breeding has largely focused on increasing palm oil yield. The effectiveness of the oil palm breeding programmes is undeniable, where improvement in breeding lines has been obvious over the past five decades, where improvements in oil yield at over 60% has been observed between 1960 and 2010 (Durand-Gasselin et al., 2010; Cochard et al., 2015). To help realize this, selection in breeding has focussed on genes of major effect that give a clear measurable phenotype such as the ratio of fruit to bunch (FTB), mesocarp to fruit (MTF) and oil to wet mesocarp (O/WM), which collectively determine the oil to bunch (OTB) ratio of a particular palm (Yong, 1992). Oil yield is thus improved by increasing OTB and the fresh fruit bunch (FFB) yield over several generations. Initially armed with a very basic and limited genetic understanding of these complex traits, breeders have managed to develop and select appropriate maternal *dura* and paternal *pisifera* lines which, when crossed together, have successfully shifted the mean progeny response for the selected traits in the desired direction.

Although effective, oil palm breeding faces several limitations, which include a long generation time of 10–12 years (Mayes et al., 1997) and large land requirement, as usually only 136–148 palms are grown per hectare of land. This makes breeding trials an expensive endeavour, especially now as resources in arable land are limited even for commercial cultivation. The advent of molecular marker technology and subsequent advances in genomics-based technologies has provided an unprecedented opportunity to improve oil palm breeding efficiency. The relationship between variation in gene sequences and phenotypic observations in the field on mature palms provides a powerful means of early selection for desired traits in juvenile plants, before the trait in question is expressed. Referred to as marker-assisted selection (MAS), this approach is especially important for oil palm indeed, where most of the phenotypic traits used for selection in conventional breeding can only be evaluated 4–7 years after field planting. Furthermore, phenotypic data collection, especially those related to bunch components, is usually carried out over for several years in order to ensure an accurate estimation of the palm productivity.

Application of markers in plant breeding was first demonstrated in the 1980s with protein-based isozyme markers (Tanksley and Rick, 1980). This approach was extended to DNA-based markers in the mid-1980s, when the applications of restriction fragment length polymorphism (RFLP) were evidenced (Beckman and Soller, 1986). With respect to oil palm, application of markers especially RFLP for DNA fingerprinting was initiated in the 1990s (Cheah et al., 1993; Mayes et al., 1996). An RFLP-based genetic map was then constructed, which revealed markers linked to a pivotal monogenic trait – the shell thickness of oil palm fruit (Mayes et al., 1997). The oil palm industry was for the first time introduced to the power of molecular markers and its potential for improving breeding efficiency. RFLP markers are cumbersome and expensive for large-scale screening, and the fact that they were also not very tightly linked to the shell thickness trait at that time did not encourage the large-scale application of such a promising tool in breeding programmes. Research was also undertaken in order to make use of modified RFLP markers in the search for the molecular determinism of the 'mantled' somaclonal variation affecting flower and fruit structure in the oil palm (Jaligot et al., 2002). The RFLP approach did however spur research in the direction towards the integration of DNA technology within breeding strategies, which gained more popularity after the development of the polymerase chain reaction (PCR) (Mullis and Faloona, 1987) and a whole generation of PCR-based molecular markers. Indeed, using PCR-based random amplified polymorphic (RAPD) markers, Moretzsohn et al. (2000) identified markers which proved to be more tightly linked to the shell thickness locus. At about the same time, Singh and Cheah (1999) described the application of amplified fragment length polymorphism (AFLP) markers for the assessment of the fidelity of controlled crosses in oil palm. Billotte et al. (2005) subsequently developed simple sequence repeat (SSR) markers for genetic mapping and produced 16 linkage groups to match the haploid chromosome number for oil palm. More importantly, these authors identified SSR markers flanking the gene influencing shell thickness, and the markers could predict the trait with over 95% accuracy. Research aimed at identifying markers linked to yield and vegetative components – which are quantitative in nature – was initially carried out using RFLP-based markers by Rance et al. (2001) and subsequently by Billotte et al. (2005), who evidenced links between SSRs markers and quantitative trait loci (QTLs) for yield and vegetative components in multi-parental linkage maps. Such pioneering works laid the basic foundation for the application of markers in oil palm breeding programmes. As the techniques for genotyping improve and the cost decreases (Poland, 2015), opportunities to incorporate MAS in oil palm breeding also increase. Such changes are also accelerated by increased expertise in basic molecular procedures across many oil palm–producing countries, especially in South East Asia and South America.

2 Application of markers for paternity testing in oil palm

Marker Assisted Selection (MAS) is yet to be widely implemented in oil palm breeding programmes. The major reasons which could explain such a situation could be that until recently, there was a lack of trained staff, appropriate genomics tools and infrastructure in many commercial plantation companies to implement MAS. The most basic application of markers in oil palm breeding is probably testing for legitimacy within the progeny of controlled crosses. This approach involves simple markers inherited in a Mendelian manner in order to ensure that the palms within a family are of the correct parentage. This is a very important information for breeding programmes, as parents are selected based on family/progeny performance, and if illegitimates are present, wrong conclusions can be drawn (Corley, 2005).

Although great care is taken to bag female flowers prior to anthesis, and in the collection, documentation and subsequent storage of pollen, controlled pollination can still run into problems due to various sources of errors as highlighted by Corley (2005). Problems with controlled pollinations spiked with the introduction of the oil palm weevil (*Elaeidobius kamerunicus*) in the early 1980s – when contamination rates as high as 20% were observed (Rao et al., 1993; Corley and Tinker, 2003a). The illegitimate palms could be noticed even without the use of molecular markers, especially in crosses involving the maternal *dura* and paternal *pisifera* palms. Although only the hybrid *tenera* palms were expected from such crosses, some contamination by *dura* palms could be observed, possibly due to uncontrollable wind and/or insect pollination (Chin, 1995). Admittedly, contamination has reduced tremendously in recent years due to improved quality control, but the problem still persists to some extent (Ooi et al., 2016).

A number of reports in recent years have highlighted the application of PCR-based markers in assessing the legitimacy of controlled crosses. Thongthawee et al. (2010) reported that 7–8 SSR loci were sufficient to detect pollination errors confidently. More recently, Hama-Ali et al. (2015) highlighted the problem with the assignment of parental palms to families in an oil palm–controlled cross involving four half sib families. The parental samples assigned to each family in their experiments were not corresponding to the SSR profiles observed, thus highlighting the errors that occur especially when a number of crosses are made at the same time. Although the authors managed to assign the correct parents to the offsprings using a minimum set of five SSR markers and appropriate statistical software, an illegitimacy rate of 1.5% was still observed. Without the use of molecular markers for parentage analysis and assignment, the breeders would have made a serious error in moving ahead with their breeding programme since family performance is essential in determining the combining ability of the parental palms. Selecting the best individual palm within a family in the presence of illegitimate palms can also have serious consequences. As such, the use of molecular markers to assess the fidelity of controlled crosses is of paramount interest, especially considering the mistakes that can happen at any stage of the crossing programme: from labelling of parental palms, pollen collection and storage to mix-up in the nursery as well as during field planting.

3 Genomic resources and tools for MAS

Research towards application of MAS in oil palm has been greatly assisted with the availability of the oil palm genome sequence (Singh et al., 2013a). Genomes of both the *Elaeis guineensis* and *Elaeis oleifera* species were uncovered, and at least ~45 000 genes were predicted. Together with the sequencing of the hypomethylated genome of eight individual palms from both species (Low et al., 2014) and the availability of expressed sequence tag (EST) sequences (Jouannic et al., 2005; Ho et al., 2007; Low et al., 2008; Chan et al., 2010), as well as transcriptome sequencing using next-generation sequencing technology (Tranbarger et al., 2011; Bourguis et al., 2011; Guerin et al., 2016), appropriate genomic resources and tools have been established for oil palm. More recently, sequencing of pooled DNA from 132 palms representing 59 different breeding lines at 35x genome coverage using Illumina Hi Seq 2000 (Teh et al., 2016) has added to the growing repertoire of genomic resources for oil palm. Such rapidly growing information has allowed the mining of SSR (Singh et al., 2008; Ting et al., 2010; Tranbarger et al., 2012; Xiao et al., 2014; Taeprayoon et al., 2016) and the highly valuable single-nucleotide polymorphism (SNP) markers (Riju et al., 2007; Ting et al., 2014; Pootokham et al., 2015; Kwong et al., 2016). Rapid advances in genotyping platforms and technologies have made SNP markers especially valuable for the application in breeding programmes. As such, the recent release of information on 4451 (Ting et al., 2014) and 200 000 SNPs (Kwong et al., 2016) printed on an Illumina array provides a powerful platform for application in oil palm genetic studies, especially in the efforts directed at identifying marker(s) for early selection of a desired trait. The recent identification and in-depth description of long terminal repeat (LTR) retrotransposons in the oil palm genome have also opened the opportunity to assess the role of these elements in genome and transcriptome variations, especially those resulting from the micropropagation of oil palm (Beule et al., 2015).

Proteomics- and metabolomics-based technologies are also now being deployed in oil palm to obtain a comprehensive understanding of gene function related to disease resistance (Syahanim et al., 2013) and yield differences in commercial material (Teh et al., 2013). Although genomics-based technologies have played a key role in improving breeding efficiency in many perennial crops, there is growing evidence that molecular mechanism underlying some agronomic traits – especially those quantitative in nature – can be complex. Regulation of traits occurs over several layers, and observed phenotype may be influenced by post-transcriptional modification, protein–protein interaction, among others and give rise to certain metabolites that may not be predicted by the genome information. As such, phenotype prediction using proteins or metabolites – small molecules that are likely to be directly associated with the phenotype – can complement and enhance breeding efficiency. Although their use for practical application in oil palm is still very premature, the promise that these technologies have for oil palm improvement is undeniable. The establishment of appropriate methods (Tahir et al., 2013; Lau et al., 2015), relevant databases for oil palm (Nur-Ain et al., 2015) and the actual profiling of proteins and metabolites associated with *Ganoderma* infection (Al-Obaidi et al., 2014), as well as fruit development (Loei et al., 2013; Neoh et al., 2013) reflect the growing prominence of proteomics and metabolomics research in oil palm.

4 MAS for predicting monogenic traits

Monogenic traits are probably the most attractive for initial improvement via MAS. The two most important monogenic traits of interest to oil palm breeders, especially in the commercial species *E. guineensis*, are those related to fruit form and colour.

4.1 Oil palm fruit colour

The fruit of the oil palm can vary considerably based on external appearance, most notably the colour of the fruit. By far, the most common type of fruit is known as *nigrescens*, in which the fruits are deep violet to black at the apex and yellow at the base when unripe and exhibit minimal change in colour upon ripening (Hartley, 1988; Singh et al., 2014a). The other major fruit type is known as *virescens*, which changes colour from green to light reddish orange at maturity. Currently, for *nigrescens* palms in commercial fields, bunches are considered ready for harvest when loose fruits are observed on the ground beneath the bunch (Corley and Tinker, 2003b). In practice, plantations prefer a minimum number of loose fruits in order to reduce the cost and time spent collecting loose fruits. Indeed, collecting loose fruits has been reported to occupy 28% of the total harvesting time (Hitam et al., 1999). Inefficient collection of loose fruits can have repercussions on the productivity of the oil palm industry. Furthermore, having a large number of loose fruits from a bunch can also affect the oil quality, as the softer exocarp is easily bruised and damaged, resulting in the oxidization and further generation of free fatty acids that can seriously affect oil quality. In this respect, *virescens* palms can be more desirable to planters as the clear difference in colour between ripe and unripe bunches makes it easier to identify ripe bunches and therefore facilitate fruit harvesting. As such, the identification of the gene responsible for oil palm fruit colour has important implications in breeding programmes. The *virescens* trait is most useful if incorporated into non-abscising palms that have been observed in plantations (Donough et al., 1996). Ongoing research to identify and characterize genes involved in oil palm fruit shedding (Roongsattham et al., 2012), combined with increased knowledge on the morphogenesis of the abscission zone

(Roongsattham et al., 2016), will undoubtedly provide additional target genes for a more effective MAS strategy.

Inheritance studies indicated that the fruit colour (*Vir)* trait is controlled by a single gene, monogenic and dominant (Corley and Tinker, 2003a). More recently, Singh et al. (2014a) assessed the inheritance of exocarp colour of oil palm fruits in a population derived from the self-pollination of the *tenera* palm, T128, from the Malaysian Palm Oil Board's (MPOB) Nigerian germplasm collection. The data, as expected, suggested that one Mendelian locus controlled the trait, and the position of the gene was determined on a genetic map derived from the selfed T128 population. Subsequently, using the reference genome of *E. guineensis*, the R2R3 MYB transcription factor was identified as the *virescens* gene (*vir*) that controls the exocarp colour of the fruit. *VIR* (R2R3-MYB transcription factor) was shown to be homozygous wild type in *nigrescens* palms but, in *virescens* palms, was heterozygous or homozygous for a nonsense mutation in the final exon of the gene (Singh et al., 2014a). In order to further validate the association of the gene with fruit colour, the entire gene was sequenced in over 500 palms from independent breeding populations as well as germplasm collection. Four additional mutations were uncovered, each resulting in the similar C-terminal truncation of the *VIR* protein giving rise to the *virescens* phenotype. Whole transcriptome gene expression profiling in *nigrescens* and *virescens* fruit demonstrated that the *VIR* truncation mutations result in a dominant-negative function linked to anthocyanin deficiencies via impaired regulation of downstream anthocyanin biosynthetic target genes (Singh et al., 2014a). A dominant-negative mutation in the same gene was also found to be regulating fruit colour in the date palm (Hazzouri et al., 2015).

The discovery of the *VIR* gene facilitates genetic testing for fruit colour prior to planting. More importantly it can be directly integrated in an MAS strategy, as it can be used as a tool in oil palm breeding to differentiate seedlings that are heterozygous for the dominant-negative mutation from those that are homozygous. This will allow breeders to develop paternal *pisifera* lines that are homozygous for *virescens* for use in introgressing this desirable trait into elite breeding materials. The use of the *VIR* gene for application in oil palm breeding via MAS was introduced by the MPOB recently (Singh et al., 2015).

4.2 Fruit form phenotype

Palm oil production is mostly based on the planting of hybrid *tenera* fruit form palms. The *tenera* hybrids are obtained by crossing thick-shelled *dura* maternal palms to *pisifera* paternal palms which are shell-less and typically female sterile. *Tenera* palms yield on average 30% more oil than *dura* (Corley and Lee, 1992), while *pisifera* palms typically do not produce any fruits. The monogenic, co-dominant inheritance of shell was first discovered in the 1940s in Zaire, Africa (Beinaert and Vanderweyen, 1941), and it reflects a classic example of single gene heterosis (or single-gene hybrid vigour). *Tenera* palms for commercial planting are produced via controlled crosses involving selected *dura* and *pisifera* palms. Although great care is taken to make the controlled crosses, error-free hybridization of the selected parental palms is often difficult to achieve in real-world settings. This can result in unintentional planting of non-*tenera* palms for commercial production. The presence of rudimentary anthers in maternal *dura* palms (Corley, 2005) can also result in self-fertilization and subsequent *dura* contamination in commercial plantings.

Over the past several decades, several attempts were made to identify markers linked to the *SHELL* gene (Mayes et al., 1997; Moretzsohn et al., 2000; Billotte et al., 2005). Despite such continuous efforts, marker(s) tightly linked to the *SHELL* gene and showing a high level of predictability for the trait over different genetic backgrounds was never made available for routine use, and the identity of the gene itself remained unknown for a long time. However, using a combination of genetic mapping, homozygosity mapping by sequencing and the oil palm whole genome sequence as a reference, Singh et al. (2013b) identified the gene responsible for the three different fruit forms of *E. guineensis*. The *SHELL* gene was revealed to be a type II MADS-box transcription factor homologous to the *Arabidopsis SEEDSTICK*, which controls ovule, seed and lignified endocarp differentiation. Singh et al. (2013b) further demonstrated that the expression level of *SHELL* was high in the outer layers of the developing kernel in *dura* fruits, consistent with its function in regulating the formation of the heavily lignified shell surrounding the kernel.

Palms bearing *dura* fruit form were found to be homozygous wild type at all nucleotide positions. Initially, two similar but independent mutations were identified in the MADS domain of the *SHELL* gene, which gave rise to the *pisifera* phenotype. These includes the Congo-derived AVROS (sh^{AVROS}) mutation which results in an asparagine substitution of a highly conserved lysine residue. The second mutation, the Nigerian T128-derived sh^{MPOB} mutation, results in a leucine to proline substitution, two amino acids from sh^{AVROS} in the N-terminal direction. Following the discovery of these two mutations, a small-scale national survey was carried out in order to determine the level of *dura* contamination among independent nurseries and smallholders in Malaysia. In the survey, additional three mutations were uncovered within the MADS domain of the *SHELL* gene. The additional mutations were sh^{MPOB2} lysine-to-glutamine substitution six amino acids N-terminal to sh^{MPOB}, the sh^{MPOB3} alanine–aspartate substitution 10 amino acids C-terminal to sh^{AVROS} and the sh^{MPOB4} lysine-to-asparagine substitution of the same amino acid altered by the sh^{MPOB2} mutation (Ooi et al., 2016). Palms that are heterozygous for any one of the five mutations are *tenera* fruit form, while palms that are homozygous mutant for any one of the five mutations (or compound heterozygous with different mutations on each of the two chromosomes) show the *pisifera* fruit form. More importantly, the discovery of the *SHELL* gene allowed the development of the first molecular diagnostic assay for oil palm which was named SureSawit™*SHELL* (Singh et al., 2014b). The assay allows the prediction of the oil palm fruit form from the leaf, and testing can be carried out at the nursery. The discovery of *SHELL* and its application into a successful diagnostic assay facilitated the measurement of non-*tenera* contamination rate in independent nurseries and plantings which was found to reach 10% (Ooi et al., 2016). Although the scope of the survey was limited to a small number of independent nurseries and plantings, it nevertheless underscored the importance of genetic testing in order to avoid uncontrolled batches of planting material derived from *dura* and *pisifera* crosses (DxP) reaching commercial fields. This quality control strategy contributes to sustainability practices as screening with the diagnostic assay will help ensure that the most appropriate materials are planted, resulting in increased productivity per unit of land area.

The discovery of the *SHELL* gene and subsequent development of a simple diagnostic assay will be most valuable to breeders in developing male parental lines for the commercial production of *tenera*. *Pisifera* palms can now be identified early in the nursery and planted separately for easier access and management. This way the planting density of selected *pisifera* palms can also be adjusted to induce the development of male inflorescence. The entire process of being able to separate out the male and female parental lines early enables a more focused breeding programme that allows considerable savings in land, labour and resources. This discovery provided a paradigm shift in assisting oil palm breeders to reduce the length and the allocation of resources to the breeding and selection cycle required to develop new and improved varieties.

5 MAS for predicting quantitative traits in oil palm breeding

Most traits of interest for oil palm breeding are polygenic in nature, as such traits are influenced by many genes, with each likely having a small effect and interacting with each other and the environment. Oil palm unlike other plants such as *Arabidopsis*, rice and maize does not have visible mutants to study the complex traits and identify the underlying genes. Instead, the 'forward-genetics' approaches such as QTL and association mapping (AM) techniques have been embraced to help understand and dissect these complex traits.

5.1 Oil palm QTL studies

One of the first QTL mapping studies in oil palm was reported by Rance et al., 2001, who revealed QTLs linked to yield components and vegetative parameters. Marker-trait association was observed for 11 of the 13 traits analysed, some explaining a large proportion of the phenotypic variance (in excess of 10%). The main setback with this initial study was that the RFLP markers linked to the QTLs (or the corresponding sequences of the markers) were not made available for further testing by other research groups and as such could not be validated in independent studies. In addition, the use in this study of a selfed *tenera* cross, that gave rise to approximately 25% *pisifera* palms, also reduced the number of palms that could be evaluated phenotypically in the mapping population. Nevertheless, this research work did lay the foundation for QTL analysis in oil palm. This was built on by Billotte et al., 2010, who used multi-parental populations involving interconnected families to identify QTLs linked to yield and vegetative parameters. The number of QTLs identified was definitely larger than those reported using the bi-parental model of Rance et al. (2001). However, many of the QTLs identified within a family by Billotte et al. (2010) did not go across families, even when sharing a common parent. The authors speculated that this could be due to the size of the mapping population within each family used in the study.

More recently, QTL studies were expanded for the commercial *E. guineensis* to include those associated with oil yield components (Jeennor and Volkaert, 2014), production traits (FFB, bunch number and bunch weight) (Tisne et al., 2015), sex ratio (Ukoskit et al., 2014) and the height-associated trait (Pootakham et al., 2015; Lee et al., 2015).

The availability of the oil palm genome sequence (Singh et al., 2013a) has made it possible to scan the QTL interval for interesting candidate genes linked to favourable traits concerned. Among the QTL studies above, interesting candidate genes were identified for the vertical growth trait, which is of interest for incorporation into commercial breeding lines in order to expand the economic life span of oil palm plantations (the slower the vertical growth, the longer the economic life). Lee et al. (2015) identified a gene known as asparagine synthase within the QTL interval for height, and this gene has been associated with a dwarf phenotype in *Arabidopsis*. Meanwhile, Pootakham et al. (2015), using a mapping population from a different genetic background, identified two interesting but separate candidate genes associated with vertical growth within their QTL interval, namely DELLA protein GAI1 and a putative gibberellin 2-oxidase 2. Both these genes have been implicated in plant height by their involvement in regulating the gibberellin signalling pathway. Although plant height is an interesting trait for improvement via MAS, the incorporation of the reduced height trait into commercial breeding lines has to be considered carefully in order to avoid any penalty in yield. This is because higher yielding palms are usually taller and hence more competitive (Corley and Tinker, 2003a).

QTL studies involving interspecific hybrids and their backcrosses have also been carried out mostly in relation to fatty acid composition. Indeed, there is a need to introgress genes related to the lower level of saturation of fatty acids observed in *E. oleifera* into *E. guineensis* through interspecific hybrid breeding. In this respect, Singh et al. (2009) initially identified QTLs linked to palmitic (C16:0) and oleic (C18:1) contents in an interspecific hybrid mapping family. Using a denser SSR-based genetic map, Montoya et al. (2013) identified 19 QTLs linked to fatty acid composition in an interspecific first-generation pseudo-backcross. More importantly, the study identified key genes related to the fatty acid biosynthesis pathway, namely acyl-ACP thioesterase type A (FATA) and Δ9 stearoyl-ACP desaturase (SAD) in the QTL intervals linked to stearic acid (C18:0) and C18:1 content. Expanding on this study, Montoya et al. (2014) compared the QTLs identified in interspecific first-generation pseudo-backcross described above to those identified in an intraspecific *E. guineensis*-based cross. The study identified only two common QTL zones and two high-confidence QTL regions specific to *E. guineensis* and *E. oleifera* in the comparison analysis. Nevertheless, Montoya et al. (2013) did conclude that the results were specific to the mapping population used in the study and on the need to further validate these findings in other independent populations segregating for fatty acid composition. Recently, Ting et al. (2016) while working on the interspecific hybrid mapping population described by Singh et al. (2009) but with a slightly bigger population size and a much denser genetic map (involving mostly SNP and SSR markers) identified ten significant QTLs for six fatty acids and iodine value (IV). Taking one of the most significant QTL regions linked to IV, C16:0 and C18:1 content, Ting et al. (2016) mapped the QTL interval back to the oil palm genome sequence. Interestingly, the corresponding genomic region contained a number of genes linked to fatty acid biosynthesis pathway such as palmitoyl-ACP thioesterase (FATB) and also an AP2-like ethylene-responsive transcription factor (WRINKLED1). Additional SSR markers were designed from intronic regions flanking these genes, and they correspondingly mapped back to the QTL interval, improving the QTL peak for the traits. Interestingly, the same QTL interval was detected in two independent second-generation interspecific backcross 2 populations (Ting et al., 2016). This was the first report on the validation of QTLs identified in independent oil palm populations, although the validation populations were somewhat linked to the original population where the QTLs were initially uncovered. Nevertheless, the study also did report major QTL regions which were specific to the three crosses evaluated and did not correspond to regions described by Montoya et al. (2013). This implies as pointed out by Ting et al. (2016) that favourable alleles for fatty acid composition can be population specific, and this has to be considered carefully when implementing MAS.

The initial studies on QTL analysis using mapping populations in oil palm have demonstrated the need for larger mapping populations, especially for traits which usually show lower heritability. Furthermore, small mapping populations also can cause an upward bias on the phenotypic effect estimated for a particular QTL linked to a trait (Butcher and Southerton, 2007). Interestingly, in order to address this issue, Cochard et al. (2015) successfully demonstrated an innovative approach for genetic mapping in oil palm using recombination found in three to four outbred pedigrees, paving the way forward for an effective pedigree-based QTL analysis. Nevertheless, the most serious limitation on most of the QTL analysis reported thus far for oil palm (with one exception) has been the lack of validation in independent populations originating from a different genetic background. It is therefore difficult to predict the accuracy or applicability of markers linked to the traits. Nevertheless, these QTL studies have revealed valuable information on the position and effect of QTLs underlying complex traits and provided confirmation that the traits were actually governed by many genes. With the availability of the genome sequence, several interesting candidate genes linked to vertical growth and fatty acid metabolism have also been identified in the QTL interval, although the specific underlying genes are yet to be identified. However, a point to take into account is that in hybrid crosses (either inter or intra), recombination rates are expected to be low, especially when compared to wild populations; the QTLs detected, especially for low heritability traits, are likely to be population specific. Furthermore, the cost factor must be considered when developing advanced bi-parental oil palm breeding lines to an extent that sufficient number of meiosis are available. As such, alternative methods have to be explored.

5.2 Association mapping studies

In recent years, owing to the limited size of the available mapping populations and the need to dissect complex traits in a more robust manner, oil palm geneticists have started to move on to AM analysis. AM was initially developed and used in human and animal genetics research, where there is also a serious limitation in terms of number of individuals available per family. Currently, AM analysis is being actively applied in a wide range of crops and is often complementary to the traditional QTL analysis (Korte and Farlow, 2013). AM analysis involves determining the genotype of wild or unadapted populations where the phenotype has been characterized, followed by the use of specific statistical software to determine population structure and subsequently marker-trait association. AM analysis can exploit natural populations where recombination events in the distant past would have selectively maintained association between a QTL and markers that are tightly linked to it (Abdurakhmonov and Abdukarimov, 2008). The so-called rapid decay in linkage disequilibrium is advantageous as once discovered and validated could be quite effective in an MAS programme. Nevertheless, for an effective discovery and implementation process, high marker density or a carefully selected set of candidate genes are required for genome-wide association (GWA) analyses (Neale and Kremer, 2011).

In the first association study published for oil palm, Teh et al. (2016) used a 200 000 SNP array to genotype over 2 000 palms, and they identified three significant QTLs on chromosome 5 linked to an important yield component trait which is the oil to dry mesocarp (O/DM) ratio. Interestingly, QTLs linked to other yield component traits such as MTF, OTB and FTB were also observed previously on the same chromosome using bi-parental families (Billotte et al., 2010; Jeennor and Volkaert, 2014). An interesting candidate gene *pyruvate kinase* was also identified near the SNP markers, and the gene had been previously linked to fatty acid biosynthesis and oil accumulation (Bourgis et al., 2011). In a powerful demonstration on the resolution of AM analysis in oil palm, Teh et al. (2016) also showed that the progeny containing the favourable allele for the most significant SNP marker in a homozygous form had elevated amounts of O/DM. It is only natural that this information is being exploited in an MAS programme to develop elite parental palms that carry the favourable allele, which in return will help generate planting materials with higher O/DM (Teh et al., 2016). Using the same SNP array which was applied in a separate GWAS analysis – involving 312 palms from four different populations – Kwong et al. (2016) identified several SNP markers around a region in chromosome 2 influencing the percentage of shell to fruit (S/F), an important determinant of shell thickness which in turn influences mesocarp and subsequently oil content. This is a convincing confirmation of the capability of AM analysis at dissecting complex traits in oil palm and making MAS for polygenic traits a tangible reality. More importantly, this technique provides the opportunity to exploit the genetic diversity available in the oil palm germplasm collections assembled (Barcelos et al., 2015). Unique traits (such as long stalk, slower vertical growth and balanced fatty acid composition, among others) that are not segregating in advanced lines could be readily associated with markers within a selected germplasm. The information will be vital to effectively introgress these desirable traits into the current commercial material. Progress in the area of genomic selection (GS) for oil palm has also been promising (Cros et al., 2014) and can further expedite the selection of palms with desirable traits. GS uses genome-wide molecular markers to estimate the breeding value of individual palms for complex traits. In contrast to MAS, GS will likely detect small-effect QTLs, especially for the quantitative traits governed by multiple loci (Spindel et al., 2015).

The results thus far obtained for QTL analysis of oil palm, either using traditional QTL or AM analysis, will likely induce significant progress in MAS programme designed to improve speed and efficiency and also reduce the cost of existing conventional breeding programmes. The markers linked to QTLs associated with yield parameters for example can be used to screen maternal *dura* lines which were preselected phenotypically, to identify those carrying favourable alleles. The *dura* palms carrying the desired genotypes can be selfed or intercrossed to develop an elite line of *dura* materials homozygous for most of the favourable alleles (Rance et al., 2001). In fact, these 'improved *dura*' palms could be selected early in the nursery for field planting. Similarly, *pisifera* palms can also be selected based on favourable alleles present in their *tenera* sibs, which may be more efficient and cost-effective than random selection (Rance et al., 2001).

In interspecific hybrid and backcross breeding programmes, the main goal is to introgress favourable alleles related to fatty acid composition from *E. oleifera* into *E. guineensis*. Although the QTLs identified thus far can allow for selection of palms with favourable alleles for higher levels of unsaturation of the oil in specific genetic backgrounds, the challenge will be to avoid unfavourable genes within the QTL confidence interval being incorporated at the same time (linkage drag). Identifying the causal gene(s) influencing fatty acid composition or using additional markers close to the QTL region to assist with *E. guineensis* genome recovery (Rance et al., 2001) will help avoid linkage drag. To help unravel the complexities of fatty acid synthesis (FAS), Guerin et al. (2016) recently revealed a gene coexpression network that regulated FAS in the oil palm. It was demonstrated that a number of genes associated with plastid biogenesis, glycolysis, starch metabolism and sugar sensing were co-ordinately transcribed with FAS genes. The most interesting finding in the study was that low level of β-ketoacyl-acyl carrier protein synthase gene (KASII) transcript is in fact responsible for the accumulation of C16: 0 in the mesocarp of oil palm. Interestingly, this is in agreement with earlier biochemical evidence by Sambathamurthi et al. (1999, 2000) that KASII in mesocarp is rate controlling and responsible for C16:0 accumulation. They also showed positive correlation between the activity of the KASII enzyme and the level of unsaturated fatty acids, in both the oil palm species and their hybrids. In fact, upregulating the expression of KASII is a major strategy employed to produce higher C18:1 palms via the genetic engineering route (Sambanthamurthi et al., 2000; Parveez et al., 2015). The combined findings from QTL analysis, biochemical analysis and gene expression profiling will facilitate the development of appropriate MAS strategies for producing palms with the desired fatty acid composition.

6 MAS in oil palm tissue culture

The production of elite planting material via the vegetative propagation of palms with superior characteristics will expedite oil palm improvement. However, genotype dependency of the cultured leaf explants results in low callogenesis rate of 19% (Corley and Tinker, 2003c) and an average embryogenesis rate of between 3 and 6% (Wooi, 1993; Ho et al., 2008). Moreover, the tissue culture process itself also induces somaclonal abnormality which results in the production of clonal palms with abnormal fruit set and flower characteristics (Rohani et al., 2000). Owing to these challenges, molecular research was initiated with a mission to unravel the molecular mechanisms controlling tissue culture amenity and the trueness-to-type of clonal regenerants.

In terms of clonal abnormality, a major breakthrough was achieved recently, when it was shown that a reduction in methylation of the retrotransposon *Karma* which lies within the *EgDEF1* gene causes the 'mantled' phenotype (Ong-Abdullah et al., 2015), a result which built on the previous work published by Jaligot et al. (2014). This discovery paves the way for the development of a diagnostic assay that will allow the culling of abnormal mantled palms prior to cultivation in the commercial fields. Now the interesting question which has to be addressed is how to correlate demethylation of *Karma* evidenced in inflorescences from mature palms with epigenetic events occurring during the various differentiation/dedifferentiation phases during in vitro micropropagation through somatic embryogenesis (SE). This putative 'gene'-assisted selection will provide confidence for large-scale commercial planting of oil palm clones, once an efficient diagnostic assay is made available. An added booster will be the ability to predict if a palm can be successfully micropropagated through SE.

The oil palm indirect SE process is initiated from leaf explants, which dedifferentiates into callus followed by somatic embryo development. The candidate gene approach was initially used to identify potential SE genes. This approach resulted in the association of a putative peroxiredoxin, *EgPER1* (Ong, 2001) and *AINTEGUMENTA-like* (*AIL*) gene, *EgAP2-1* (Morcillo et al., 2007) with oil palm SE. In order to have an improved understanding of genes expressed during tissue culture, ESTs were generated from suspension cell cultures (Ho et al., 2007; Lin et al., 2009; Roowi et al., 2010), leaf-derived embryogenic callus (EC), non-embryogenic callus (NEC) and embryoid (EMB) (Low et al., 2008; Chan et al., 2010). Potential gene expression markers for embryogenesis such as *EgPK1*, a member of serine/threonine kinase family (Ong-Abdullah and Ooi, 2007; Ooi et al., 2008); *EgHOX1*, a HD-Zip II subfamily member (Ong-Abdullah and Ooi, 2007; Ooi et al., 2010); *OPSC10*, coding for aminocyclopropane-1-carboxylate oxidase (Ong-Abdullah and Ooi, 2007; Ooi et al., 2010); *EgLSD*, a putative lignostilbene-α,β-dioxygenase (Roowi et al., 2010); *EgER6*, a putative ethylene-responsive 6 (Roowi et al., 2010); and *Eg*707, an unknown protein (Thuc et al., 2011) were identified from these ESTs.

Transcript profiling of ESTs during the tissue culture process was made possible by the DNA microarray technology. The EST DNAs were printed on a glass slide, and relative gene expression data were obtained through differential co-hybridization between test and reference samples. Preliminary work focused on the identification of differentially expressed genes across EC, NEC, EMB and shoot (Ooi et al., 2001; Chan et al., 2005; Low et al., 2006) prior to early stages of tissue culture process. A total of 76 unigenes whose expression was upregulated in NEC as compared to EC/EMB and shoot were identified (Low, 2009). Research was subsequently directed at targeting a group of hormone-responsive genes such as *EgIAA9*, a putative *Aux/IAA* to predict for embryogenic competency (Ooi et al.,

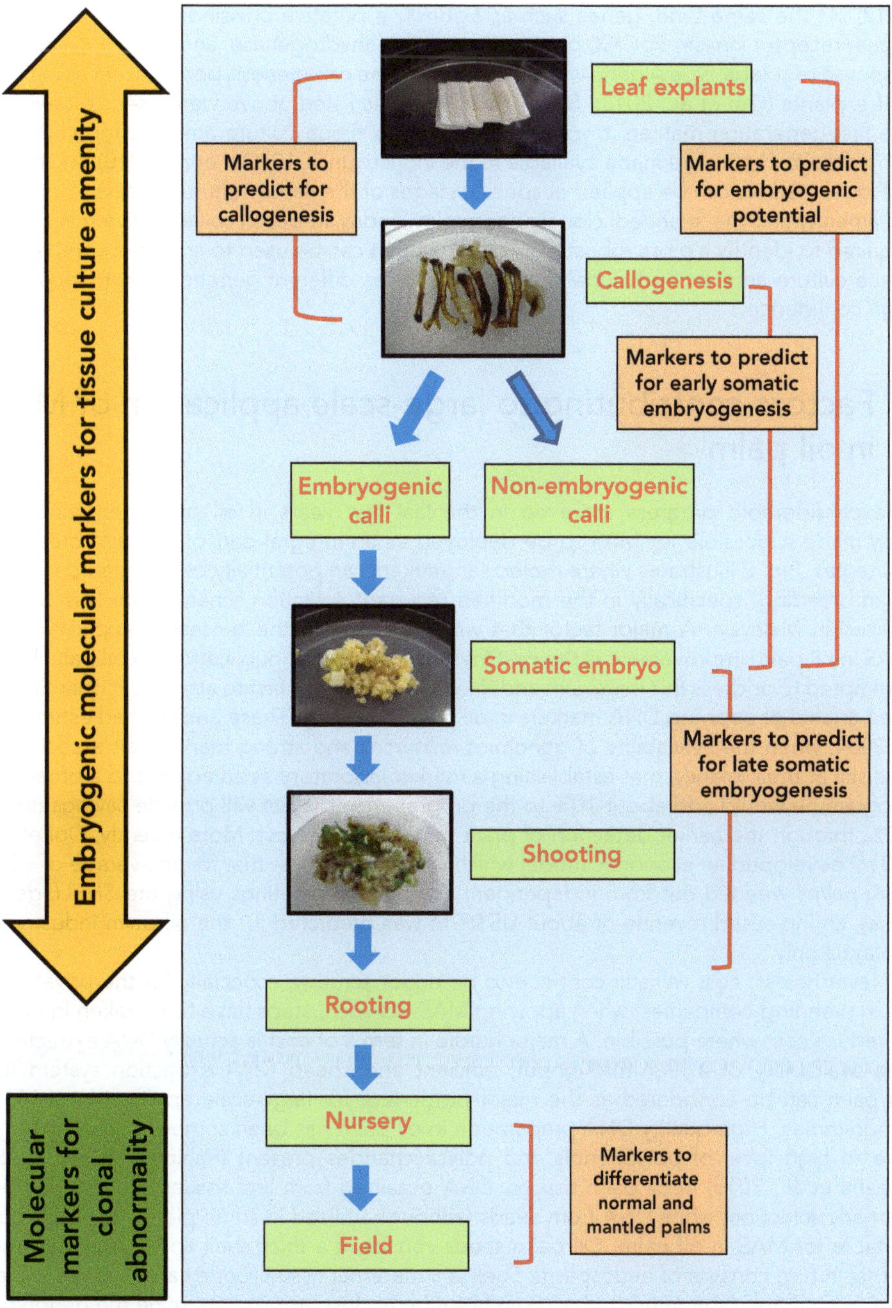

Figure 1 Application of molecular markers during the oil palm tissue culture process. Molecular markers can be used at specific stages to predict for amenity and clonal abnormality.

2012). At the same time, genes such as *EgBrRK*, a putative brassinosteroid leucine-rich repeat receptor kinase; *EgCKX*, putative cytokinin dehydrogenase; and *EgRR1*, a putative response regulator type A gene were also linked to the callogenesis potential of uncultured leaf explants (Ooi et al., 2013). Several of the genes listed above were used to develop the first-generation markers to predict for oil palm tissue culture amenity, for which 15 candidate markers were made available to the industry in 2010 (Ooi et al., 2010). As shown in Fig. 1, markers can be applied at specific stages of the tissue culture process to predict for amenity and the 'mantled' clonal abnormality. Today, in-depth collaborative research is required to identify a more robust set of markers that can be used to accurately predict for tissue culture amenity and clonal abnormality across different genetic backgrounds with high confidence.

7 Factors contributing to large-scale application of MAS in oil palm

The considerable progress achieved in the last few years in oil palm genomics has now made it possible for MAS to be deployed as an integral part of oil palm breeding strategies. Fig. 2 illustrates where molecular markers can potentially be integrated into oil palm breeding, specifically in the modified recurrent selection scheme, which is widely utilized in Malaysia. A major factor that will contribute to the successful application of MAS in oil palm improvement is the cost involved. Not many publications in oil palm have attempted to address this issue. Lim and Rao (2004) were the first to attempt at quantifying the benefits of applying DNA markers in oil palm breeding. These authors had estimated in 2004, when the availability of genomics resources and strong marker-trait association was still at their infancy, that establishing a marker laboratory as an adjunct to a breeding programme would add about 10% to the programme cost but will provide savings up to 30%, through the earlier detection of plant material of interest. More recently, Ooi et al. (2016) developed an economic model which was able to show that for an average of ~8% *dura* palms weeded out from independent commercial plantings using the *SHELL* gene assay, an increased revenue of about US$89M was predicted for the oil palm industry in Malaysia only.

Nevertheless, cost will still continue to be under scrutiny, especially for the private oil palm breeding companies, when applying MAS. As such, steps have to be taken in order to reduce cost where possible. A major hurdle in terms of cost is actually DNA extraction. The availability of a high-throughput, efficient and cheap DNA extraction system for oil palm can be considered as the major bottleneck for large-scale application of MAS programmes. High-quality DNA preparation in oil palm has been somewhat challenging, due to high level of polyphenols and polysaccharides present (Rahimah et al., 2006; Suzana et al., 2015) in oil palm tissues. DNA obtained from leaf tissues is most feasible for early selection, while DNA from seeds (although utilized in some plant species) is not suitable for MAS in oil palm. Oil palm seeds consist of a thick shell and kernel, and the kernel in turn consists of endosperm. Shell is a maternal tissue (Corley and Tinker, 2003a) and as such, the sample has to be taken from the endosperm to determine the genotype of the palm that will eventually emerge. High cost of penetrating the thick shell to get into the endosperm without damaging the embryo makes the process impractical. In addition, up to 20% of seeds may eventually be culled as off-types at the various developmental

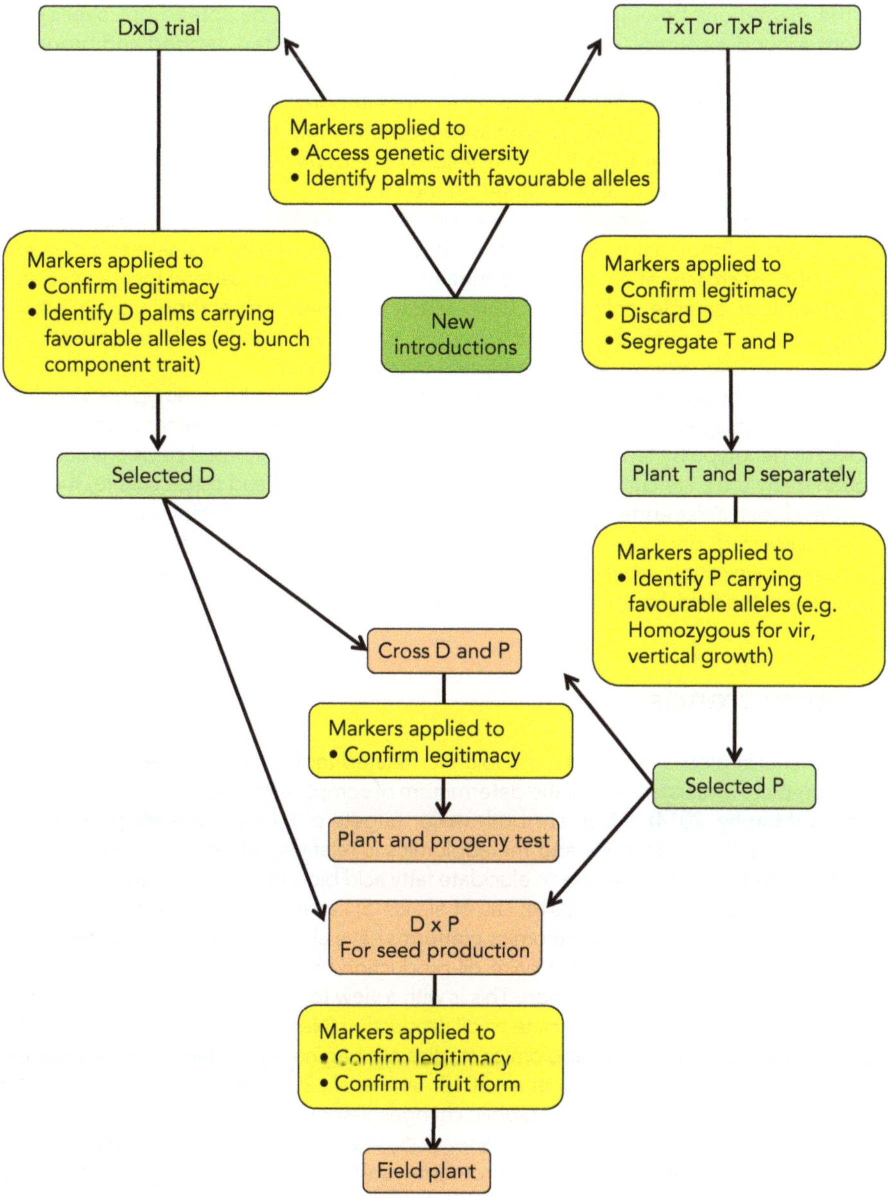

Figure 2 A simplified diagram describing integration of markers in the modified recurrent selection breeding scheme (D: *dura*, T: *tenera*, P: *pisifera*).

stages, making testing at this stage not cost-effective. Furthermore, seeds are also not appropriate for DNA-based quality control in the tissue culture process, as clonal materials are not seed derived.

As the plantations belonging to an oil palm company can be spread out over several locations around a country, samples must be efficiently collected, tagged and brought to a central DNA testing facility. An efficient sample collection kit, allowing sample tracking, is essential for a large-scale MAS programme. More importantly, investment in developing an efficient sample tracking system similar to laboratory information management system is important. Tracking samples from the field to the DNA testing lab and subsequently correlating the genotype profile obtained in the laboratory back to the original sample is time consuming and can be error-prone if not managed well (Xu and Crouch, 2008). This is especially true for oil palm, for which breeders tend to establish, evaluate and manage a large number of crosses at any one time that are spread out over several locations. Careful management of sample collection, DNA extraction, data generation via an efficient tracking system at a high-throughput level will contribute significantly to cost-effective and successful MAS programmes for oil palm in the long run.

Undeniably, a huge amount of data is going to be generated by genotyping samples of interest in MAS programmes, and this dataset must be integrated with other databases, [e.g. breeding or genotype by environment (GxE) database] to make informed decisions accurately. Although many tools are currently available in oil palm breeding and tissue culture, they were almost developed independently, with separate data formats and in different operating systems. If not synchronized, decision-making in a timely manner will be hampered and may restrict effective communication between geneticists working in the laboratory as well as breeders and agronomists in the field, hampering efforts at implementing MAS for oil palm improvement.

8 Future trends

Advancements in proteomics- and metabolomics-related technologies are essential to obtain a more complete understanding of the determinism of complex traits such as yield and disease resistance (Murphy, 2014), which are both major targets of oil palm breeding. Admittedly, research in oil palm proteomics and metabolomics is relatively at its infancy. Nevertheless, pioneering efforts have been made to elucidate fatty acid biosynthesis and lipid metabolism in oil palm mesocarp (Hasliza et al., 2014; Lau et al., 2015; Lau et al., 2016) using gel-based and nanoLC–MS/MS technologies. Proteomics platforms have also been utilized in understanding the biological processes involved in palm oil production by analysing the fruits harvested at different time points during maturation. This is with a view to identify key temporal changes in the mesocarp proteome that contribute to oil production (Neoh et al., 2013; Teh et al., 2013). At present, proteomics and metabolomics studies are beginning to identify various group of metabolites from the oil palm (Tahir et al., 2012; Tahir et al., 2013) and reveal the relationships between specific proteins and/or metabolic changes associated with *Ganoderma boninense* infection, which causes the basal stem rot disease (Syahanim et al., 2013; Nurazah et al., 2013). Efforts are already underway to develop proteome- and metabolome-related databases (Nur-Ain et al., 2015) in order to have a centralized data repository system, which will facilitate access to data and dissemination of relevant information.

9 Conclusion

In oil palm breeding and micropropagation, the adoption and utilization of MAS have not yet met expectations. Even if markers for DNA fingerprinting to assess the fidelity of controlled crosses have been available for at least a decade, widespread application is lacking. Most of the efforts had actually been channelled at developing appropriate genomic resources and finding marker-trait association for both monogenic and quantitative traits of interest. The significant achievements which occurred in the last 3–4 years, especially in deciphering the oil palm genome and in identifying genes linked to the two important monogenic traits of oil palm, namely fruit form and colour, will likely pave the way for MAS in oil palm. The identification of these genes and subsequent development of a diagnostic assay for *SHELL* has opened up the prospects for error-free early selection in oil palm breeding. The identification of the gene linked to the expression of the 'mantled' clonal abnormality has also opened up the prospects for the use of markers for quality control during the oil palm micropropagation process. In the medium term, the advances made in conventional QTL analysis and association genetics in identifying markers linked to economically important polygenic traits will facilitate improvement of quantitative traits via MAS. MAS undoubtedly will be most effective where phenotyping is difficult or expensive for the polygenic traits. The emerging technologies of proteomics and metabolomics will also likely contribute to the long-term goal of routine selection via MAS. With the right support structure in place, in terms of sample collection, processing and tracking, MAS will facilitate the exploitation of the genetic variation available in exotic germplasms to improve yields and provide considerable value addition for sustainable development of the industry.

10 Where to look for further information

Additional information on oil palm genome assembly, transcripts, gene annotation and genetic markers is available at the Genomsawit website 'http://genomsawit.mpob.gov.my'.

11 References

Abdullah, R. and Wahid, M. B. (2010), World palm oil supply, demand, price and prospects: Focus on the Malaysian and Indonesian palm oil industry. Malaysian Palm Oil Council (MPOC) Full report (http://www.mpoc.org.my/upload/WorldPalmOil_SupplyDemandPriceProspects_MalaysianIndonesianIndustry_FullReport.pdf).

Abdurakhmonov, I. Y. and Abdukarimov, A. (2008), 'Application of association mapping to understanding genetic diversity of plant germplasm resources', *International Journal of Genomics*, 208, 18.

Al-Obaidi, J. R., Mohd-Yusuf, Y., Razali, N., Jayapalan, J. J., Tey, C. C., Md-Noh, N., Junit, S. M., Othman, R. Y. and Hashim, O. H. (2014), 'Identification of proteins altered abundance in oil palm infected with *Ganoderma boninense*', *International Journal of Molecular Sciences*, 15, 5175–92.

Barcelos, E. Rios, S. D. A., Cunha, R. N. V., Lopes, R., Motoike, S. Y., Babiychuk, E., Skirycz, A. and Kushnir, S. (2015), 'Oil palm natural diversity and the potential for yield improvement', *Frontiers In Plant Science*, 6, 1.

Basiron, Y. (2000), Techno-economic aspects of research and development in the Malaysian oil palm industry. In: Basiron, Y., Jalani, B. S. and Chan, K. W. (eds), *Advances in Oil Palm Research*. Volume 1. Kuala Lumpur: Malaysian Palm Oil Board, pp. 1–18.

Beckmann, J. S., and Soller, M. (1986), 'Restriction fragment length polymorphisms and genetic improvement of agricultural species', *Euphytica*, 35, 111–24.

Beirnaert, A. and Vanderweyen, R. (1941), 'Contribution a l'etude genetique et biometrique des varieties *d'Elaeis guineensis* Jacq', *Publ. Inst Nat. Etude agron. Congo Belge, Ser. Sci.*, 27, 1–101.

Billotte, N., Marseillac, N., Risterucci, A. M., Adon, B., Brotteir, P., Baurens, F. C., Singh, R., Herran, A., Asmady, H., Billot, C., Amblard, P., Durrand-Gasselin, T., Courtois, B., Asmono, D., Cheah, S. C., Rohde, W. and Charrier, A. (2005), 'Microsatellite-based high density linkage map in oil palm (*Elaeis guineensis* Jacq.)', *Theoretical and Applied Genetics*, 110, 754–65.

Billotte, N., Jourjon, M F., Marseillac, N., Berger, A., Flori, A., Asmady, H., Adon, B., Singh, R., Nouy, B., Potier, F., Cheah, S. C., Rohde, W., Ritter, E., Courtois, B., Charrier, A. and Mangin, B. (2010), 'QTL detection by multi-parent linkage mapping in oil palm (*Elaeis guineensis* Jacq.)', *Theoretical and Applied Genetics*, 120, 1673–87.

Beule, T., Agbessi, M. D., Dussert, S., Jaligot, E. and Guyot, R. (2015). 'Genome-wide analysis of LTR-retrotransposons in oil palm', *BMC Genomics*, 16(1).

Bourgis, F., Kilaru, A., Cao, X., Ngando-Ebongue, G.-F., Drira, N., Ohlrogge, J. B. and Arondel, V. (2011), 'Comparative transcriptome and metabolite analysis of oil palm and date palm mesocarp that differ dramatically in carbon partitioning', *Proceedings of the National Academy of Sciences*, 108, 12527–32.

Butcher, P. and Southerton, S. (2007), Marker assisted selection in forestry species, In: Guimaraes, E. P., Ruane, J., Scherf, B. D., Sonnino, A. and Dargie, J. D. (eds), *Marker assisted selection – Current Status and future perspectives in crops, livestock, forestry and fish*. Rome: Food and Agriculture Organization of the United Nations.

Chan, P. L., Low, E. T. L., Boon, S. H., Rozana, R., Ooi, L. C.-L.,., Rajinder, S. and Cheah, S. C. (2005), *Monitoring gene expression changes in different stages of oil palm tissue culture via cDNA microarray*, In 2005 Conference on Biotechnology of Plantation Commodities, MIECC, Mines Resort City, Selangor, 9th to 11th August 2005. Kuala Lumpur: Malaysian Palm Oil Board. pp. 445–53.

Chan, P. L., Ma, L. S., Low, E. T. L., Elyana, M. S., Ooi, L. C.-L., Cheah, S. C. and Rajinder, S. (2010), 'Normalized embryoid cDNA library of oil palm (*Elaeis guineensis*)', *Electronic Journal of Biotechnology*, 13(1), 1–17.

Cheah, S. C., Siti Nor Akmar, A., Ooi, L. C. L., Rahimah, A. R. and Maria, M. (1993), Detection of DNA variability in the oil palm using RFLP probes. In 1991 PORIM International Palm oil Conference-Agriculture (Module 1). Kuala Lumpur, 20th to 25th September 1993. Kuala Lumpur: Palm Oil Research Institute of Malaysia, pp. 144–50.

Chin, C. W. (1995), *Oil Palm planting materials and quality control*. In 1995 PORIM National Oil Palm Conference-Tech In Plant. Kuala Lumpur, 11th to 12th July 1995. Kuala Lumpur: Palm Oil Research Institute of Malaysia, pp. 38–47.

Cochard, B., Carasco-Lacombe, C., Pomies, V., Dufayard, J. F., Suryana, E., Omore, A., Durand-Gasselin, T. and Tisne, S. (2015), 'Pedigree-based linkage map in two genetic groups of oil palm', *Tree Genetics and Genomes*, 11, 68. Available from: doi:10.1007/s11295–015-0893–7.

Corley, R. H. V. and Lee, C. H. (1992), 'The physiological basis for genetic improvement of oil palm in Malaysia', *Euphytica*, 60, 179–84.

Corley, R. H. V. and Tinker, P. B. (2003a), 'Selection and Breeding', *The Oil Palm*. Fourth edition. Oxford: Blackwell Science Ltd.

Corley, R. H. V. and Tinker, P. B. (2003b), 'Care and maintenance of oil palms', *The Oil Palm*. Fourth edition. Oxford: Blackwell Science Ltd.

Corley, R. H. V. and Tinker, P. B. (2003c), Vegetative propagation and biotechnology, *The Oil Palm*. Fourth edition. Oxford: Blackwell Science Ltd.

Corley, R. H. V. (2005), 'Illegitimacy in oil palm breeding- A review', *Journal of Oil Palm Research*, 17, pp. 64–9.

Cros, D., Denis, M., Sanchez, L., Cochard, B., Flori, A., Durand-Gasselin, T., Nouy, B., Omore, A., Pomies, V., Riou, V., Suryana, E. and Bouvet, J. M. (2015), 'Genomic selection prediction accuracy in a perennial crop: case study of oil palm (*Elaeis guineensis* Jacq.)', *Theoretical and Applied Genetics*, 128(3), 397–410.

Donough, C. R., Corley, R. H. V. and Law, I. H. (1996), '"Non-shedding" mutants of the oil palm', *International Society of Oil Palm Breeders Newsletter*, 11(2), 10–11.

Durand-Gasselin, T., Blangy, L., Picasso, C., De Franqueville, H., Breton, F., Amblard, P., Cochard, B., Louise, C. and Nouy, B. (2010), 'Sélection du palmier à huile pour une huile de palme durable et responsabilité sociale', *Oléagineux, Corps Gras, Lipides*, 17(6). 385–92.

Godswill, N-N., Frank, N-E. G., Walter, A-N., Edson, M-Y. J., Kingsley, T-M., Arondel, V., Martin, B.J. and Emmanuel, Y. (2016), Oil palm, In: Gupta, S. K. (ed) *Breeding Oil Seed Crops for Sustainable Production: Opportunities and Constraints*. USA: Elsevier Inc.

Guerin, C., Joet, T., Serret, J., Lashermes, P., Vaissayre, V., Agbessi, M. D. T., Beule, T., Severac, D., Amblard, P., Tregear, J., Durand-Gasselin, T., Morcillo, F. and Dussert, S. (2016), 'Gene coexpression network analysis of oil biosynthesis in an interspecific backcross of oil palm', *The Plant Journal*, 87(5), 423–41.

Hama-Ali, E. O., Syed Alwee, S. S. R., Tan, S. G., Panandam, J. M., Ho, C.-L., Namasivayam, P. and Peng, H. B. (2015), 'Illegitimacy and sibship assignments in oil palm (*Elaeis guineensis* Jacq.) half-sib families using single locus DNA microsatellite markers', *Molecular Biology Reports*, 42, 917–25.

Hartley, C. W. S. (1988), The origin and development of the oil palm industry, *The Oil Palm*. 3rd edition. London: Longman.

Hasliza, H., Lau, B. Y. C. and Ramli, U. S. (2014), 'Extraction method for analysis of oil palm leaf and root proteins by Two-dimensional Gel Electrophoresis', *Journal of Oil Palm Research*, 26, 54–61.

Hazzouri, K. M., Flowers, J. M., Visser, H. J., Khierallah, H. S. M., Rosas, U., Pham, G. M., Meyer, R. S., Johansen, C. K., Fresquez, Z. A., Masmoudi, K., Haider, N., Kadri, N. E., Idaghdour, Y., Malek, J. A., Thirkhill, D., Markhand, G. S., Krueger, R. R., Zaid, A. and Purugganan, M. D. (2015), 'Whole genome re-sequencing of date palms yields insights into diversification of a fruit tree crop', *Nature Communications*, 6, 8824. Available from: doi:10.1038/ncomms9824 (2015).

Hitam, A., Darus, A. and Weng, C. K. (1999), *Systems Approach to Mechanization in Oil Palm Plantations*, In 1999 PORIM International Palm Oil Congress (PIPOC)– Emerging Technologies and Opportunities in the Next Millennium. 1st to 6th February 1999. Kuala Lumpur: Palm Oil Research Institute of Malaysia, pp. 325–36.

Ho, C. L., Kwan, Y. Y., Choi, M. C., Tee, S. S., Ng, W. H., Lim, K. A., Lee, Y. P., Ooi, S. E., Lee, W. W., Tee, J. M., Tan, S. H., Kulaveerasingam, H., Syed Alwee, S. S. R. and Ong Abdullah, M. (2007), 'Analysis and functional annotation of expressed sequence tags (ESTs) from multiple tissues of oil palm (*Elaeis guineensis* Jacq.)', *BMC Genomics*. 8, 381.

Ho, Y. W., Tan, C. C., Soh, A. C., Wong, G., Chong, S. P., Choo, C. N. and Norazura, A. (2008), Biotechnological approaches in producing oil palm planting material-a success story, *AAR News*, 1–5 December.

Jaligot, E., Beule, T. and Rival, A. (2002), 'Methylation-sensitive RFLPs: characterisation of two oil palm markers showing somaclonal variation-associated polymorphism', *Theoretical and Applied Genetics*, 104, 1263–9.

Joonnor, S. and Volkaert, H. (2014), 'Mapping of Quantitative Trait Loci (QTLs) for oil yield using SSRs and gene-based markers in African Oil Palm (*Elaeis guineensis* Jacq.)', *Tree Genetics and Genomes*, 10, 1–14.

Jouannic, S., Argout, X., Lechauve, F., Fizames, C., Borgel, A., Morcillo, F., Aberlenc-Bertossi, F., Duval, Y. and Tregear, J. (2005), 'Analysis of expressed sequence tags from oil palm (*Elaeis guineensis*)', *FEBS letters*, 579(12), 2709–14.

Korte, A. and Farlow, A. (2013), 'The advantages and limitations of trait analysis with GWAS: a review', *Plant methods*, 9(1), 1.

Kwong, Q. B., TeH, C. K., Ong, A. L., Heng, H. Y., Lee, H. L., Mohamed, M., Low, J. Z.-B., Apparow, S., Chew, F. T., Mayes, S., Kulaveerasingam, H., Tammi, M. and Appleton D. R. (2016), 'Development and validation of a high-density SNP genotyping array for African Oil Palm', *Molecular Plant*, 9, 1132–41.

Lau, B. Y. C., Clerens, S., Morton, J. D., Dyer, J. M., Deb-Choudhury, S. and Ramli, U. S. (2015), 'Method developments to extract proteins from oil palm chromoplast for proteomic analysis', *Springer Plus* 4, 791.

Lau, B. Y. C., Clerens, S., Morton, J. D., Dyer, J. M., Deb-Choudhury, S. and Ramli, U. S. (2016), 'Application of Mass Spectrometry approach to detect the presence of fatty acid biosynthetic phosphopeptides', *Protein Journal*, 35, 163–70.

Lee, M., Xia, J. H., Zou, Z., Ye, J., Alfiko, Y., Jin, J., Lieando, J. V., Purnamasari, M. I., Lim, C. H., Suwanto, A., Wong, L., Chua, N. M. and Yue, G. H. (2015), 'A consensus linkage map of oil palm and a major QTL for stem height', *Scientific reports*, 5, 8232. Available from: doi:10.1038/srep08232.

Lim, C. C. and Rao, V. (2004), 'DNA marker technology and private sector oil palm breeding', *The Planter*, 80(943), 611–28.

Lin, H.-C., Morcillo, F., Dussert, S., Tranchant-Dubreuil, C., Tregear, J. W. and Tranbarger, T. J. (2009), 'Transcriptome analysis during somatic embryogenesis of the tropical monocot *Elaeis guineensis*: evidence for conserved gene functions in early development', *Plant Molecular Biology*, 70, 173–92.

Loei, H., Lim, J., Tan, M., Lim, T. K., Lin, Q. S., Chew, F. T., Harikrishna, K. and Cung, M. C. (2013), 'Proteomic Analysis of the Oil Palm Fruit Mesocarp Reveals Elevated Oxidative Phosphorylation Activity is Critical for Increased Storage Oil Production', *Journal of proteome research*, 12, 5096–109.

Low, E. T. L. (2009), Development and application of expressed sequenced tags and DNA microarray for somatic embryogenesis in oil palm. A Thesis Submitted in partial fulfilment of the Requirements of Universiti Putra Malaysia for the Degree of Doctor of Philosophy. Malaysia: Universiti Putra Malaysia.

Low, E. T. L., Halimah, A., Boon, S. H., Elyana, M. S., Tan, C. Y., Ooi, L. C. L., Cheah, S. C., Abdul-Rahim, R., Wan, K. L. and Singh, R. (2008), 'Oil palm (*Elaeis guineensis* Jacq.) tissue culture ESTs: identifying genes associated with callogenesis and embryogenesis', *BMC Plant Biology*, 8, 62.

Low, E. T. L., Rosli, R., Jayanthi, N., Mohd-Amin, A. B., Azizi, N., Chan, K. L., Maqbool, N. J., Maclean, P., Brauning, R., Mcculloch, A., Moraga, R., Ong-Abdullah, M. and Singh, R. (2014), 'Analyses of hypomethylated oil palm gene space', *PLoS ONE*, 9(1), e86728.

Low, E. T. L., Tan, J. S., Chan, P. L., Boon, S. H., Wong, Y. L., Rozana, R., Ooi, L. C.-L., Ma, L. S., Ong-Abdullah, M., Cheah, S. C., and Rajinder, S. (2006), 'Developments toward the application of DNA chip technology in oil palm tissue culture', *Journal of Oil Palm Research*, Special issue, 87–98.

Mayes, S., James, C. M., Horner, S. F., Jack, P. L., and Corley, R. H. V. (1996), 'The application of restriction fragment length polymorphism for the genetic fingerprinting of oil palm (*Elaeis guineensis* Jacq)', *Molecular Breeding*, 2, 175–80.

Mayes, S., Jack, P. L., Marshall, D. and Corley, R. H. V. (1997), 'Construction of a RFLP genetic linkage map for oil palm (*Elaeis guineensis* Jacq.)', *Genome*, 40, 116–22.

Montoya, C., Cochard, B., Flori, A., Cros, D., Lopes, R., Cuellar, T., Espeout, S., Syaputra, I., Villeneuve, P., Pina, M., Ritter, E., LeroY, T. and Billotte, N. (2014), 'Genetic architecture of palm oil fatty acid composition in cultivated oil palm (*Elaeis guineensis* Jacq.) compared to its wild relative *E.oleifera* (H. B. K) Corte´s', *PLoS ONE*, 9(5), e95412. Available from: doi:10.1371/journal.pone.0095412.

Montoya, C., Lopes, R., Flori, A., Cros, D., Cuellar, T., Summo, M., Espeout, S., Rivallan, R., Risterucci, A. M., Bittencourt, D., Zambrano, J. R., Alarcon, G. W. H., Villeneuve, P., PINA, M., Nouy, B., Amblard, P., Ritter, E., Leroy, T. and Billotte, N. (2013), 'Quantitative Trait loci (QTLs) Analysis of Palm Oil Fatty Acid Composition in an Interspecific Pseudo-backcross from *Elaeis oleifera* (H. B. K.) Cortés and Oil Palm (*Elaeis guineensis* Jacq.)', *Tree Genetics and Genomes*, 9, 1207–25.

Morcillo, F., Gallard, A., Pillot, M., Jouannic, S., Aberlenc-Bertossi, S., Collin, M., Verdeil, J. L. and Tregear, J. W. (2007), '*EgAP2–1*, an *AINTEGUMENTA-like* (*AIL*) gene expressed in meristematic and proliferating tissues of embryos in oil palm', *Planta*, 226, 1353–62.

Moretzsohn, M. C., Nunes, C. D. M., Ferreira, M. E. and Grattapaglia, D. (2000), 'RAPD linkage mapping of the shell thickness locus in oil palm (*Elaeis guineensis* Jacq.)', *Theoretical and Applied Genetics*, 100, 63–70.

Mullis, K. and Faloona, F. (1987), 'Specific synthesis of DNA in vitro via a polymerase-catalyzed chain reaction', *Methods Enzymol*, 155, 335–50. https://doi.org/10.1016/0076-6879(87)55023-6

Murphy, D. J. 2014, 'The future of oil palm as a major global crop: opportunities and challenges', *Journal of Oil Palm Research*, 26, 1–24.

Neale, D. B. and Kremer, A. (2011), 'Forest tree genomics: growing resources and applications', *Nature Reviews Genetics*, 12, 111–22.

Neoh, B. K., Teh, H. F., Ng, T. L. M., Tiong, S. H., Thang, Y. M., Ersad, M. A., Mohamed, M., Chew, F. T., Kulaveerasingam, H. and Appleton, D. R. (2013), 'Profiling of metabolites in oil palm mesocarp at different stages of oil biosynthesis', *Journal of Agricultural and Food Chemistry*, 61, 1920–7.

Nurazah, Z., Idris, A. S., Kushairi, A. D., and Ramli, U. S. (2013), 'Metabolite profiling of oil palm towards understanding Basal Stem Rot (BSR) disease', *Journal of Oil Palm Research*, 25(4), 58–71.

Nur-Ain, I., Nurazah, Z., Nurul Liyana, R., Halim, M. A., Rozana, R., Abrizah, O., Ramli, U. S. and Tahir, N. I. (2015), *Construction of integrated oil palm metabolome chemical resources library*, In Palm Oil International Conference 2015. Kuala Lumpur, 6th to 8th October 2015. Kuala Lumpur: Malaysian Palm Oil Board, pp. 381–5.

Ong-Abdullah M. (2001), An examination of embryogenic and non-embryogenic cultures of oil palm (*Elaeis guineensis)*, A Thesis Submitted in partial fulfilment of the Requirements of Universiti Putra Malaysia for the Degree of Doctor of Philosophy. Malaysia: Universiti Putra Malaysia.

Ong-Abdullah, M. and Ooi, S. E. (2007), 'Biomarkers: finding a niche in oil palm tissue culture. Part 2-Targeting the transcriptome', *Oil Palm Bulletin*, 54, 68–88.

Ong-Abdullah, M., Ordway, J. M., Jiang, N., Ooi, S. E., Kok, S. Y., Sarpan, N., Azimi,N., Hashim, A. T., Ishak, Z., Kamal Rosli, S., Ahmad Malike, F., Abu Bakar, N A., Marjuni, M., Abdullah, N., Yaakub, Z., Amiruddin, M. D., Nookiah, R., Singh, R., Leslie Low, E-T., Chan, K.-L., Azizi, N., Smith, S. W., Bacher, B., Budiman, M. A., Van Brunt, A., Wishchmeyer, C., Beil, M., Hogan, M., Lakey, N., Lim, C.-C., Arulando, X., Wong, C.-K., Choo, C.-N., Wong, W.-C., Kwan, Y.-Y., Syes Alwee, S. S. R., Sambanthamurthi, R. and Martienssen, R. A. (2015), 'Loss of *Karma* transposon methylation underlies the mantled somaclonal variant of oil palm', *Nature*, 525, 533–7.

Ooi, L. C.-L., Cheah, S. C., Rahimah, A. R. and Bresnahan, P. (2001), Preliminary microarray gene profiling and differential gene expression studies in oil palm, Paper presented at the Microarray User Group Meeting (MUGV) meeting. Victoria, B. C., Canada, 12–14 August.

Ooi, S.-E., Choo, C.-N., Ishak, Z. and Ong-Abdullah, M. (2012), 'A candidate auxin-responsive expression marker gene, EgIAA9, for somatic embryogenesis in oil palm (*Elaeis guineensis* Jacq.)', *Plant Cell Tissue and Organ Culture*, 110, 201–12.

Ooi, S.-E., Choo, C.-N., Ishak, Z. and Ong-Abdullah, M. (2013), 'Candidate hormone-responsive marker for callogenesis of oil palm (*Elaeis guineensis* Jacq.)', *Journal of Oil Palm Research*, 25(1), 9–21.

Ooi, S.-E., Harikrishna, K. and Ong-Abdullah, M. (2008), 'Isolation and characterization of a putative serine/threonine kinase expressed during oil palm tissue culture', *Journal of Oil Palm Research, Special Issue Malaysia-MIT Biotechnology Partnership Programme 1-Oil Palm Tissue Culture*, 14–22.

Ooi, S. E., Low, E. T. L., Chan, P. L., Ooi, C. L. L., Ishak, F., Kadri, F., Singh, R. and Ong-Abdullah, M. (2010), First generation embryogenic markers (GEM) for tissue culture amenity, MPOB Information Series No. 508, MPOB TT No. 452, Malaysia: Malaysian Palm Oil Board.

Ooi, L. C.-L., Low, E.-T. L., Abdullah, M. O., Nookiah, R., Ting, N. C., Nagappan, J., Manaf, M. A. A., Chan, K. L., Halim, M. A., Azizi, N., Omar, W., Murad, A ., Lakey, N., Ordway, J M., Favello, A., Budiman, M. A., Van Brunt, A., Beil, M., Leininger, M. T., Jiang, N., Smith, S. W., Brown, C. R., Kuek, A. C. S., Bahrain, S., Hoynes-O'Connor, A., Nguyen, A. Y., Chaudhari, H. G., Shah, S. A., Choo, Y.-M., Sambanthamurthi, R. and Singh, R. (2016), 'Non-*tenera* contamination and the economic impact of *SHELL* genetic testing in the Malaysian independent oil palm industry', *Frontiers in Plant Sciences*, 7, 771. Available from: doi:10.3389/fpls.2016.00771.

Parveez, G. K. A., Rasid, O. A., Masani, M. Y. A. and Sambanthamurthi, R. (2015), 'Biotechnology of oil palm: strategies towards manipulation of lipid content and composition', *Plant Cell Reports*, 34, 533–43.

Poland, J. (2015), 'Breeding-assisted genomics', *Current Opinion In Plant Biology*, 24, 119–24.

Pootakham, W., Jomchai, N., Ruang-Areerate, P., Shearman, J. R., Sonthirod, C., Sangsrakru, D., Tragoonrung, S., and Tangphatsornruang, S. (2015), 'Genome-wide SNP Discovery and identification of QTL Associated with Agronomic Traits in Oil Palm Using Genotyping-by-Sequencing (GBS)', *Genomics*, 105, 288–95.

Rahimah, A. B., Cheah, S. C. and Rajinder, S. (2006), 'Freeze-drying of oil palm (*Elaeis guineensis*) leaf and its effect on the quality of extractable DNA', *Journal of Oil Palm Research*, 18, 296–304.

Rance, K. A., Mayes, S., Price, Z., Jack, P. L. and Corley, R. H. V. (2001), 'Quantitative trait loci for yield components in oil palm (*Elaeis guineensis* Jacq.)', *Theoretical and Applied Genetics*, 103, 1302–10.

Rao, V., Jalani, B. S. and Rajanaidu, N. (1993), Effect of dura contamination on oil extraction rate (OER), In National Seminar on 'Palm oil extraction Rate; problem and issues'. Kuala Lumpur, 21st to 22nd December 1993. Kuala Lumpur: Palm Oil Research Institute, pp. 58–60.

Riju, A., Chandrasekar, A and Arunachalam, V. (2007), 'Mining for single nucleotide polymorphisms and insertions/deletions in expressed sequence tag libraries of oil palm', *Bioinformation*, 2(4), 128–31.

Rohani, O., Sharifah, S. A., Mohd Rafii, Y., Ong, M., Tarmizi, A H and Zamzuri, I (2000), Tissue culture of oil palm, In: Basiron. Y., Jalani, B. S. and Chan, K. W. (eds), *Advances in Oil Palm Research*. Kuala Lumpur: Malaysian Palm Oil Board, pp. 238–83.

Roongsattham, P., Morcillo, F., Jantasuriyarat, C., Pizot, M., Moussu, S., Jayaweera, D., Collin, M., Gonzalez-Carranza, Z. H., Amblard, P., Tregear, J. W., Tragoonrung, S., Verdeil, J.-L. and Tranbarger, T. J. (2012), 'Temporal and spatial expression of polygalacturonase gene family members reveals divergent regulation during fleshy fruit ripening and abscission in the monocot species oil palm', *BMC Plant Biology*, 12, 1.

Roongsattham, P., Morcillo, F., Fooyontphanich, K., Jantasuriyarat, C., Tragoonrung, S., Amblard, P., Collin, M., Mouille, G., Verdeil, J. L. and Tranbarger, T. J. (2016), 'Cellular and pectin dynamics during abscission zone development and ripe fruit abscission of the monocot oil palm', *Frontiers In Plant Science*, 7, 1.

Roowi, S. H., Ho, C.-L., Syed Alwee, S. S. R., Ong-Abdullah, M. and Napis, S. (2010), 'Isolation and characterization of differentially expressed transcripts from the suspension cells of oil palm (*Elaeis guineensis* Jacq.) in response to different concentration of auxins', *Molecular Biotechnology*, 46, 1–19.

Sambanthamurthi, R., Abrizah, O. and Umi Salamah, R. (1999), 'Biochemical factors that control oil composition in the oil palm', *Journal of Oil Palm Research (Special Issue)*, 24–33.

Sambanthamurthi, R., Parveez G. K. and Cheah S. C. (2000), Genetic Engineering of the Oil Palm, In: Basiron. Y., Jalani, B. S. and Chan, K. W. (eds), *Advances in Oil Palm Research*. Kuala Lumpur: Malaysian Palm Oil Board, pp. 284–331.

Sharma, M. and Tan, Y. (1997), 'Oil palm breeding programme and the performance of D × P planting materials at United Plantations Berhad', *Planter*, 73, 860.

Singh, R. and Cheah, S. C. (1999), 'Analysis of the Inheritance of AFLP markers in an interspecific cross of oil palm using the pseudo-testcross strategy', *Journal of Oil Palm Research (Special Issue)*, 64–73.

Singh, R., Zaki, N. M., Ting, N. C., Rosli, R., Tan, S.-G., Low, E.-T. L., Ithnin, M. and Cheah, S. C. (2008), 'Exploiting an Oil Palm EST Database for the Development of Gene-derived SSR Markers and Their Exploitation for Assessment of Genetic Diversity', *Biologia*, 63(2), 227–35.

Singh, R., Tan, S. G., Panandam, J. M., Rahimah, A. R., Ooi, L. C., Low, E.-T. L., Sharma, M., Jansen, J., and Cheah, S. C. (2009), 'Mapping quantitative trait loci (QTLs) for fatty acid composition in an interspecific cross of oil palm', *BMC Plant Biology*, 9(1), 1.

Singh, R., Ong-Abdullah, M., Low, E. T. L., Abdul Manaf, M. A., Rosli, R., Rajanaidu, N., Ooi, L. C.-L., Ooi, S. E., Chan, K. L., Halim, M. A., Azizi, N., Jayanthi, N., Bacher, B., Lakey, N., Smith, S. W., He, D., Hogan, M., Budiman, M. A., Lee, E. K., Desalle, R., Kudrna, D., Goicoechea, J. L., Wing, R. A., Willson, R. K., Fulton, R. S., Ordway, J. M., Martienssen, R. A. and Sambanthamurthi, R. (2013a), 'Oil palm genome sequences reveals divergence of interfertile species in old and new worlds', *Nature*, 500(7462), 335–9.

Singh, R., Low, E.-T. L., Ooi, L. C.-L., Ong-Abdullah, M., Ting, N.-C., Jayanthi, N., Rajanaidu, N., Amiruddin, M. D., Rosli, R., Abdul Manaf, M. A., Chan, K. L., Halim, M. A., Azizi, N., Lakey, N., Smith, S. W., Budiman, M. A, Hogan, M., Bacher, B., Brunt, A. V., Wang, C., Ordway, J. M., Sambanthamurthi, R. and Martienssen, R. A. (2013b), 'The oil palm SHELL gene controls oil yield and encodes a homologue of SEEDSTICK', *Nature*, 500(7462), 340–4.

Singh, R., Low, E. T. L., Ooi, L. C. L., Ong-Abdullah, M., Nookiah, R., Ting, N. C., Marjuni, M., Chan, P. L., Ithnin, M., Abdul Manaf, M. A. A., Nagappan, J., Chan, K. L., Rosli, R., Halim, M. A., Azizi, N., Budiman, M. A., Lakey, N., Bacher, B., Van Brunt, A., Wang, C., Hogan, M., He, D., Macdonald, J. D., Smith, S. W., Ordway, J. M., Martienssen, R. A. and Sambanthamurthi, R. (2014a), 'The oil palm VIRESCENS gene controls fruit colour and encodes a R2R3-MYB', *Nature Communications*, 5, 4106.

Singh, R., Low, E.-T. L., Ooi, L.C.-L., Abdullah, M. O., Rahimah, A. R., Manaf, M. A. A., Nookiah, R. and Sambanthamurthi, R. (2014b), *SureSawit*™ SHELL- A diagnostic assay to predict oil palm fruit forms, MPOB T. T No. 548, *MPOB Information Series*, June 2014.

Singh, R., Ooi, L.C.-L., Low, E.-T. L., Abdullah, M. O., Nagappan, J., Chin, T. N., Manaf, M. A. A., Nookiah, R. and Sambanthamurthi, R. (2015), SureSawit™*VIR* - A Diagnostic assay to predict colour of oil palm fruits, MPOB T. T No. 568, *MPOB Information Series*, June 2015.

Spindel, J., Begum, H., Akdemir, D., Virk, P., Collard, B., Redona, E., Atlin, G., Jannink, J. L. and McCouch, S. R. (2015), 'Genomic selection and association mapping in rice (*Oryza sativa*): effect of trait genetic architecture, training population composition, marker number and statistical model on accuracy of rice genomic selection in elite, tropical rice breeding lines', *PLoS Genetics*, 11(2), e1004982.

Suzana, M., Rahimah, A. R., Maizura, I. and Singh, R. (2015), 'A simple and rapid protocol for isolation of genomic DNA from oil palm leaf tissue', *Journal of Oil Palm Research*, 27 282–7.

Syahanim, S., Abrizah, O., Mohamad Arif, A. M., Idris, A. S., and Mohd Din, A (2013), 'Identification of differentially expressed proteins in oil palm seedlings artificially infected with G*anoderma*: a proteomics approach', *Journal of Oil Palm Research*, 25, 298–304.

Taeprayoon, P., Tanya, P., Kang, Y. J., Limsrivilai, A., Lee, S. H. and Srinives, P. (2016), 'Genome-wide SSR marker development in oil palm by Illumina HiSeq for parental selection', *Plant Genetic Resources*, 14(02), 157–60.

Tahir, N. I., Shaari, K., Abas, F., Ahmad Parveez, G. K., Ishak, Z., and Ramli, U. S. (2012), 'Characterization of apigenin and luteolin derivatives from oil palm (*Elaeis guineensis* Jacq.) leaf using LC-ESI-MS/MS', *Journal of Agricultural and Food Chemistry*, 60, 11201–10.

Tahir, N. I., Shaari, K., Abas, F., Ahmad Parveez, G. K., Ahmad Tarmizi, H. and Ramli, U. S. (2013), 'Identification of oil palm (*Elaeis guineensis*) spear leaf metabolites using mass spectrometry and neutral loss analysis', *Journal of Oil Palm Research*, 25, 72–83.

Tanksley, S. D. and Rick, C. M. (1980), 'Isozyme gene linkage map of the tomato: Applications in genetics and breeding', *Theoretical and Applied Genetics*, 57, 161–70.

Teh, H. F., Neoh, B. K., Hong, M. P. L., Low, J. Y. S., Ng, T. L. M., Ithnin, N., Thang, Y. M., Mohamed, M., Chew, F. T., and Yusof, H. M. (2013), 'Differential metabolite profiles during fruit development in high-yielding oil palm mesocarp', *PloS one*, 8, e61344.

Teh, C. K., Ong, A. L., Kwong, Q. B., Apparow, S., Chew, F. T., Mayes, S., Mohamed, M., Appleton, D. and Kulaveersasingham, H. (2016), 'Genome-wide association study identifies three key loci for high mesocarp oil content in perennial crop oil palm', *Scientific reports* 6, Article No. 19075, Available from: doi 10.1038/sreport19075.

Thongthawee, S., Tittinutchanon, P. and Volkaert, H. (2010), 'Microsatellites for parentage analysis in an oil palm breeding population, *Thai Journal of Genetics*, 3(2), 172–81.

Thuc, L. V., Sarpan, N., Ky, H., Ooi, S. E., Napis, S., Ho, C.-L., Ong-Abdullah, M., Chin, C.-F and Namasivayam, P. (2011), 'A novel transcript of oil palm (*Elaeis guineensis* Jacq.), Eg707, is specifically upregulated in tissues related to totipotency', *Molecular Biotechnology*, 48, 156–64.

Ting, N. C., Zaki, N. M., Rosli, R., Low, E.-T. L., Ithnin, M., Cheah, S. C., Tan, S. G. and Singh, R. (2010), 'SSR mining in oil palm EST database: application in oil palm germplasm diversity studies', *Journal of Genetics*, 89(2), 135–45.

Ting, N. C., Jansen, J., Mayes, S., Massawe, F., Sambanthamurthi, R., Ooi, L. C. L., Chin, C. W., Arulandoo, X., Seng, T. Y., Syed Alwee, S. S. R., Ithnin, M. and Singh, R. (2014), 'High density SNP and SSR-based genetic maps of two independent oil palm hybrids', *BMC Genomics*, 15, 309.

Ting, N. C., Yaakub, Z., Kamaruddin, K., Mayes, S., Massawe, F., Sambanthamurthi, R., Jansen, J., Low, E. T. L., Ithnin, M., Kushairi, A., Arulandoo, X., Rosli, R., Chan, K. L., Amiruddin, N., Kandha Sritharan, K., Lim, C. C., Nookiah, R., Amiruddin, M. A. and Singh, R. (2016), 'Fine-mapping and cross-validation of QTLs linked to fatty acid composition in multiple independent interspecific crosses of oil palm', *BMC Genomics*, 17, 289.

Tisné, S., Denis, M., Cros, D., Pomiès, V., Riou, V., Syahputra, I., Omoré, A., Durand-Gasselin, T., Bouvet, J. M., and Cochard, B. (2015), 'Mixed model approach for IBD-based QTL mapping in a complex oil palm pedigree', BMC Genomics, 16, 798. Available from: doi:10.1186/s12864–015-1985–3.

Tranbarger, T. J., Dussert, S., Joet, T., Argout, X., Summo, M., Champion, A., Cros, D., Omore, A., Nouy, B. and Morcillo, F. (2011), 'Regulatory mechanisms underlying oil palm fruit mesocarp maturation, ripening, and functional specialization in lipid and carotenoid metabolism', *Plant Physiology*, 156, 564–84.

Tranbarger, T. J., Kluabmongkol, W., Sangsrakru, D., Morcillo, F., Tregear, J. W., Tragoonrung, S. and Billotte, N. (2012), 'SSR markers in transcripts of genes linked to post-transcriptional and transcriptional regulatory functions during vegetative and reproductive development of *Elaeis guineensis*, *BMC Plant Biology*, 12, 1.

Ukoskit, K., Chanroj, V., Bhusudsawang, G., Pipatchartlearnwong, K., Tangphatsornruang, S., and Tragoonrung, S. (2014), 'Oil palm (*Elaeis guineensis* Jacq.) linkage map and quantitative trait locus analysis for sex ratio and related traits', *Molecular Breeding*, 33, 415–24.

Wooi, K. C. (1993), *Oil palm tissue culture-current practice and constraints*, In ISOPB International Symposium on Recent Developments in Oil Palm Tissue Culture and Biotechnology. Kuala Lumpur: 24th to 25th September 1993. Kuala Lumpur: Palm Oil Research Institute of Malaysia, pp. 21–32.

Xiao, Y., Zhou, L., Xia, W., Mason, A. S., Yang, Y., Ma, Z. and Peng, M. (2014), 'Exploiting transcriptome data for the development and characterization of gene-based SSR markers related to cold tolerance in oil palm (*Elaeis guineensis*)', *BMC Plant Biology*, 14, 384.

Xu, Y and Crouch, J. (2008), 'Marker-Assisted Selection in plant breeding: from publications to practice', *Crop Science*, 48, 391–407.

Yong, Y. Y. (1992), *The oil palm breeding programme in Malaysia*, In Symposium on the Science of Oil Palm Breeding. Kuala Lumpur, 30th June To 2nd July 1992. Kuala Lumpur: Palm Oil Research Institute of Malaysia, pp. 141–64.

Chapter 9

Advances in the genetic modification of oil palm

Denis J. Murphy, Head of Genomics and Computational Biology Research Group, University of South Wales, United Kingdom

1 Introduction

Transgenic crop modification uses recombinant DNA methods to alter gene expression in order to create new varieties for breeders that may be either difficult or impossible to produce using conventional approaches. Over the past few decades, transgenic methods have been successfully applied to develop genetically modified (GM) varieties of the major oilseed crops, soybean, rapeseed and sunflower. In this section, the research efforts for production of GM oil palm plants over the past 20 years will be reviewed. Although these efforts have yet to result in stable lines of commercially useful GM varieties of oil palm, there are good prospects that the greatly improved knowledge of genomics, coupled with advanced technologies such as genome editing, will be successful in the future (Murphy, 2015).

Since the advent of scientific breeding in the early twentieth century, many novel methods have been applied to agricultural improvement (Murphy, 2011). The term 'genetic modification' is popularly applied to the use of recombinant DNA methods for the creation of new varieties of plants, animals or microbes. Strictly speaking, however, humans have been using genetic modification as part of crop and livestock breeding for over 10 000 years. Initially, such breeding efforts involved controlled reproduction followed by rigorous selection of progeny by farmers so that new types of domesticated plants and animals

http://dx.doi.org/10.19103/AS.2017.0018.09

were created. In many cases the human-selected versions of their crops and livestock were radically different from their wild ancestors that had only been subject to Darwinian natural selection. Indeed, most domesticated organisms are unable to compete in the 'natural' world and would soon become extinct without their human guardians.

During the twentieth century, breeding methods became increasingly scientifically based and even further removed from so-called natural selection. Such methods include the use of DNA-damaging chemicals, for example alkylating agents and azides, or gamma irradiation, for instance from cobalt-60 or caesium-137, to create large numbers of mutations in plant or animal genomes. While the vast majority of such mutations have adverse consequences, and are often lethal, a few may be useful to the breeder if they result in phenotypes that are more adapted to agriculture. In other cases, breeders have forced different species to breed together via wide crosses. The progeny of such interspecific sexual crosses are normally non-viable, but they can sometimes be grown to maturity using *in vitro* cell and tissue culture methods referred to as embryo rescue. These and other highly intrusive breeding technologies have been widely used in plant breeding since the 1930s with little or no controversy (Murphy, 2011). Indeed, they are often collectively referred to as 'conventional breeding' and are not classified as genetic modification – despite the fact that their express purpose is to use laboratory techniques to modify crop genomes.

2 Early and current genetically modified (GM) crop varieties

The first GM bacteria were created in the 1970s, and the first GM plants in the early 1980s. In many cases, GM methods are faster, more specific and safer than alternative breeding techniques such as mutagenesis. GM methods also have the advantage for commercial breeders that the resultant crop varieties can be patented under more rigorous legislation than is available for the protection of non-GM varieties (Murphy, 2007). Interestingly, GM methods are now used routinely and without controversy in a host of medical applications, such as the production of life-saving recombinant therapeutic products including human insulin and blood-clotting factors. In contrast, however, the use of GM technologies in agriculture has caused substantial (if arguably unwarranted) public concern. The concern about GM crops has resulted in an effective ban on their cultivation in most of Europe although they are increasingly being grown in the Americas, Africa and Asia (Murphy, 2016).

The earliest commercial transgenic oilseed crop was a variety of rapeseed (canola) produced by the biotech start-up company, Calgene, and first grown under licence by farmers in the USA in 1995 (Murphy, 2006). Interestingly, this transgenic rapeseed, termed Laurical™, had been modified to produce a high-lauric oil with the aim of replacing imported palm kernel oil in US markets. Laurical™ was produced by transferring a gene encoding a C12-specific thioesterase from the California Bay plant, *Umbelluria californica*, into rapeseed. Although yields of lauric acid were relatively low in the early versions of Laurical™, the addition of additional transgenes eventually enabled Calgene breeders to produce a seed oil that contained in excess of 60% w/w lauric (Voelker et al., 1996).

However, even this relatively lauric-rich oil crop was unable to compete commercially with the standard commodity lauric oil, which was (and still is) obtained from the far cheaper and more plentiful (and non-GM) palm kernel oil. Therefore, the Laurical™ brand of transgenic rapeseed was a commercial failure and was only grown for a few seasons during the mid-1990s in the southern United States. This demonstrates some of the major challenges involved in producing transgenic crop varieties that also apply to oil palm. In the case of Laurical™, the technical challenges to produce a high-lauric oil were eventually resolved but the unfavourable commercial realities of trying to compete with palm kernel oil prevented the crop from succeeding in the marketplace.

Remarkably, since the mid-1990s there have been no further releases of GM crops specifically modified to produce novel oils, and none of the major 'first generation' GM crops released between 1996 and 2010 were modified for output traits such as oil composition. The reasons for this include the technical challenges in producing GM oil crops with acceptable fatty acid contents, plus the lack of markets for such modified oils. Moreover, even if these technical and commercial hurdles could be overcome, new transgenic varieties would still face problems due to the public opposition to GM crops that has existed in some parts of the world since 1999. Such opposition has been particularly intense in Europe right up to the present day. Hence, in late 2015, many local regions in the EU took advantage of a change in regulations and instituted formal bans on the cultivation of all GM crops, this even included varieties that had been approved by the European Commission (Nelson, 2015). In the case of a long-lived perennial tree crop like oil palm, all three of these challenges, namely technical feasibility, commercial prospects and anti-GM sentiment, are potentially more severe when compared with annual oil crops such as rapeseed and soybean.

In 2015, GM crops were grown on 180 million hectares or about 10% of the global arable crop area (James, 2015). Over half of this area was in developing countries, with Brazil, Argentina, India and South Africa emerging as increasingly important centres of GM crop breeding as well as cultivation. Interestingly an increasing number of GM crop varieties are being developed by public-sector organizations, such as EMBRAPA (Murphy, 2016) in Brazil. This is beginning to challenge the dominance of the major agbiotech multinationals, such as Monsanto, Dupont and Syngenta, that have hitherto been the major sources of seeds for GM crops. Indeed, in 2016, a Chinese government-owned company called ChemChina acquired Syngenta in a development that many analysts regarded as signalling the beginning of a large-scale deployment of GM (and other biotech-derived) crops in China, while several other large agbiotech companies also announced merger plans (Clapp, 2016).

3 GM oil palm in Malaysia

During the 1980s and 1990s, the oil palm industry in Malaysia was expanding rapidly with a trebling of oil output between 1980 and 1995 (Abdullah, 2005). During this period Malaysia was also aspiring to develop into a leading high-tech economy as outlined by then prime minister, Dr Mahatir, as part of the '2020 vision' for the country (Murphy 2003). This coincided with the development of transgenic (GM) methods for oil modification in annual crops by several biotech companies, including Calgene, Dupont and Monsanto. Production of the first experimental transgenic plants was reported in 1983 and over the next two decades many hundreds of transgenes were inserted into dozens of plant species. The late 1990s were a relatively optimistic time for GM technology, with the first large-scale crop releases occurring in 1998 and with the backlash from anti-GM groups was still in the future. At this time there was considerable pressure against imported palm oil from activist groups, such as the US soybean industry. The major criticism of what was referred to as 'tropical oils' was based on the perceived nutritional disadvantages of relatively highly saturated oil.

Existing accessions of oil palm were from a relatively restricted genetic base and did not display much variation in fatty acid profile, which ruled out a conventional breeding programme. A mutagenesis approach would also be complicated by the need to manipulate several independent genes (see Fig. 1). Given the success of Calgene in using GM methods to modify high-oleic rapeseed into a high-lauric variety after a few years of research and development, the prospect of creating a transgenic oil palm with a lower amount of saturated fatty acids was very enticing (see below). Moreover, if it was possible to manipulate acyl content at will, other more niche novel oil varieties could be created, such as high stearate for solid fat markets or ultra-high laurate for cosmetic/cleaning products (Barcelos et al., 2015; Murphy, 2014; Sambanthamurthi et al., 2009).

This led to the launching of a research programme to produce GM oil palm with target traits that included oil composition plus several additional key agronomic characters. This research was carried out at the Palm Oil Research Institute of Malaysia (PORIM) – and later the Malaysian Palm Oil Board (MPOB) (Cheah et al., 1995; Sambanthamurthi, et al., 2009). Even at this early stage, it was recognized that the choice of target traits was important and that disruption of the existing highly successful commodity palm oil products should be avoided. It was also recognized that oil palm transformation was a multi-decade endeavour that was unlikely to bear fruit before the 2020s. This is illustrated by the following quote about the goals of the GM oil palm programme from the leader of the PORIM/MPOB transformation group, GKA Parveez:

> Among these targets are high-oleate and high-stearate oils, and the production of industrial feedstock such as biodegradable plastics. The efforts in oil palm genetic engineering are thus not targeted as commodity palm oil. Due to the long life cycle of the palm and the time taken to regenerate plants in tissue culture, it is envisaged that commercial planting of transgenic palms will not occur any earlier than the year 2020. (Parveez et al., 2000)

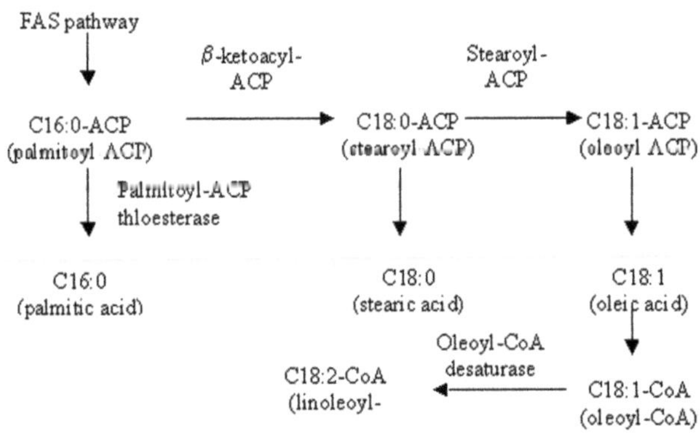

Figure 1 Metabolic pathway for the biosynthesis of C16 and C18 fatty acids in plants.

4 Improving the fatty acid composition of palm oil

By far the most important target trait for transgenic oil palm is to have a significantly increased oleic acid content in the mesocarp oil. Existing commercial varieties of oil palm produce a mesocarp oil that typically contains only about 40–45% oleate whereas competitor crops such as rapeseed, olive, soybean and sunflower have varieties containing well above 60% and in some cases as high as >90% oleate (see Table 1). The main reason for the low oleate content of palm oil is that the mesocarp has a highly active palmitoyl-ACP thioesterase that results in C16 palmitate being channelled towards TAG accumulation instead of being elongated to stearate and desaturated to oleate (see Fig. 1).

In order to change the mesocarp oil into a high-oleate profile, it is therefore necessary to manipulate at least three, and possibly more, genes. First, palmitoyl-ACP thioesterase needs to be downregulated to stop palmitate being diverted to TAG. Secondly, the key elongation gene, KASII needs to be upregulated so that the palmitate is efficiently converted to C18 stearate. Thirdly, the stearoyl-ACP desaturase needs to be upregulated so that stearate is efficiently desaturated to oleate. Experience with other crops, such as the transgenic Laurical™ rapeseed described above, has shown that it may also be necessary to transfer additional acyltransferase genes to ensure that the new fatty acids are efficiently assembled onto triacylglycerols. For example, evidence from several plant species suggests that the type-2 acyl-CoA:diacylglycerol acyltransferase (DGAT2) can stimulate accumulation of exotic fatty acids in storage TAG (Kroon et al., 2006; Shockey, 2006).

Due to these complications, until very recently, the levels of the novel fatty acids in most transgenic plants have been relatively modest and were far from achieving commercial viability (see reviews by McKeon, 2005; Murphy, 2006; Cahoon et al., 2007; Dyer and Mullen, 2008). This means that it is likely to be some time before an ultra-high oleate trait can be created in oil palm, but this goal is still important due to the usefulness of oleate-rich feedstocks for both food and non-food applications (Murphy, 2014). It also means that the engineering of oil palm for more minor fatty acid traits, such as high palmitate or palmitoleate, or for acyl derivatives such as bioplastics (Parveez et al., 2015; Yunus et al., 2008) is going to be even more problematic. Given that these will be niche products with much more limited markets, it may be appropriate to question whether large R&D investments in a transgenic strategy aimed at such targets are justified at this stage.

Table 1 The 'big four' major global oil crops and their potential for ultra-high oleate varieties

Type of oil	Global production, Mt	Global production, %	Ease of producing GM varieties	Highest % oleate achieved
Palm oil	65	40	+	50
Palm kernel oil	7	4	+	15
Soybean oil	50	30.5	++++++	75
Rapeseed oil	27	16.5	+++++++	75
Sunflower oil	15	9	+++	>90

Data from USDA website, 2015.

In addition to oil modification, transgenic approaches can, in principle, be applied to any other trait where suitable genes have been identified. This means that there is a long list of possible GM varieties of oil palm, but high research and development costs are likely to limit their development to traits with a guaranteed agronomic and commercial potential. For example, expression of Bt insecticidal proteins has been proven to be an effective strategy to control certain insect pests in a range of annual crops and is also being investigated in oil palm (Lee et al., 2006). A much more important threat to oil palm yields is the fungal pathogen, *Ganoderma boninense*, and the possibility of modifying lignin to make oil palm plants more resistant is being examined (Paterson et al., 2008). As of 2016, progress in addressing any of these traits has been limited and the high costs of existing transgene technology mean that scarce resources would arguably be better targeted to addressing the major key trait of high-oleate mesocarp oil where the biochemical basis of the trait is reasonably well understood and there is a guaranteed global market for the resulting oil (Murphy, 2014).

5 Progress to date in oil palm transformation

Early work on oil palm transformation was initially focused on the assessment of suitable tissue culture systems transformation and suitable vectors for transgene delivery. This work was based largely at PORIM/MPOB with additional input from several university-based groups in Malaysia, as outlined in Abdullah (2005). During the 1990s, *Agrobacterium*-mediated gene transfer was still largely restricted to dicot plant species. At this time a few monocots such as rice and maize had been transformed, but oil palm was well outside the host range of *Agrobacterium* so initial strategies for oil palm transformation used the newer method of biolistics, which could in principle be applied to any plant species (Parveez 1998; Parveez and Christou, 1998; Parveez et al., 2000). A disadvantage of biolistics is that it often results in the integration of multiple transgene copies or even fragments of transgenes, whereas *Agrobacterium*-mediated transformation is more likely to introduce either single-copy or low-copy-number transgenes. Nevertheless, it was possible to bombard oil palm cultures with marker genes such as GUS and achieve transient expression of the transgene (Parveez et al., 1997). It was also possible to experiment with different gene promoters in order to investigate levels of gene expression and tissue specificity (Chowdhury et al., 1997).

In order to improve transformation efficiency, new tissue culture systems were developed in order to have a suitable starting material for transformation. One of the best systems was found to be immature oil palm embryos and these were used as the target tissue for both biolistics and *Agrobacterium* mediated transformation experiments (Ruslan et al., 2005; Bhore and Shah, 2012). This resulted in transformed tissues that expressed marker transgenes such as those encoding *Bacillus thuringiensis* (Bt) insecticidal proteins (Lee et al, 2006) and the cowpea trypsin inhibitor (CpTI) (Ismail et al, 2010). Unfortunately, these experiments required the transformation of thousands of tissue-culture samples. Even after several stages of selection and regeneration, there were many hundreds of plants that required housing in containment facilities and subsequent analysis. It took 3–5 years to generate transgenic oil palm plants by either particle bombardment or *Agrobacterium*-mediated transformation and that both processes were highly inefficient. The frequency of escapes and chimeric plants was high because of the long selection process during callus formation and the fact that somatic embryogenesis encourages the growth of non-transformed cells (Masani et al., 2014). This meant that the logistics of establishing and maintaining such large numbers of tissue explants and regenerating plantlets was extremely challenging.

Despite several decades of intensive research, the development of transgenic oil palm remains a relatively young science and is fraught with many technical and logistical challenges. For example, it is still unclear whether biolistics or *Agrobacterium*-mediated gene transfer will be the gene delivery method of choice (Izawati et al., 2012; Parveez and Bahariah, 2012). The choice of plant material is also crucial with some of the best options including various types of callus and embryo culture. Then there is the issue of which gene promoters to use. Unlike most existing commercial transgenic crops where constitutive viral promoters are used, the manipulation of palm oil composition will probably require deployment of strong mesocarp-specific or kernel-specific gene promoters, ideally sourced from the oil palm genome itself.

Also, in order to achieve high oleic oil phenotypes, it will probably be necessary to downregulate some genes while adding or upregulating other genes. Even when the primary transgenic plantlets have been produced they will still need to be grown on for three to five years to obtain fruits that can be screened for oil content. Finally, these primary transformants will need to be taken through several sexual generations, backcrossed with existing elite lines and then multiplied via micropropagation before any new commercial transgenic varieties can be released. Some recent strategies for achieving transgenic oil palm are listed below:

- Improved transformation vectors for oil palm (Yunus, and Parveez, 2008)
- Development of positive selectable marker systems such as phosphomannose isomerase (Bahariaha et al., 2013) and 2-deoxyglucose-6-phosphate phosphatase (Izawati et al., 2015)
- Transforming oil palm protoplasts by PEG-mediated transfection and microinjection (Masani et al., 2014)
- Using more rigorous molecular analytical methods to detect transgene presence (Nurfahisza et al., 2014)
- Improved tissue-specific gene promoters for mesocarp expression (Taha et al., 2012)

Given the gradual rate of progress to date, it seems likely that the development of routine methods for transforming oil palm with gene constructs optimized for commercial phenotypes will take more than a decade from the present time. The initial target of having commercial transgenic oil palm varieties under cultivation in the 2020s is now looking rather optimistic. However, a target date in the late 2020s is not necessarily totally unrealistic, especially if new genome-editing technologies prove useful, as discussed below. Meanwhile it is important that efforts at using classical and/or more modern genome manipulation technologies are continued for oil palm. This is because there are likely to be several important genetic traits that cannot be created using non-transgenic approaches. For example, although it may be possible to create a medium-high (55–65%) oleic oil phenotype via conventional methods (e.g. by using identified accessions obtained from Africa (Murphy, 2014)), a more desirable ultra-high (80–90%) oleic trait might only be possible via transgenic technology. Another example where the use of transgenic methods is essential is the production of completely novel compounds, such as biopolymers like polyhydroxyalkanoates, in oil palm.

6 New technologies for genome editing – an alternative to 'classical GM'

During recent years, biological advances have started to make it much easier to manipulate genomes in more radical and precise ways than were possible with traditional late twentieth-century GM technologies (Murphy, 2016). For example, several new forms of plant and animal (including humans) GM technologies known as 'genome editing' have recently been developed. Probably the most powerful is the CRISPR (Clustered, Regularly Interspaced, Short Palindromic Repeats) system (Hsu et al., 2014; Mao et al., 2013). In 2015, the CRISPR/Cas9 system was described in a Nature article as 'the biggest game changer to hit biology since PCR' (Ledford, 2015a). Applying this method to crop improvement has opened up many new possibilities for radical genome modifications (Belhaj et al., 2015; Zhang et al., 2014). Genome editing will greatly accelerate crop breeding by enabling precise and predictable genetic modifications directly in elite individuals as well as simultaneous modification of multiple traits.

This means that breeders will progress well beyond the random insertion of single or small numbers of genes into a genome (as in traditional GM) to the highly precise insertion into a defined location of large numbers of genes, chromosome segments or pseudo-segments encoding entire metabolic pathways into virtually any plant species. Methods such as TALEN (Transcription Activator-Like Effector Nucleases) and CRISPR/Cas9 can be used for gene knockouts, for example, to eliminate unwanted genes that adversely affect food quality or confer susceptibility to pathogens or that divert metabolism away from valuable end-products. An example reported in 2014 was the use of both TALEN and CRISPR/Cas9 to target the genes of the mildew resistance locus in wheat. This resulted in the production of plants resistant to powdery mildew disease, which is a serious crop disease (Wang et al. 2014). Throughout 2015 and 2016, improved versions of CRISPR have been reported that will enable the technology to be used more widely and effectively to create new crop phenotypes (Ledford, 2015b; Zetsche et al., 2015).

In addition to transgenic approaches, modern breeding is able to profit from a host of advanced technologies such as association genetics (Rafalski, 2010) and molecular mutagenesis, including TILLING (TargetIng Local Lesions IN Genomes) (Shu, 2009) and marker-assisted selection (Ribaut et al., 2010). Deployment of these technologies is already enabling breeders to begin to address complex traits such as drought tolerance and disease resistance (Tuberosa and Salvi, 2006; Xu, 2010). Such technologies could be used together with GM methods to develop additional useful traits such as dwarf trees, increased oil content in mesocarp and kernel tissues, and wider resistance/tolerance profiles to a range of pests and pathogens (Murphy, 2014).

7 Conclusions and future prospects

Despite a lack of clear scientific evidence to support adverse claims from anti-GM activists, many politicians in Europe remain wary about allowing GM crops or products to be grown or imported into the region (Nelson, 2015). Given that oil palm already suffers from an image problem (especially in Europe) due to its perceived environmental impacts, the industry is rightly cautious about possible future GM developments. Indeed, in many cases palm oil producers actually stress that the crop is entirely non-GM at the present time.

However, the current status of GM crops may change radically over the next decade as the technology becomes ever more refined and globalized and is applied to more crops with traits that have better consumer appeal than current GM varieties (Murphy, 2016). In line with this prospect, it was announced in mid-2016 that CRISPR-edited crops (which some would regard as being GMOs) will be able to be cultivated and sold in the USA without regulatory oversight by the USDA (Waltz, 2016).

By 2016, over half of all GM crops were being grown in developing countries, including several countries in Africa. There has been an increasing tendency for transgenic crops to be developed by public-sector institutions in developing countries and then distributed at low cost to poorer farmers. An example is golden rice developed by IRRI in the Philippines and scheduled for release to farmers in 2016–17 (Kowalski, 2015; Moghissi et al., 2015). As developing countries produce new varieties of GM crops with traits such as improved nutritional quality drought tolerance and disease resistance, it will become increasingly difficult to defend the 1990s-era argument that GM crops represent a tool for the imposition of global food hegemony by Western governments and multinationals (Engdahl, 2007). These factors of improved genome-editing technologies, ability to alter many more traits, more public-sector and public-good research and development, and increased attractiveness to consumers around the world mean that the coming decades are more likely to see an expansion in the number and range of GM crops and traits. For these reasons, one can be cautiously optimistic about the future of transgenic oil palm and the possibility that commercial releases might be carried out before the end of the 2020s.

8 Where to look for further information

The most reliable source of information about this topic for researchers is the peer reviewed literature and especially the articles cited below. In some cases there may also be useful web-based information from bodies such as FAO and national scientific academies such as US National Academy of Science and the Royal Society in the UK.

9 References

Abdullah, R. 2005. A Decade of Oil Palm Gene Manipulation. Where are we now? 9th International Conference on Agricultural Biotechnology: Ten Years After, CEIS-University of Rome.

Bahariaha, B., Parveez, G. K. A., Masani, M. Y. A., et al. 2013. Biolistic transformation of oil palm using the phosphomannose isomerase (pmi) gene as a positive selectable marker. *Biocatal. Agric. Biotechnol.* 2: 295–304.

Barcelos, E. et al. 2015. Oil palm natural biodiversity and the potential for yield improvement. *Front. Plant Sci.* 2015, 6: 190.

Belhaj. K., Chaparro-Garcia, A., Kamoun, S., et al. 2014. *Curr. Opin. Biotechnol.* 32: 76–84.

Bhore, S. J. and Shah, F. H. 2012. Genetic transformation of the American oil palm (*Elaeis oleifera*) immature zygotic embryos with antisense palmitoyl-acyl carrier protein thioesterase (PATE) gene. *World Appl. Sci. J.* 16: 362–9.

Cheah, S. C., Sambanthamurthi, R., Siti Nor Akmar, A. et al. 1995. Towards genetic engineering oil palm (*Elaeis guineensis* Jacq.). In: Ed Kader, J. C., Mazliak, P. Netherlands, Kluwer. *Plant Lipid Metabolism*, pp. 570–2.

Cahoon, E. B., Shockey, J. M., Dietrich C. R., et al. 2007. Engineering oilseeds for sustainable production of industrial and nutritional feedstocks: solving bottlenecks in fatty acid flux. *Curr. Biol.* 10: 236–44.

Chowdhury, M. K. U., Parveez, G. K. A. and Saleh, N. M. 1997. Evaluation of five promoters for use in transformation of oil palm (*Elaeis guineensis* Jacq.). *Plant Cell Rep.* 5: 277–81.

Clapp, J. 2016. Monsanto, Dow, Syngenta: rush for mega-mergers puts food security at risk, *Guardian*, 5 May 2016, available online: http://www.theguardian.com/sustainable-business/2016/may/05/monsanto-dow-syngenta-rush-for-mega-mergers-puts-food-security-at-risk

Dyer, J. M. and Mullen, R. T. 2008. Engineering plant oils as high-value industrial feedstocks for biorefining: the need for underpinning cell biology research, *Physiol. Plant.* 132: 11–22.

Engdahl, W. F. 2007. *Seeds of destruction – the hidden agenda of genetic manipulation.* Montreal: Global Research.

Hsu, P. D., Lander, E. S. and Zhang, F. 2014. Development and applications of CRISPR-Cas9 for genome engineering, *Cell* 157: 1262–78.

Ismail, I., Lee, F. S., Abdullah, R. et al. 2010. Molecular and expression analysis of cowpea trypsin inhibitor (CpTI) gene in transgenic *Elaeis guineensis* Jacq leaves. *Aust. J. Crop Sci.* 4: 37–48.

Izawati A. M., Parveez G. K. and Masani, M. Y. 2012. Transformation of oil palm using *Agrobacterium tumefaciens. Methods Mol Biol.* 847: 177–88.

Izawati, A. M. D., Masani, M. Y. A., Ismanizan, I., and Parveez, G. K. A. 2015. Evaluation on the effectiveness of 2-deoxyglucose-6-phosphate phosphatase (*DOGR1*) gene as a selectable marker for oil palm (*Elaeis guineensis* Jacq.) embryogenic calli transformation mediated by *Agrobacterium tumefaciens. Front. Plant Sci.* 2015, 6: 727.

James, C. 2015. Global Status of Commercialized Biotech/GM Crops: 2015. ISAAA Brief No. 51. ISAAA: Ithaca, NY, available online: http://isaaa.org/resources/publications/briefs/51/executivesummary/default.asp

Kowalski, S. P. 2015. Golden rice, open innovation, and sustainable global food security, *Ind. Biotechnol.* 11: 84–90.

Kroon, J. T., Wei W., Simon, W. J. and Slabas, A. R. 2006. Identification and functional expression of a type 2 acyl-CoA:diacylglycerol acyltransferase (DGAT2) in developing castor bean seeds which has high homology to the major triglyceride biosynthetic enzyme of fungi and animals. *Phytochemistry* 67, 2541–9.

Ledford, H. 2015a. CRISPR, the disruptor, *Nature* 522: 20–4.

Ledford, H. 2015b. Alternative CRISPR system could improve genome editing. Smaller enzyme may make process simpler and more exact. *Nature* 526: 17.

Lee, M. P., Yuen, L. H. and Abdullah, R. 2006. Expression of *Bacillus thuringiensis* insecticidal protein gene in transgenic oil palm. *Electron. J. Biotechnol.* 9(2).

Mao, Y., Zhang, H., Xu, N., et al. 2013. Application of the CRISPR-Cas system for efficient genome engineering in plants, *Mol. Plant* 6: 2008–11.

Masani, M. Y. A., Noll, G. A., Parveez, G. K. A. et al. 2014. Efficient transformation of oil palm protoplasts by PEG-mediated transfection and dna microinjection, PLOS ONE 0096831.

McKeon, T. A. 2005. Genetic Modification of Seed Oils for Industrial Applications, In: Erhan, S. Z. (ed.), *Industrial Uses of Vegetable Oils*, AOCS Press, Urbana, pp. 1–13.

Moghissi, A. A., Pei, S. and Liu, Y. 2015. Golden rice: scientific, regulatory and public information processes of a genetically modified organism, *Crit. Rev. Biotechnol.* doi:10.3109/07388551.2014.993586.

Murphy, D. J. 2003. Working to improve the oil palm crop, *Int. News Fats, Oils Relat. Mater.* 14: 670–1.

Murphy, D. J. 2006. Molecular breeding strategies for the modification of lipid composition, In Vitro Cell. *Dev. Biol. Plant* 42: 89–99.

Murphy, D. J. 2007. *Plant Breeding and Biotechnology*, Cambridge University Press, UK.

Murphy, D. J. 2011. *Plants, Biotechnology, and Agriculture*, CABI, UK.

Murphy, D. J. 2014a. The future of oil palm as a major global crop: opportunities and challenges, *J. Oil Palm Res.* 26: 1–24, http://jopr.mpob.gov.my/

Murphy, D. J. 2015. What is the future for oil palm as a global crop? *World Agric.* 5: 24–34 http://issuu.com/wharncliffe/docs/world_agriculture_winter_2015/1

Murphy, D. J. 2016. Has agricultural biotechnology finally turned a corner? *World Agric.* 6, available online: https://issuu.com/wharncliffe/docs/world_agriculture_vol_6_issue_1_201

Nelson, A. 2015. Half of Europe opts out of new GM crop scheme, *Guardian* 1 October 2015, www.guardian.com

Nurfahisza, A. R. Rafiqah M. A. Masani M. Y. A. et al. 2014. Molecular analysis of transgenic oil palm to detect the presence of transgenes, *J. Oil Palm Res.* 26: 96–103.

Parveez, G. K. A. 1998. Optimization of parameters involved in the transformation of oil palm using the biolistics method. PhD Thesis, Universiti Putra Malaysia.

Parveez, G. K. A. and Bahariah, B. 2012. Biolistic-mediated production of transgenic oil palm. *Methods Mol. Biol.* 847:163–75.

Parveez, G. K. A., Chowdhury, M. K. U. and Saleh, N. M. 1997. Physical parameters affecting transient GUS gene expression in oil palm (*Elaeis guineensis* Jacq.) using the biolistic device. *Ind. Crops Prod.* 6: 41–50.

Parveez, G. K. A. and Christou, P. 1998. Biolistic-mediated DNA delivery and isolation of transgenic oil palm (*Elaeis guineensis* Jacq) embryogenic callus cultures. *J. Oil Palm Res.* 10: 29–38.

Parveez, G. K. A., Masri, M. M., Zainal, A., et al. 2000. Transgenic oil palm: production and projection. *Biochem. Soc. Trans.* 28: 969–72.

Parveez, G. K. A., Bahariah, B., Ayub, N. H., et al. 2015. Production of polyhydroxybutyrate in oil palm (*Elaeis guineensis* Jacq.) mediated by microprojectile bombardment of PHB biosynthesis genes into embryogenic calli. *Front. Plant Sci.* 6: 598.

Paterson, R. R. M., Moen, S. and Lima, N. 2009. The feasibility of producing oil palm with altered lignin content to control ganoderma disease. *J. Phytopathol.* 157: 649–56.

Rafalski, J. A. 2010. Association genetics in crop improvement, *Curr. Opin. Plant Biol.* 13: 174–80.

Ribaut, J. M., deVicente, M.C. and Delannay, X. 2010. Molecular breeding in developing countries: challenges and perspectives, *Curr. Opin. Plant Biol.* 13: 213–18.

Ruslan, A., Zainal, A., Heng, W. Y. et al. 2005. Immature embryo: A useful tool for oil palm (*Elaeis guineensis* Jacq.) genetic transformation studies. *Electron. J. Biotechnol.* 8: 25–34.

Sambanthamurthi, R., Singh, R., Parveez, G. K. A. et al. 2009. Opportunities for the Oil Palm via Breeding and Biotechnology, In: Mohan Jain, M. and Priyadarshan, P. M. (eds), *Breeding Plantation Crops*, Springer, New York, pp. 377–421.

Shockey, J. M., Gidda, S. K., Chapital, D. C. et al. 2006. Tung tree DGAT1 and DGAT2 have nonredundant functions in triacylglycerol biosynthesis and are localized to different subdomains of the endoplasmic reticulum. *Plant Cell* 18: 2294–313.

Shu, Q. Y., ed. 2009. *Induced Plant Mutations in the Genomics Era*, FAO, Rome, http://www.fao.org/docrep/012/i0956e/i0956e00.htm

Taha, R. S., Ismail, I., Zainal, Z. and Abdullah, S. N. A. 2012. The stearoyl-acyl-carrier-protein desaturase promoter (Des) from oil palm confers fruit-specific GUS expression in transgenic tomato, *J. Plant Physiol.* 169: 1290–300.

Tuberosa, R. and Salvi, S. 2006. Genomics-based approaches to improve drought tolerance of crops, *Trends Plant Sci.* 11: 405–12.

Voelker, T. A., Hayes, T. R., Cranmer, A. M., et al. 1996. Genetic engineering of a quantitative trait: metabolic and genetic parameters influencing the accumulation of laurate in rapeseed. *Plant J.* 9: 229–41.

Waltz, E. 2016. CRISPR-edited crops free to enter market, skip regulation, *Nat. Biotechnol.* 43: 582.

Wang, Y., Cheng, X., Shan, Q., et al. 2014. Simultaneous editing of three homoeoalleles in hexaploid breadwheat confers heritable resistance to powdery mildew, *Nat. Biotechnol.* 32: 947–51.

Xu, Y. 2010. Molecular Plant Breeding, CABI, Oxford, UK.

Yunus, A. M. M., Chai-Ling, H., Parveez, G. K. A. 2008. Construction of PHB and PHVB transformation vectors for bioplastics production in oil palm. *J. Oil Palm Res.* Special Issue on Malaysia-MIT Biotechnology Partnership Programme, Volume 2 - Oil Palm Metabolic Engineering.

Yunus, A. M. M. and Parveez, G. K. A. 2008. Development of transformation vectors for the production of potentially high oleate transgenic oil palm. *Electron. J. Biotechnol.* 11: 3.

Zetsche B., Gootenberg, J. G. and Abudayyeh, O. O. et al. 2015. Cpf1 Is a single RNA-guided endonuclease of a class 2 CRISPR-Cas system. *Cell* in press, doi:10.1016/j.cell.2015.09.038

Zhang, H., Zhang, J., Wei, P., et al. 2014. The CRISPR/Cas9 system produces specific and homozygous targeted gene editing in rice in one generation. *Plant Biotechnol. J.* 12: 797–807.

Part 3

Cultivation techniques

Chapter 10

Modelling crop growth and yield in palm oil cultivation

Christopher Teh Boon Sung, Universiti Putra Malaysia, Malaysia; and Cheah See Siang, Sime Darby Research Sdn. Bhd., Malaysia

1 Introduction

The first semi-mechanistic oil palm model was OPSIM (van Kraalingen, 1985). It was developed by taking into account some aspects of oil palm physiology and the physical processes and causal relationships between the environment and crop. OPSIM was modified by Gerristma (1988) in order to propose a more rigorous approach to estimate oil palm photosynthesis and later by Dufrêne et al. (1990), whose model (SIMPALM) relied much more on measured data of oil palm vegetative parameters. GPHOT, GPHOT2, OPLFSIM3 and later OPRODSIM (Henson, 1989, 2000, 2004, 2009) were the most comprehensive models at that period, as Henson's models progressively improved by including increasingly more effects and factors of oil palm growth and yield, including modelling the seasonal cycles of oil palm yield. Supplementary factors such as the effects of air vapour pressure deficit and available soil water on oil palm photosynthesis were also modelled. Fitted relationships, based

http://dx.doi.org/10.19103/AS.2017.0018.10

on measured data collected from various oil palm studies, were later included to better estimate oil palm root turnover, dry matter partitioning and flower sex ratios. Since then, the development of new oil palm models, such as WaNuLCAS (van Noordwijk et al., 2011), ECOPALM (Combres et al., 2013) and PALMSIM (Hoffmann et al., 2014) have increased in frequency. More recent oil palm models such as APSIM-Oil Palm (Huth et al., 2014), CLM-Palm (Fan et al., 2015) and CLIMEX-Oil Palm (Paterson et al., 2015) are actually components of a larger, more general modelling framework. The APSIM-Oil Palm, for instance, is one of the 30 sub-models for various crops, trees and pastures under the APSIM model framework.

The purpose of this chapter is to discuss the development of a new oil palm growth and yield model called PySawit. There are several key aspects that differentiate PySawit from other models. First, PySawit attempts to model oil palm photosynthesis in a more rigorous manner based on the biochemical photosynthesis model by Collatz et al. (1991), a modification of the original photosynthesis model by Farquhar et al. (1980). Other oil palm models tend to use oil palm's radiation use efficiency (based on Beer's law) to convert the intercepted radiation into gross photosynthesis. In our study, actual leaf measurements were done on young to mature palms to determine oil palm's Rubisco maximum capacity rate (V_{cmax}) and intercellular concentration of carbon dioxide (CO_2), both of which are needed to determine the leaf photosynthesis rate.

Second, the microclimate environment within and under the oil palm canopies is modelled by describing the soil–plant–atmosphere system as a network of resistances in which latent and sensible heat fluxes from the soil and crop must traverse, akin to the flow of electrical current, driven by potential differences and impeded by a series of resistances (Shuttleworth and Wallace, 1985). The Shuttleworth–Wallace (SW) evapotranspiration model is an important extension of the Penman–Monteith model (Monteith, 1965) as it allows the simultaneous heat fluxes from both soil and crop, unlike its predecessor that would only allow fluxes from either soil or crop but not both simultaneously. We further measured the relationship between oil palm's leaf stomatal conductance with photosynthetically active radiation (PAR) and air vapour pressure deficit (VPD), as these are needed to determine fluxes of latent heat and, thus, the amount of water loss from oil palm.

And third, PySawit is developed specifically for oil palm planted on a wide range of densities, from about 120 to 300 palms ha^{-1}. The recent oil palm models such as APSIM-Oil Palm, CLM-Palm and PALMSIM were validated only over a narrow planting density range of 127–156 palms ha^{-1}. The accuracy of the CLIMEX-Oil Palm, on the other hand, was not even validated. Moreover, most oil palm models, especially recent ones, have only been validated over a limited period of the oil palm growth stage (usually only when the palms have fully matured).

Consequently, the second purpose of this chapter is to discuss the evaluation of PySawit's accuracy when its predictions were compared with several measured parameters of growth and yield in oil palm aged 1–19 yrs. These oil palm were planted with 10 different planting densities, ranging from 122 to 296 palms ha^{-1}, in an oil palm estate at Merlimau, Malaysia. The degree of agreement between PySawit's predictions and observations was evaluated by visual inspection of scatterplots between predictions and observations and by the use of statistical goodness-of-fit indexes.

2 Theory and model development

PySawit simulations are for crop production level 2, where the growth and yield of oil palm are limited only by weather conditions and water availability (de Witt and Penning de Vries, 1982). The model assumes that oil palm grows under optimal nutrient levels and without any detrimental effects from pests, diseases or weeds, and is managed by following standard practices (such as pruning and mulching) by the oil palm industry.

PySawit comprises five core components: 1) meteorology, 2) photosynthesis, 3) energy balance, 4) soil water and 5) crop growth. The meteorology component contains calculations pertaining to both daily and hourly meteorological elements such as air temperature, wind speed, humidity, solar irradiance and rain. The photosynthesis component calculates the oil palm photosynthesis rate, and the energy balance component deals with the various heat fluxes with the aim of determining the potential evapotranspiration and canopy temperature. The soil water component determines the soil water balance and the various water fluxes in the soil layers to estimate the amount of water available to the plant and the level of crop water stress, if any. The last component, crop growth, is aimed at estimating the crop maintenance, vegetative growth respirations and the bunch production and yield.

3 Modelling meteorology

Simulations begin with the meteorology component. PySawit requires only four daily meteorological parameters: minimum and maximum air temperature (°C), wind speed (m s^{-1}) and rain (mm) to estimate the various instantaneous meteorological properties such as solar irradiance, wind speed, air temperature, vapour pressure and humidity. Routine information on the site's local solar hours of sunrise t_{sr} and sunset t_{ss}, day length *DL* (hours), solar inclination θ (solar angle from vertical) (radians), extraterrestrial solar irradiance I_{et} (W m^{-2}), relative humidity *RH* (%) and saturated vapour pressure (mbar) was obtained from the work of Goudriaan and van Laar (1994).

Instantaneous total solar irradiance comprises two components: direct and diffuse components, where the direct component arrives from a single direction (the solar position), whereas the diffuse component arrives from all directions due to scattering and reflection from various surfaces in the environment. Calculations from Liu and Jordan (1960) were used to determine these solar radiation components as

$$I_t = I_{dr} + I_{df} \tag{1a}$$

$$I_{dr} = I_{et} \cdot \tau^m \tag{1b}$$

$$I_{df} = 0.3\left(1 - \tau^m\right)I_{et} \tag{1c}$$

where I_t, I_{dr} and I_{df} are the instantaneous total, direct and diffuse solar irradiance, respectively (all in W m^{-2}); τ is the sky clearness index (or atmospheric transmittance, the ratio between I_t and I_{et}); and *m* is the optical mass number. The optical mass number *m* is determined from Campbell and Norman (1998) as

$$m = P_a / (101.3 cos\theta) \tag{2}$$

where P_a is the atmospheric pressure (kPa), assumed constant at 101 kPa.

Dimas et al. (2011) showed that τ can be estimated from relative humidity (*RH*) and in turn be used to calculate I_t and its components: I_{dr} and I_{df}. Hourly meteorological data for years 2015 and 2016 were obtained from weather stations located at five Malaysian

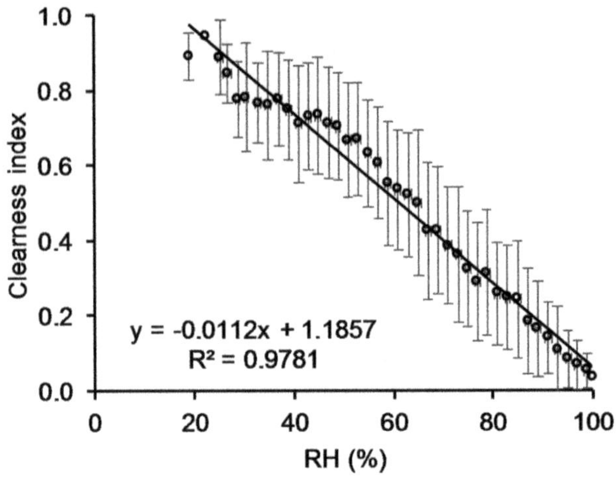

Figure 1 Linear relationship between measured relative humidity (*RH*) and sky clearness index between 2015 and 2016 for five stations located in Malaysia (N = 25548). Note: the bars represent one standard deviation.

oil palm estates: Bukit Selarong, Kedah (5.462824°N, 100.597084°E); Diamond Jubilee, Melaka (2.33333°N, 102.483333°E); Imam, Sabah (4.333333°N, 117.833333°E); Seri Intan, Perak (3.976583°N, 100.9739°E); and Ulu Remis, Johor (1.827778°N, 103.461944°E), where the following linear relationship between τ and relative humidity (RH) was derived:

$$\tau = 1.1857 - 0.0112RH \tag{3}$$

where *RH* (%) (Fig. 1).

The daily trend of instantaneous air temperature for the considered stations was also observed to vary sinuously with time during the period between 1.5 hours after the time of minimum air temperature and time of sunset. Outside this period, air temperature was observed to change linearly with time. This trend can be accurately depicted by the following mathematical relationships according to Wilkerson et al. (1983) as

$$T_a = \begin{cases} T_{set} + \dfrac{(T_{min} - T_{set})(24 + t_h - t_{ss})}{(t_{sr} + 1.5) + (24 - t_{ss})} & t_h < (t_{sr} + 1.5) \\ T_{set} + \dfrac{(T_{min} - T_{set})(t_h - t_{ss})}{(t_{sr} + 1.5) + (24 - t_{ss})} & t_h > t_{ss} \\ T_{min} + (T_{max} - T_{min}) sin\left[\dfrac{\pi(t_h - t_{sr} - 1.5)}{t_{ss} - t_{sr}}\right] & otherwise \end{cases} \tag{4a}$$

$$T_{set} = T_{min} + (T_{max} - T_{min}) sin\left[\frac{\pi(t_{ss} - t_{sr} - 1.5)}{t_{ss} - t_{sr}}\right] \tag{4b}$$

where T_a is the air temperature; T_{min} and T_{max} are the minimum and maximum air temperature, respectively; and T_{set} is the air temperature at the time of sunset (t_{ss}). All temperatures are in °C.

The dew point temperature T_{dew} for these five stations were rather constant throughout the year, with a mean of 23°C. This value is within the range reported by Mohd. Desa and Rakhecha (2006), who also observed a rather constant dew point temperature between 20°C and 24°C for 24 towns across Malaysia for the 1994 and 2003 periods.

Knowing the dew point temperature, T_{dew}, enables the determination of air vapour pressure according to Ephrath et al. (1996) as follows:

$$e_a = 6.1078 exp\left(\frac{17.269 T_{dew,cal}}{T_{dew,cal} + 237.3}\right) \tag{5}$$

where e_a is the air vapour pressure (mbar) and $T_{dew,cal}$ is the calibrated dew point temperature (°C), where it should be lower than the current air temperature, T_a:

$$T_{dew,cal} = MIN[\![T_a,\ T_{dew}]\!] \tag{6}$$

where $MIN[\![n_1, n_2, \ldots, n_n]\!]$ is the minimum function, returning the smallest of the enclosed values.

The instantaneous wind speed typically varies sinuously within the day as

$$u = MAX\left[\!\!\left[u_{min},\ u_{min} + (u_{max} - u_{min}) sin\left[\frac{\pi}{DL}(t_h - t_{sr} - 1.5)\right]\right]\!\!\right] \tag{7}$$

where u is the wind speed (m s^{-1}); u_{max} and u_{min} are the maximum and minimum wind speed for the day, respectively (m s^{-1}); and $MAX[\![n_1, n_2, \ldots, n_n]\!]$ is the maximum function, returning the largest of the enclosed values. The mean daily minimum and maximum wind speed can be estimated by the following equations, developed by fitting the best function to the measured hourly wind speed data for several towns in Malaysia from 1982 to 1991 (Sopian et al., 1995):

$$u_{min} = 0.5591 u_d^{1.25} \tag{8}$$

$$u_{max} = 1.7976 u_d^{0.75} \tag{9}$$

where u_d is the mean daily wind speed (m s^{-1}).

4 Modelling photosynthesis

The photosynthesis model by Collatz et al. (1991), which is based on Farquhar et al. (1980), was adapted in order to determine the amount of CO_2 assimilated by the oil palm canopy, where gross leaf CO_2 assimilation is limited by the lowest of the three potential assimilation rates due to Rubisco, light and sink:

$$\Lambda_{sl/sh} = MIN[\![v_c,\ v_{q,sl/sh},\ v_s]\!] \tag{10}$$

where $\Lambda_{sl/sh}$ is the gross leaf CO_2 assimilation rate for either sunlit (subscript *sl*) or shaded (*sh*) leaves, and v_c, v_q and v_s are the Rubisco-, light- and sink-limited assimilation rates, respectively. All assimilation rates are in μmol CO_2 m^{-2} leaf s^{-1}.

Leaf CO_2 assimilation rate is scaled up to canopy level, following Campbell and Norman (1998), by

$$\Lambda_{canopy} = \Lambda_{sl}L_{sl} + \Lambda_{sh}L_{sh} \tag{11}$$

where Λ_{canopy} is the gross canopy CO_2 assimilation rate (μmol CO_2 m^{-2} ground s^{-1}) and L_{sl} and L_{sh} are the sunlit and shaded leaf area indexes, respectively (both in m^2 leaf m^{-2} ground).

Equation 11 is integrated over sunrise (t_{sr}) to sunset (t_{ss}) to determine the daily gross canopy photosynthesis rate as

$$\Lambda_{canopy,d} = \frac{1.08}{PD}\int_{t_{sr}}^{t_{ss}}\Lambda_{canopy}dt \tag{12}$$

where $\Lambda_{canopy,d}$ is the daily gross photosynthesis rate (converted to kg CH_2O $palm^{-1}$ day^{-1}); and *PD* is the planting density (palms ha^{-1}). Integration is performed by the numerical Gaussian integration method, where five points over the diurnal period (from sunrise to sunset) are selected, and for each selected hour, the gross canopy CO_2 assimilation rate is calculated (Teh, 2006).

The following sub-sections describe the meaning and calculations of the various parameters in Eq. 10 and 11.

4.1 Rubisco-limited assimilation (v_c)

Rubisco-limited leaf CO_2 assimilation rate v_c is determined by

$$v_c = \frac{V_{cmax}\left(C_i - \Gamma^*\right)}{K_c\left(1 + O_a / K_o\right) + C_i} \tag{13}$$

where K_c is the Michaelis–Menten constant for CO_2 (μmol CO_2 mol^{-1} air); K_o is the Michaelis–Menten constant for O_2 (μmol O_2 mol^{-1} air); Γ^* is the CO_2 compensation point (μmol CO_2 mol^{-1} air); O_a is the ambient O_2 concentration in air (210000 μmol O_2 mol^{-1} air); C_i is the intercellular CO_2 concentration (μmol CO_2 mol^{-1} air); and V_{cmax} is the maximum Rubisco capacity rate (μmol CO_2 m^{-2} leaf s^{-1}). The CO_2 compensation point Γ^* (CO_2 concentration at the point where assimilation is zero) is determined by

$$\Gamma^* = 0.5O_a / \tau \tag{14}$$

where τ is the CO_2/O_2 specificity factor (μmol O_2 μmol^{-1} CO_2). For C3 plants like oil palm, τ indicates the level of competition between O_2 and CO_2 substrates for the Rubisco enzyme.

Most of the parameters in Eq. 13 are temperature-dependent and must be corrected for foliage temperature (T_f). Correction is made by multiplying their values at 25°C with their corresponding Q_{10} temperature coefficients (Table 1). Q_{10} is a measure of the rate of change of a given biochemical activity due to the increase in the temperature by 10°C.

Temperature correction for these parameters follows this general form:

$$\xi = \xi_{(25)} \times Q_{10,\xi}^{(T_f - 25)/10} \tag{15}$$

where $\xi_{(25)}$ is the parameter value at 25 °C, T_f is the canopy temperature (°C) and $Q_{10,\xi}$ is the relative change in parameter ξ for every 10 °C change (Table 1). However, V_{cmax} has to be additionally corrected for foliage temperatures greater than 40°C, after which the Rubisco enzyme degrades, causing a rapid decline in CO_2 assimilation:

Table 1 Values of temperature-dependent photosynthesis parameters $\xi_{(25)}$ at 25 °C. Except for $V_{cmax(25)}$, all values are from Bernacchi et al. (2001, 2002), and they are general values for C3 plants

$\xi_{(25)}$	Description	Unit	Value	$Q_{10,\xi}$
$K_{c(25)}$	Michaelis–Menten constant for CO_2	μmol CO_2 mol^{-1} air	270	2.786
$K_{o(25)}$	Michaelis–Menten constant for O_2	μmol O_2 mol^{-1} air	165000	1.355
$\tau_{(25)}$	CO_2/O_2 specificity factor	μmol O_2 μmol^{-1} CO_2	2800	0.703
$V_{cmax(25)}$	Maximum Rubisco capacity rate	μmol CO_2 m^{-2} leaf s^{-1}	79	2.573

$$V_{cmax} = \frac{V_{cmax(25)} \times 2.573^{(T_f - 25)/10}}{1 + exp[0.29(T_f - 40)]} \tag{16}$$

$V_{cmax(25)}$ for oil palm is determined to decline with tree age, from a mean (± s.e.) of 91.1 (±3.7) to 69.8 (±2.6) μmol CO_2 $m-^2$ leaf s^{-1} between palm ages 1 and 19 yrs (Fig. 2). The relationship between oil palm $V_{cmax(25)}$ and tree age can be described by the following linear equation:

$$V_{cmax(25)} = 87.935 - 0.0026 age \tag{17}$$

where *age* refers to the tree age (in days).

Our measured $V_{cmax(25)}$ is between 10% and 30% smaller than that used by Fan et al. (2015) for their oil palm growth simulations. Fan et al. (2015) used a constant $V_{cmax(25)}$ of

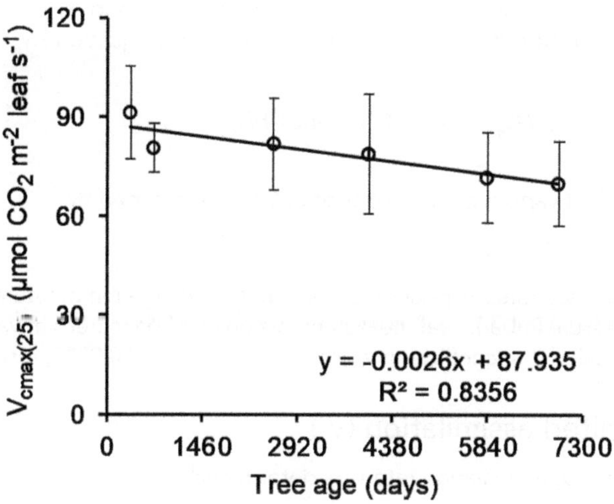

Figure 2 Changes in the maximum Rubisco capacity rate $V_{cmax(25)}$ with oil palm tree age. Leaf gas exchanges for 1- to 19-year-old palm trees were measured (N = 138) (Long and Bernacchi, 2003) using the TPS-2 Portable Photosynthesis System (PP Systems International, Inc., Amesbury, US), and based on the gas exchange data, $V_{cmax(25)}$ was estimated based on Sharkey et al. (2007) and Sharkey (2005). Note: the bars represent one standard deviation.

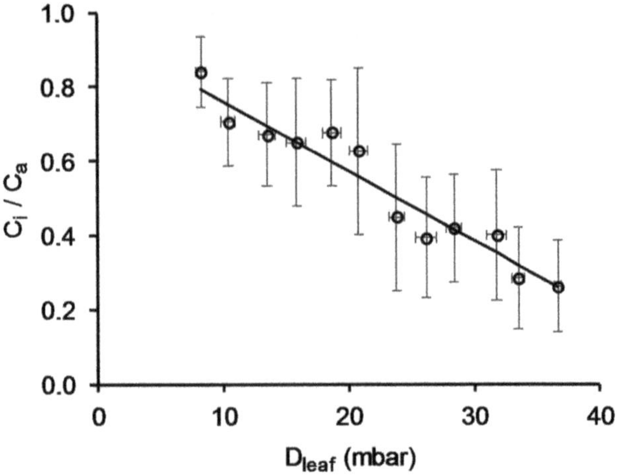

Figure 3 Decline in the ratio between intercellular CO_2 concentration (C_i) and ambient CO_2 concentration (C_a) with increasing leaf vapour pressure deficit (D_{leaf}) for 1- to 19-year-old oil palm trees (N = 654). Leaf gas exchanges were measured based on Long and Bernacchi (2003) using the TPS-2 Portable Photosynthesis System (PP Systems International, Inc., Amesbury, US). Note: the bars represent one standard deviation.

100.7 μmol CO_2 m^{-2} leaf s^{-1}, a value they had calculated based on the leaf N content and specific leaf area for crops in general (not specifically based on oil palm leaves).

The intercellular CO_2 concentration (C_i) can be determined from its relationship with ambient CO_2 concentration (C_a) (both concentrations expressed as μmol CO_2 mol^{-1} air) and vapour pressure deficit in the leaves by the following equation by Yin and van Laar (2005):

$$C_i / C_a = 1 - \left(1 - \Gamma^* / C_a\right)\left(a + bD_{leaf}\right) \tag{18}$$

where D_{leaf} is the leaf vapour pressure deficit (mbar), determined by

$$D_{leaf} = e_s\left[T_f\right] - e_a \tag{19}$$

where $e_s\left[T_f\right]$ is the saturated vapour pressure (mbar) at foliage temperature T_f (°C) and e_a is the air vapour pressure (mbar). Leaf measurements on of oil palm trees at various ages (from 1 to 19 years) revealed that coefficients a and b are 0.0615 and 0.0213, respectively (Fig. 3).

4.2 Light-limited assimilation (v_q)

Light-limited leaf CO_2 assimilation rate v_q is determined by

$$v_{q,sl/sh} = 0.8e_m Q_{sl/sh}\left(C_i - \Gamma^*\right) / \left(C_i + \Gamma^*\right) \tag{20}$$

where $v_{q,sl/sh}$ is the light-limited leaf CO_2 assimilation rate (μmol CO_2 m^{-2} leaf s^{-1}) for either sunlit (subscript *sl*) or shaded (*sh*) leaves; $Q_{sl/sh}$ is the photosynthetically active radiation

(PAR) flux density absorbed by either sunlit or shaded leaves (μmol m^{-2} leaf s^{-1}); and e_m is the intrinsic quantum efficiency or yield (μmol CO_2 μmol^{-1} photons), taken as 0.051 for oil palm (Dufrêne and Saugier, 1993). Skillman (2008) reviewed 10 studies that measured e_m of various C3 plants and the mean value (± s.e.) was found to be 0.052 ± 0.003 (N = 61).

According to Campbell and Norman (1998), there are four PAR flux components within the canopies, namely: 1) PAR irradiance for unintercepted beam *with* scattering $Q_{p,dr}$ (μmol photons m^{-2} ground s^{-1}), 2) PAR irradiance for unintercepted beam *without* scattering $Q_{p,dr,dr}$ (μmol photons m^{-2} ground s^{-1}), 3) PAR irradiance of *only* the scattered component $Q_{p,dr,\alpha}$ (μmol photons m^{-2} leaf s^{-1}) and 4) mean diffuse PAR irradiance $\overline{Q_{p,df}}$ (μmol photons m^{-2} leaf s^{-1}). Note that both $Q_{p,dr,\alpha}$ and $\overline{Q_{p,df}}$ are expressed based on per unit leaf area (not per unit ground area). The scattered and diffuse irradiance on a unit horizontal ground area are assumed equal to their irradiance on a unit leaf area. These four PAR flux components can be calculated by

$$Q_{p,dr} = (1 - p_{dr})Q_{dr}exp\left(-k_{dr}\omega\sqrt{0.8}L\right) \tag{21}$$

$$Q_{p,dr,dr} = (1 - p_{dr})Q_{dr}exp(-k_{dr}\omega L) \tag{22}$$

$$Q_{p,dr,\alpha} = \left(Q_{p,dr} - Q_{p,dr,dr}\right)/2 \tag{23}$$

$$\overline{Q_{p,df}} = (1 - p_{df})Q_{df}\left[1 - exp\left(-k_{df}\sqrt{0.8}L\right)\right]/k_{df}L \tag{24}$$

where Q_{dr} and Q_{df} are the instantaneous direct and diffuse PAR irradiance, respectively (both in μmol photons m^{-2} ground s^{-1}); p_{dr} and p_{df} are the canopy reflection coefficients for direct and diffuse PAR, respectively (both unitless); k_{dr} and k_{df} are the canopy extinction coefficients for direct and diffuse PAR, respectively (both unitless); ω is the canopy clustering coefficient for discontinuous canopies (0 to 1); and L is the total leaf area index (m^2 leaf m^{-2} ground).

The PAR irradiance components Q_{dr} and Q_{df} are determined by

$$Q_{dr} = I_{dr} \times 4.55 \times 0.5 \tag{25}$$

$$Q_{df} = I_{df} \times 4.55 \times 0.5 \tag{26}$$

where I_{dr} and I_{df} are the instantaneous direct and diffuse solar irradiance, respectively (W m^{-2} ground). Total PAR irradiance is assumed to be 50% of total solar irradiance, and 1 W m^{-2} ground is taken as equal to 4.55 μmol photons m^{-2} ground s^{-1} (Goudriaan and van Laar, 1994).

The reflection coefficients p_{dr} and p_{df} are determined according to Goudriaan (1977) by

$$p_{dr} = MAX\left[0.04, p_s exp\left(-2k_{dr}\omega\sqrt{0.8}L\right)\right] \tag{27}$$

$$p_{df} = MAX\left[0.04, p_s exp\left(-2k_{df}\sqrt{0.8}L\right)\right] \tag{28}$$

where p_s is the soil reflection coefficient, taken as 0.15.

Assuming the oil palm canopies are randomly distributed in the aerial space, the canopy extinction coefficient for direct PAR k_{dr} is determined by

$$k_{dr} = 0.5 / cos\theta \tag{29}$$

where θ is the solar inclination (radians). Equation 29, however, is valid only for fully closed canopies. For discontinuous or partially closed canopies, k_{dr} must be multiplied by a clump factor ω, which ranges from 0 (no canopy) to 1 (fully closed canopies), and which can be determined according to Kustas and Norman (1999) as

$$\omega = \omega_0 + 6.6557(1-\omega_0)exp\left[exp(-\theta + 2.2103)\right] \tag{30a}$$

$$\omega_0 = \frac{1}{k_{dr}L} ln\left\{\tau_b + (1-\tau_b)exp\left[-k_{dr}\frac{L}{1-\tau_b}\right]\right\} \tag{30b}$$

where ω_0 is the canopy clustering coefficient when the sun is at zenith (highest point in the sky) and τ_b is the canopy gap fraction (0 to 1). Awal (2008) measured the canopy gap fraction for various oil palm tree ages, and reanalysing and fitting the best function to his measured data, τ_b can be estimated from L by

$$\tau_b = \left(1 + 1.33\sqrt{L}\right)^{-1} \tag{31}$$

Following the method by Teh (2006), the canopy extinction coefficient for diffuse PAR k_{df} is determined by integrating the light penetration function based on Beer's law (together with Eq. 29–31) over the whole sky (assuming a uniform bright sky) and for several leaf area indexes L to finally obtain the following estimate of k_{df}:

$$k_{df} = exp\left(0.038042 - 0.38845\sqrt{L}\right) \tag{32}$$

Note that this k_{df} estimation already accounts for discontinuous canopies by the use of Eq. 30 in the integration.

The PAR flux densities absorbed by the oil palm canopy are made up of two components: that absorbed by the canopy directly exposed to direct PAR (sunlit portion of L) and that absorbed by the portion of canopy shaded from direct PAR (shaded L). Dividing the oil palm canopies in this way, rather than treating the canopies as one whole, would provide a more accurate estimation of canopy photosynthesis because the CO_2 assimilation by the sunlit and shaded portions of the canopies would often be different from each other.

L_{sl} and L_{sh} are the sunlit and shaded leaf area indexes, respectively (both in m^2 leaf m^{-2} ground), and they can be estimated according to Goudriaan and van Laar (1994) as

$$L_{sl} = \frac{1-exp(-k_{dr}\omega L)}{k_{dr}\omega} \tag{33}$$

$$L_{sh} = L - L_{sl} \tag{34}$$

Q_{sl} and Q_{sh} are the PAR flux densities absorbed by the sunlit leaves and shaded leaves, respectively (both in μmol photons m^{-2} leaf s^{-1}), and they can be determined according to Campbell and Norman (1998) as

$$Q_{sl} = 0.8\left(k_{dr}\omega Q_{dr} + \overline{Q_{p,df}} + Q_{p,dr,\alpha}\right) \tag{35}$$

$$Q_{sh} = 0.8\left(\overline{Q_{p,df}} + Q_{p,dr,\alpha}\right)p1 \tag{36}$$

Note that the PAR flux densities absorbed (Q_{sl} and Q_{sh}) are on a per unit leaf area (not per unit ground area basis).

4.3 Sink-limited assimilation (v_s)

Besides limitations by Rubisco and light, photosynthesis could also be limited by the amount of sink in the oil palm, as the greater the storage or sink, the slower the photosynthetic rate, akin to chemical reactions which slow down due to the buildup of assimilates. In a plant, the most likely limiting sink is the storage of sucrose, and Collatz et al. (1991) assumed that v_s is merely half of the maximum Rubisco capacity rate:

$$v_s = 0.5V_{cmax} \tag{37}$$

5 Modelling energy balance

The energy balance of the soil–plant–atmosphere system is described as a network of resistances in which latent and sensible heat fluxes must traverse within the system to reach some reference level above the canopy (Fig. 4) (Shuttleworth and Wallace, 1985). Solving the energy balance equation gives the potential water loss by soil evaporation and plant transpiration.

The energy balance in Fig. 4 can be described in eight independent equations, and as shown by Teh (2006) and Shuttleworth and Wallace (1985), these equations can be solved and summarized by the following series of equations to determine the total latent heat flux:

$$\lambda ET = C_c PM_c + C_s PM_s \tag{38a}$$

$$PM_c = \frac{\Delta A + \left(\rho c_p D - \Delta r_a^c A_s\right) / \left(r_a^a + r_a^c\right)}{\Delta + \gamma\left[1 + r_s^c / \left(r_a^a + r_a^c\right)\right]} \tag{38b}$$

$$PM_s = \frac{\Delta A + \left(\rho c_p D - \Delta r_a^s A_c\right) / \left(r_a^a + r_a^s\right)}{\Delta + \gamma\left[1 + r_s^s / \left(r_a^a + r_a^s\right)\right]} \tag{38c}$$

$$C_c = \left\{1 + R_c R_a / \left[R_s\left(R_c + R_a\right)\right]\right\}^{-1} \tag{38d}$$

$$C_s = \left\{1 + R_s R_a / \left[R_c\left(R_s + R_a\right)\right]\right\}^{-1} \tag{38e}$$

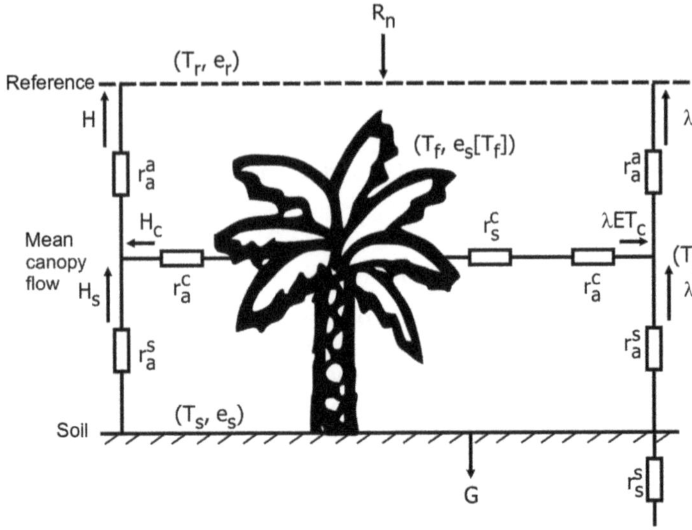

Figure 4 Energy balance of the oil palm system described as a network of resistances. Key: λET and H are components of the latent and sensible heat fluxes, respectively; R_n is the net radiation flux; G is the heat flux into the soil; T and e are the temperature and vapour pressure components; $e_s[T]$ is the saturated vapour pressure at temperature T; r_a^a and r_a^s are the aerodynamic resistances; and r_a^c, r_s^c and r_s^s are the surface resistances.

$$R_a = (\Delta + \gamma) r_a^a \tag{38f}$$

$$R_c = (\Delta + \gamma) r_a^c + \gamma r_s^c \tag{38g}$$

$$R_s = (\Delta + \gamma) r_a^s + \gamma r_s^s \tag{38h}$$

where λET is the total latent heat flux (W m^{-2}); r_a^a is the aerodynamic resistance between the mean canopy flow and reference height (s m^{-1}); r_a^s is the aerodynamic resistance between the soil and mean canopy flow (s m^{-1}); r_a^c is the bulk boundary layer resistance (s m^{-1}); r_s^c and r_s^s are the canopy and soil surface resistance, respectively (both in s m^{-1}); A, A_s and A_c are energy available to the system (total), soil and crop, respectively (all in W m^{-2}); Δ is the slope of the saturated vapour pressure curve (mbar K^{-1}); γ is the psychometric constant (0.658 mbar K^{-1}); D is the air vapour pressure deficit (mbar) and ρc_p is the volumetric heat capacity for air (1221.09 J m^{-3} K^{-1}).

Determination of total latent heat flux λET allows the determination of the vapour pressure deficit D_0 (mbar) at the theoretical mean canopy flow height as

$$D_0 = D + \frac{r_a^a}{\rho c_p}\left[\Delta A - (\Delta + \gamma)\lambda ET\right] \tag{39}$$

so that D_0 can in turn be used to determine the partitioning of the total latent heat flux λET and total sensible heat flux H into their respective soil (λET_s and H_s) and crop (λET_c and H_c) flux components:

$$\lambda ET_s = \frac{\Delta A_s + \rho c_p D_0 / r_a^s}{\Delta + \gamma\left(r_s^s + r_a^s\right) / r_a^s} \tag{40}$$

$$\lambda ET_c = \frac{\Delta A_c + \rho c_p D_0 / r_a^c}{\Delta + \gamma\left(r_s^c + r_a^c\right) / r_a^c} \tag{41}$$

$$H_s = \frac{\gamma A_s\left(r_s^s + r_a^s\right) - \rho c_p D_0}{\Delta r_a^s + \gamma\left(r_s^s + r_a^s\right)} \tag{42}$$

$$H_c = \frac{\gamma A_c\left(r_s^c + r_a^c\right) - \rho c_p D_0}{\Delta r_a^c + \gamma\left(r_s^c + r_a^c\right)} \tag{43}$$

where all the soil and crop flux components are in W m^{-2}.

Numerical Gaussian integration method is used to obtain the daily latent heat and sensible fluxes (Teh, 2006). The daily amount of water transpired (by plants) and evaporated (by soil) are estimated by such integration methods. Five points over 24 hours in a day were selected, and for each selected hour, the heat fluxes were calculated.

5.1 Available energy

The balance of net radiation R_n after subtracting soil heat flux G is the energy available to the plant A_c and soil A_s (all in W m^{-2}), or

$$A = A_s + A_c = R_n - G \tag{44a}$$

$$A_s = R_n \tau_{dr,\alpha} - G \tag{44b}$$

$$A_c = R_n\left(1 - \tau_{dr,\alpha}\right) \tag{44c}$$

$$G = R_n\left[t_c + \tau_{dr,\alpha}\left(t_s - t_c\right)\right] \tag{44d}$$

$$\tau_{dr,\alpha} = \exp\left(-k_{dr}\omega\sqrt{0.5L}\right) \tag{44e}$$

where L is the total leaf area index (m^2 leaf m^{-2} ground); k_{dr} is the canopy extinction coefficient for direct solar radiation; ω is the canopy clustering coefficient; and t_c and t_s are the fraction of net radiation available as soil heat flux under full canopies and for bare soil (no canopies), respectively. Parameters t_c and t_s are taken as 0.05 (Monteith, 1973) and 0.315 (Kustas and Daughtry, 1990), respectively.

Following Brutsaert (1982), net radiation R_n (W m^{-2}) is determined as the difference between the incoming shortwave radiation I_t (W m^{-2}) and outgoing net long-wave radiation R_{nL} (W m^{-2}):

$$R_n = (1-p)I_t + R_{nL} \tag{45a}$$

$$R_{nL} = 0.98\sigma\left(T_a + 273.15\right)^4\left[1.31\left(\frac{e_a}{T_a + 273.15}\right)^{1/7} - 1\right] \tag{45b}$$

where p is the surface albedo (0.15) and σ is the Stefan-Boltzmann constant (5.67×10^{-8} W m^{-2} K^{-4}).

5.2 Vertical profile of wind speed and eddy diffusivity

Wind speed decreases exponentially with decreasing height due to increasing drag until wind speed theoretically reaches zero at a height equal to the total height of zero plane displacement d and crop roughness length z_0. The methods by Su et al. (2001) and Massman (1997) were adapted to determine d and z_0 (both in m) as

$$d = h\Delta H \tag{46a}$$

$$z_0 = h(1-\Delta H)exp(-k/0.32) \tag{46b}$$

$$\Delta H = 1 - \frac{exp(-2k_w)\left[exp(-2k_w) - 1\right]}{2k_w} \quad \text{for } 0.30 \leq \Delta H \leq 0.95 \tag{46c}$$

where k_w is the vertical wind speed extinction coefficient (unitless); h is the tree height (m) and k is the von Karman constant (0.4).

Vertical wind speed extinction coefficient k_w (assumed equal to eddy diffusivity k_e) increases with increasing leaf area index. Simulations by Massman (1987) for hypothetical canopies with uniform foliage density were used by Nikolov and Zeller (2003) to approximate the relationship between k_w and total leaf area index L (m^2 leaf m^{-2} ground) simply as

$$k_w = k_e = 3\left[1 - \exp(-L)\right] \tag{47}$$

The wind speed at the canopy height u_h (m s^{-1}) is determined by

$$u_h = \frac{u_*}{k} ln\left(\frac{h-d}{z_0}\right) \tag{48}$$

where u_* is the friction velocity (m s^{-1}), determined by

$$u_* = \frac{ku}{ln\left(\frac{z_r - d}{z_0}\right)} \tag{49}$$

where u is the wind speed (m s^{-1}) measured at the reference height z_r (m).

5.3 Resistances

Aerodynamic resistances are calculated by

$$r_a^s = \frac{\exp(k_e)}{k_e k u_*}\left[\exp\left(-k_e\frac{z_{s0}}{h}\right) - \exp\left(-k_e\frac{z_0 + d}{h}\right)\right] \tag{50}$$

$$r_a^a = \frac{1}{k_e u_*} ln\left(\frac{z_r - d}{h - d}\right) + \frac{1}{k_e k u_*}\left\{\exp\left[k_e\left(1 - \frac{z_0 + d}{h}\right)\right] - 1\right\} \tag{51}$$

where r_a^s and r_a^a are the aerodynamic resistance between the mean canopy flow and reference height and between the soil and mean canopy flow, respectively (both in s m^{-1}); and z_{s0} is the soil surface roughness length (m), taken as 0.004, the value for a flat, tilled land (Hansen, 1993).

Calculations are adapted from Campbell and Norman (1998) to determine the bulk boundary layer resistance r_a^c (s m^{-1}) as

$$r_a^c = \frac{k_w}{0.01 L_{eff}\left[1 - \exp(-0.5 k_w)\right]\sqrt{u_h / w}} \tag{52}$$

where L_{eff} is the effective leaf area index (m^2 leaf m^{-2} ground); and w is the mean leaf width (m), which, for oil palm pinnae, can be determined from the work by Rao et al. (1992) as

$$w = 0.0165 + 0.0152 ln(age / 365) \tag{53}$$

where *age* refers to the tree age (days).

Not all leaves, especially at near or full canopy, would contribute equally to affect fluxes. This is due to self-shading of leaves. Consequently, Szeicz and Long (1969) recommended that an effective leaf area index L_{eff} be used instead of total L, where L_{eff} is taken as the smaller value of the the current leaf area index L and half of the maximum possible leaf area index; that is,

$$L_{eff} = MIN\left[\!\left[L,\ 0.5 L_{max,PD}\right]\!\right] \tag{54}$$

where $L_{max,PD}$ is the maximum leaf area index (m^2 leaf m^{-2} ground) for a given PD planting density (palms ha^{-1}), and for oil palm, $L_{max,PD}$ is determined by

$$L_{max,PD} = 0.0274 PD^{1/A} \tag{55}$$

where A is 0.935. Equation 55 was derived by fitting the best function to the maximum L obtained in various oil palm planting densities from Foong (1999), Kwan (1994), Rao et al. (1992) and Tan and Ng (1977).

Canopy resistance is determined by scaling up leaf stomatal resistance (the inverse of leaf stomatal conductance) to the canopy level by

$$r_s^c = 1/(gst \times L_{eff}) \tag{56}$$

where r_s^c is the canopy resistance (s m^{-1}) and gst is the leaf stomatal conductance (m s^{-1}).

Leaf stomatal conductance gst is at maximum conductance gst_{max}, but water stress, low PAR irradiance, and vapour pressure deficit will reduce gst by the following relationships:

$$gst = gst_{max} \times f_{water} \times f_{PAR} \times f_D \tag{57a}$$

$$f_{water} = ET_c / PET_c \tag{57b}$$

$$f_{PAR} = gst_{PAR}[[PAR]] / gst_{PAR}[[PAR_{max}]] \tag{57c}$$

$$gst_{PAR}[[PAR]] = 0.014614[1 - exp(-0.008740PAR)] \tag{57d}$$

$$f_D = gst_D[[D]] / gst_D[[D_{min}]] \tag{57e}$$

$$gst_D[[D]] = 0.031970 - 0.007516 ln(D) \tag{57f}$$

where gst_{max} for oil palm is 500 mmol H_2O m^{-2} leaf s^{-1} or 0.0125 m s^{-1} (Kallarackal et al., 2004); and f_{PAR}, f_{water} and f_D are scaled reductions from 0 to 1 to gst_{max} due to PAR irradiance, water stress and vapour pressure deficit, respectively.

Note that crop water stress is described in Eq. 57b, where the level of stress is determined, following Kropff (1993), as the ratio of the actual to potential transpiration (ET_c and PET_c, respectively; both in m day^{-1}).

Equations 57d and 57f were derived by fitting the best functions to measured leaf stomatal conductance at several levels of PAR and air vapour pressure deficit D (Fig. 5). Leaf stomatal conductance for palm aged 1–20 years was measured using the AP4 dynamic diffusion porometer (Delta-T Devices Ltd., Cambridge, UK). From Fig. 5a, PAR_{max}, which is the maximum PAR irradiance (W m^{-2}) for maximum leaf conductance, is taken at 330 W m^{-2} (or 1500 μmol photons m^{-2} ground s^{-1}) (Dufrêne and Saugier, 1993). Similarly, maximum conductance occurs when air vapour pressure deficit is minimum (D_{min}), taken at 10 mbar (Fig. 5b).

Finally, soil surface resistance r_s^s (s m^{-1}) was determined solely from the first (topmost) soil layer as this is the layer in direct contact with the atmosphere. Equations from Farahani and Ahuja (1996) and Choudhury and Monteith (1988) are used:

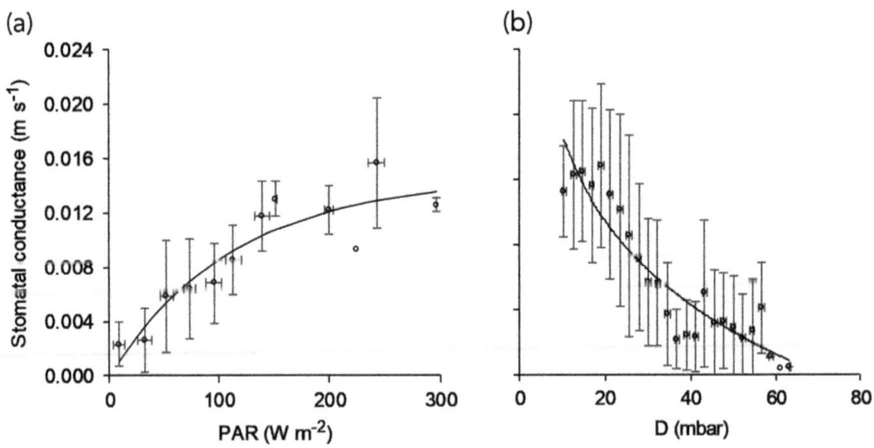

Figure 5 Relationship between measured leaf stomatal conductance with: (a) photosynthetically active radiation (PAR) irradiance (N = 381) and (b) air vapour pressure deficit D (N = 2591). Leaf measurements were made on 1- to 20-year-old oil palm trees. Note: the bars represent one standard deviation.

$$r_s^s = \frac{\tau l}{\phi_p D_{m,v}} \exp\left(-\frac{1}{\lambda} \times \frac{\theta}{\theta_0}\right) \tag{58}$$

where $D_{m,v}$ is the vapour diffusion coefficient in air (24.7×10^{-6} m² s⁻¹); ϕ_p is the soil porosity (m³ m⁻³); l is the soil layer thickness of the first soil layer (m); θ is the volumetric soil water content (m³ m⁻³); θ_0 is the saturated soil water content (m³ m⁻³); λ is the soil pore-size distribution index or the slope of the logarithmic suction–soil moisture curve (Bittelli et al. 2015) and τ is the soil tortuosity (unitless), determined from Shen and Chen (2007) as

$$\tau = \sqrt{\phi_p + 3.79(1-\phi_p)}, \tag{59}$$

Soil properties and water flow in oil palm plantations are discussed next.

6 Modelling soil water flow

Rather than treating the whole soil profile as one large and homogeneous soil layer, it is more accurate to divide the soil profile into two or more soil layers and then use Darcy's law to describe the flow of water from one soil layer to the next, taking into account the physical properties of each soil layer (Fig. 6). Water flow is modelled following the 'tipping bucket' system, where water flow is treated in a sequential manner, beginning in the first soil layer, then moving successively down through the soil profile until the last soil layer (Hillel, 1977).

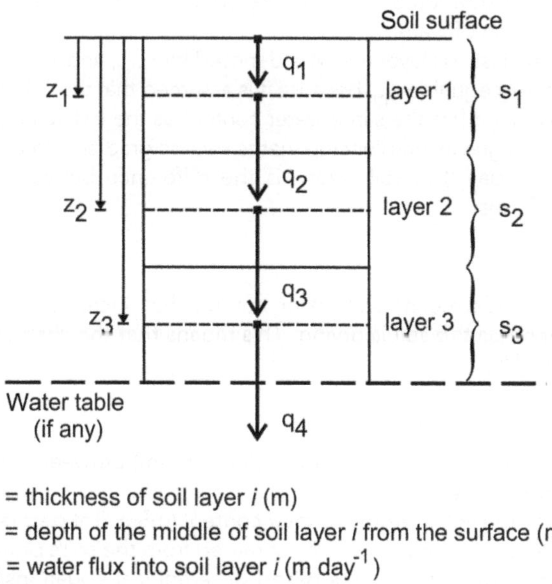

Figure 6 Water flow in a soil profile is divided into three successive layers, with the presence of a water table, if any, always just beneath the last (in this case, third) soil layer.

It is recommended to divide a soil profile into two or more consecutive layers, with the first soil layer as a thin layer and the second layer covering up to at least the entire rooting depth. Soil layer i (i = 1 to N) has a thickness of s_i (m), and the depth from the soil surface to the middle of layer i is z_i.

Water flux into soil layer i is denoted as q_i (m day^{-1}). Water flow follows the downward positive coordinate system, where the downward and upward directions of flow of water are taken as a positive and negative value, respectively, and the reference level is taken as the soil surface level. Darcy's law is used to describe the water flow in the soil. Water flow is taken to occur from the middle of layer i – 1 to the middle of layer i. The method based on Hillel (1977) is followed. Water flux into soil layer i is:

$$q_i = \begin{cases} P_{net} - ET_s - ET_{c,i} & i = 1 \\ \overline{K_{\theta,i}} \dfrac{H_i - H_{i-1}}{z_i - z_{i-1}} - ET_{c,i} & 1 < i \leq N \\ K_{\theta,N} & i = N+1 \end{cases} \tag{60a}$$

$$\overline{K_{\theta,i}} = \frac{K_{\theta,i-1} - K_{\theta,i}}{lnK_{\theta,i-1} - lnK_{\theta,i}} \tag{60b}$$

where q is the water flux (m day^{-1}); P_{net} is the net daily rainfall (m day^{-1}); ET_s is the actual daily soil evaporation (which occurs only from the first soil layer) (m day^{-1}); ET_c is the daily extraction of water by roots (actual plant transpiration) (m day^{-1}); $\overline{K_{\theta,i}}$ is the logarithmic mean of the hydraulic conductivities of layer i and i – 1 (m day^{-1}); z is the soil layer's depth (m) and H is the total head (matric suction and gravity heads) (m). Note that water table, if present, is treated as an additional but saturated soil layer N+1, and Eq. 60 is used to describe the capillary rise of water.

Water flux out of the last soil layer (i = N) is denoted by q_{N+1}, and without the presence of a water table, it is merely equal to $K_{\theta,N}$ because it is assumed that the soil below the last layer is uniformly wet and that it has the same water content as the last soil layer. Consequently, water flux is only due to gravity gradient (no matric suction gradient). In this case, $q_{N+1} = K_{\theta,N}$.

The net flux $\widehat{q_i}$ (m day^{-1}) in soil layer i is the difference between incoming q_i and outgoing water q_{i+1} fluxes:

$$\widehat{q_i} = q_{i-1} - q_{i+1} \tag{61}$$

where a positive net flux means soil water content has increased, and in contrast, a negative net flux denotes the soil is drying. This means that the change in the soil water content is determined by:

$$\Theta_{i,t+1} = \Theta_{i,t} + \widehat{q_i} \tag{62}$$

where $\Theta_{i,t}$ and $\Theta_{i,t+1}$ are the water content in soil layer i (m) between two successive time steps t and t+1, respectively.

For each soil layer i, the volumetric soil water content (m^3 m^{-3}) at permanent wilting point θ_{1500}, field capacity θ_{33} and saturation θ_0 are estimated from the soil's primary particles (clay and sand) and organic matter contents based on empirical equations from Saxton and Rawls (2006). Matric suction head and soil hydraulic conductivity (m day^{-1}) for saturated $K_{s,i}$ and unsaturated $K_{\theta,i}$ flows are estimated based on Bittelli et al. (2015) and Saxton and Rawls (2006), using the geometric mean distribution of the soil's primary particles.

Actual soil evaporation ET_s (taken as m day^{-1}) is calculated from Teh (2006) and van Keulen and Seligman (1987) as

$$ET_s = PET_s \times R_{Ds} \tag{63a}$$

$$R_{Ds} = \frac{1}{1+\left(3.607\theta_1 / \theta_{s,1}\right)^{-9.3172}} \tag{63b}$$

where PET_s is the potential evaporation (m day^{-1}); R_{Ds} is the reduction factor for evaporation (ranging from 0 to 1); and θ_1 and $\theta_{s,1}$ are the current and saturated soil water content, respectively, for the first soil layer (i = 1) (both in m^3 m^{-3}). Note: λET_s (W m^{-2}) from Eq. 40 is divided by λ (2454000 J kg^{-1}) and multiplied by 86.4 to obtain PET_s in m day^{-1}.

Actual transpiration (ET_c, m day^{-1}) is calculated from Kropff (1993) as

$$ET_c = PET_c \times R_{Dc} \tag{64a}$$

$$R_{Dc} = \begin{cases} 1 & \theta_{root} \geq \theta_{cr,root} \\ \dfrac{\theta_{root} - \theta_{1500,root}}{\theta_{cr,root} - \theta_{1500,root}} & \theta_{1500,root} < \theta_{root} < \theta_{cr,root} \\ 0 & \theta_{root} \leq \theta_{1500,root} \end{cases} \tag{64b}$$

$$\theta_{cr,root} = \theta_{1500,root} + p\left(\theta_{0,root} - \theta_{1500,root}\right) \tag{64c}$$

where PET_c is the potential transpiration (after converting λET_c from Eq. 41 into m day^{-1}); R_{Dc} is the reduction factor for transpiration (0 to 1); θ_{root} is the soil water content currently in the root zone; $\theta_{1500,root}$ and $\theta_{0,root}$ are the soil root zone's permanent wilting and saturation points, respectively; and $\theta_{cr,root}$ is the volumetric water content in the root zone below which water stress occurs. All soil water contents are expressed in m^3 m^{-3}.

For C3 plants in general, p in Eq. 64c is often taken as 0.5. However, a comparison of the soil water content between irrigated and non-irrigated oil palm trials from 1983 to 1990 by Foong (1999) suggested that oil palm is more sensitive to water stress because p is more likely 0.6 than 0.5 of $\left(\theta_{0,root} - \theta_{1500,root}\right)$. This 0.6 critical point corresponds to about 45% of the available soil water content (AWC) of Munchong soil series (Typic Hapludox), the type of soil in the oil palm trials by Foong (1999). Incidentally, Rey et al. (1998) also observed that oil palm stomatal conductance would begin to decline only when the soil water content fell below the level of about 50% of their soil's AWC (Fig. 7).

The amount of water extracted by roots in each soil layer is based on the measured data for oil palm by Nelson et al. (2006) and on the root water uptake algorithm by Miyazaki (2005):

$$ET_{c,i} = ET_c\left(\varphi_i - \varphi_{i-1}\right) \tag{65a}$$

$$\varphi_i = 1.8c_j - 0.8c_j^2 \tag{65b}$$

$$c_j = MIN\left[1,\ S_j / d_{root}\right] \tag{65c}$$

where S_j is the cumulative thickness of soil layer j (summation of thickness of the current soil layer and all its preceding soil layers).

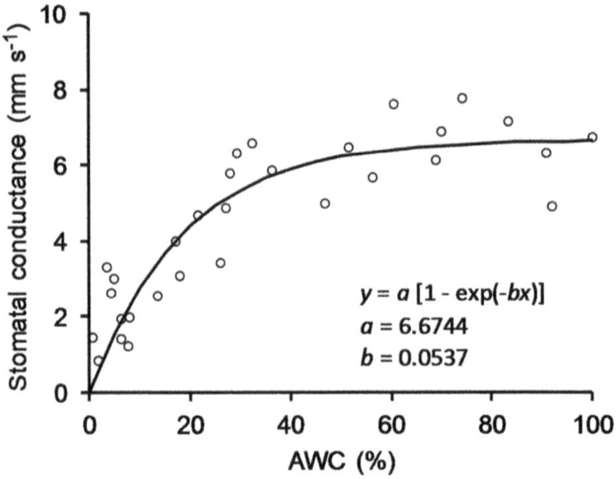

Figure 7 Fitting the best function to the relationship between oil palm leaf stomatal conductance and available soil water content (AWC), as measured by Rey et al. (1998). Stomatal conductance declined only when AWC was about 50% or less.

Lastly, net rainfall P_{net} refers to the amount of rain reaching the ground as both throughfall and stemflow. The larger the canopy cover or leaf area index, the larger the fraction of intercepted gross rainfall by the canopies and the smaller the net rainfall. Net rainfall studies on closed oil palm canopies by Lubis (2016), Chong (2012), Bentley (2007), Zulkifli et al. (2006) and Damih (1995) showed that throughfall and stemflow account for on average (± s.e.) 61.3 ± 2.1 and 8.4 ± 1.0% of P_g, respectively (N = 430 rain events). P_{net} (m day^{-1}) is related to oil palm leaf area index L (m^2 leaf m^{-2} ground) and P_g (m day^1) as follows:

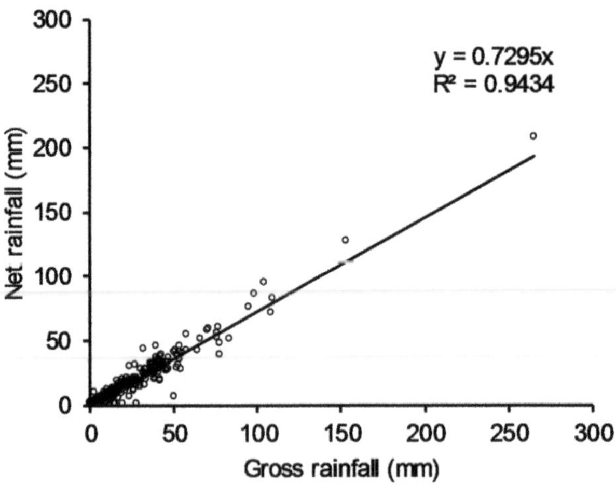

Figure 8 Strong linear relationship between net rainfall and gross rainfall under closed oil palm canopies (Lubis 2016; Chong 2012; Bentley, 2007; Zulkifli et al. 2006; Damih 1995) (N = 430).

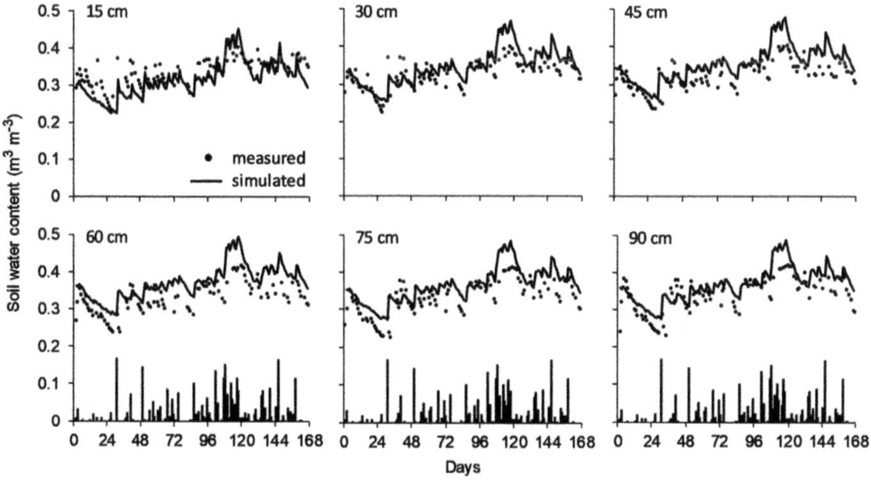

Figure 9 Comparisons between simulated and measured soil water content at several soil depths (from 15 to 90 cm) under fully matured oil palms. Note: the bar charts in the lower panel represent the daily rainfall amount (mm) (total rainfall during this period was 1890 mm, with a maximum daily rainfall of 100 mm).

$$P_{net} = P_g \times MAX\left[0.7295,\ 1 - 0.0541L\right] \tag{66}$$

where it is assumed that P_{net} / P_g decreases linearly with L until closed canopies are reached, after which P_{net} never exceeds 72.95% of P_g (Fig. 8). Figure 9 shows the accuracy of this soil water model component when it was tested against field measurements of soil water content at several soil depths (0.15–0.90 m) under fully matured oil palms (20 years old; 148 palms ha 1) in an oil palm estate at Serdang, Malaysia (2.98053403°N, 101.72884214°E). Daily soil water measurements were made for 167 days from 17 July 2012. The soil type was Munchong series (Typic Hapludox), with a rather homogeneous soil profile (54% clay and 37% sand). The average absolute difference between soil water predictions and measurements was 0.03 $m^3\ m^{-3}$.

7 Modelling crop growth

Plant assimilates (expressed as amount of CH_2O) produced via photosynthesis are first used for plant maintenance respiration (supporting processes for continual plant survival) and vegetative growth respiration (synthesis of new cells) (Thornley 1970), after which the balance is then used for generative growth (development of flowers, bunches and yield). The method by van Kraalingen et al. (1989) for oil palm was adapted.

7.1 Maintenance respiration

The maintenance respiration for pinnae M_{pinnae}, rachis M_{rachis}, trunk M_{trunk} and roots M_{roots} (all in kg CH_2O palm^{-1} day^{-1}) was calculated as follows:

$$M_{pinnae} = W_{pinnae} \times M_{c,pinnae} \times (24 - DL) / 24 \tag{67a}$$

$$M_{c,pinnae} = (N_{pinnae} \times 0.036 \times 6.25) + (X_{pinnae} \times 0.072 \times X_c) \quad (67b)$$

$$M_{rachis} = W_{pinnae} \times M_{c,rachis} \quad (68a)$$

$$M_{c,rachis} = (N_{rachis} \times 0.036 \times 6.25) + (X_{rachis} \times 0.072 \times X_c) \quad (68b)$$

$$M_{trunk} = (W_{top.trunk} + 0.06W_{bottom,trunk}) \times M_{c,trunk} \quad (69a)$$

$$M_{c,trunk} = (N_{trunk} \times 0.036 \times 6.25) + (X_{trunk} \times 0.072 \times X_c) \quad (69b)$$

$$W_{top,trunk} = MIN[\![W_{trunk}, 45.0]\!]$$

$$W_{bottom,trunk} = W_{trunk} - W_{top,trunk} \quad (69c)$$

$$M_{roots} = W_{roots} \times M_{c,roots} \quad (70a)$$

$$M_{c,rachis} = (N_{roots} \times 0.036 \times 6.25) + (X_{roots} \times 0.072 \times X_c) \quad (70b)$$

where $W_{(part)}$ represents the dry weight (DM) of an individual plant part (*e.g.*, pinnae, rachis, trunk, or roots) (kg DM palm^{-1}); *DL* is the day length (hour) $M_{c,(part)}$ is the maintenance coefficient for each specific plant part (kg CH_2O kg^{-1} DM); $N_{(part)}$ and $X_{(part)}$ are the nitrogen and mineral fractions (by weight) in a given plant part, respectively; and X_c is the correction factor for mineral content (unitless), taken as 2 by van Kraalingen et al. (1989).

Maintenance respiration to generate new organs (bunches and flowers) M_{organs} (kg CH_2O kg^1 DM) is determined as follows:

$$M_{organs} = 0.0027W_{matbunch} + M_{c,rachis}(W_{immbunch} + W_{maleflo}) \quad (71)$$

where $W_{matbunch}$, $W_{immbunch}$ and $W_{maleflo}$ are the dry weights of the mature bunches, immature bunches and male flowers, respectively (kg DM palm^{-1}), and it is assumed that the maintenance coefficients for immature bunches and male flowers are the same as that for rachis.

Total maintenance is determined by:

$$M_{total} = M_{metabolic} + \sum_{plant\ parts} M_{(part)} \quad (72a)$$

$$M_{metabolic} = \frac{0.16\Lambda_{canopy,d}}{\sum_{plant\ parts} W_{(part)}} \quad (72b)$$

where M_{total} is the total maintenance requirement for the whole tree, which is the summation of the maintenance requirement for all plant parts, as well as that to support the metabolic activity $M_{metabolic}$ (all units in kg CH_2O palm^{-1} day^{-1}); $\Lambda_{canopy,d}$ is the daily canopy photosynthesis (Eq. 12) (kg CH_2O palm^{-1} day^{-1}) and $\sum_{plant\ parts} W_{(part)}$ is the total dry weight of all plant parts (kg DM palm^{-1}).

Maintenance respiration increases with increasing air temperature, so M_{total} must be corrected for air temperature. Correction was performed using Eq. 15 and taking the Q_{10} coefficient as 2. Growth of oil palm is completely inhibited when air temperature is below

15°C (Henry, 1955). The maximum air temperature for oil palm growth is taken as 45°C, a general maximum value for most plants (Hasanuzzaman et al., 2013). Consequently, if the current air temperature is beyond oil palm's growing temperature range of 15–45°C, all assimilates if any will solely be for maintenance respiration.

Any balance of assimilates after maintenance respiration is available for vegetative growth respiration G_{growth} (kg CH_2O palm^{-1} day^{-1}) as:

$$G_{growth} = MAX\left[\!\left[0,\ \Lambda_{canopy,d} - M_{total}\right]\!\right] \tag{73}$$

7.2 Vegetative growth respiration

The annual vegetative dry matter requirement VDM (kg DM palm^{-1} year^{-1}) is determined based on leaf area index L (m^2 leaf m^{-2} ground), following a rectangular hyperbola relationship, as

$$VDM = MIN\left[\!\left[20,\ \left(a + b / L^{1.5}\right)^{-1}\right]\!\right] \tag{74a}$$

$$a = A / VDM_{max,PD} \tag{74b}$$

$$b = 0.1(1/A - 1)(PD/100)^{1/A} \tag{74c}$$

$$VDM_{max,PD} = 231PD^{1-1/A} \tag{74d}$$

where $VDM_{max,PD}$ is the maximum VDM for the given planting density PD (palms ha^{-1}) and A is 0.935. Coefficients a, b and A, as well as $VDM_{max,PD}$, were obtained by fitting the best functions to measured data from van Kraalingen et al. (1989).

The growth rates $G_{growth,(part)}$ (kg DM palm^{-1} day^{-1}) for the individual plant parts (pinnae, rachis, trunk and roots) are determined by

$$G_{growth,\ (part)} = DM_{(part)} \times A_{growth} \times CVF \tag{75a}$$

$$A_{growth} = MIN\left[\!\left[\frac{VDM/365}{CVF},\ G_{growth}\right]\!\right] \tag{75b}$$

$$CVF = 0.70\left(DM_{pinnae} + DM_{rachis}\right) + 0.66DM_{trunk} + 0.65DM_{roots} \tag{75c}$$

where A_{growth} is the actual amount of assimilates available for vegetative growth respiration (kg CH_2O palm^{-1} day^{-1}); CVF is a factor (kg DM kg^{-1} CH_2O) to convert a weight expressed on a CH_2O per weight basis to that on per dry matter (DM) basis. Dry matter partitioning $DM_{(part)}$ for pinnae, rachis, trunk and roots are 0.24, 0.46, 0.14 and 0.16, respectively (their mean values are from Henson and Mohd Tayeb 2003; Henson 1995; Corley et al. 1971). Dry matter partitioning would likely be affected by planting densities, nutritional status and water stress, but during model development, PySawit's simulations showed better fit with training data sets when constant dry matter partitioning was used instead.

The death rates (kg DM palm^{-1} day^{-1}) for leaves (pinnae and rachis) and roots are calculated according to Dr. Ian E. Henson (then at Malaysian Palm Oil Board; personal communication) as

$$G_{death,leaves} = 0.0016W_{leaves} \begin{cases} 0 & age \leq 600 \\ \dfrac{age - 600}{2500 - 600} & 600 < age \leq 2500 \\ 1 & age > 2500 \end{cases} \tag{76}$$

$$G_{death,roots} = \frac{W_{roots}}{365} \times \begin{cases} 0 & age \leq 1200 \\ 0.0000959210^{-5} age - 0.115 & 1200 < age \leq 3285 \\ 0.2 & age > 3285 \end{cases} \tag{77}$$

where *age* refers to the tree age (days), and W_{leaves} and W_{roots} are the dry weights of leaves and roots, respectively (both in kg DM palm^{-1}).

The net increase in dry weight of a plant part is the difference between its growth and death rates between two successive time steps *t* and *t*+1:

$$W_{(part),t+1} = W_{(part),t} + G_{growth,(part)} - G_{death,(part)} \tag{78}$$

where $W_{(part)}$ is the dry weight of a given plant part (kg DM palm^{-1}). Note that there is no death rate for trunk: $G_{death,trunk}$ = 0.

Leaf area index *L* (m^2 leaf m^{-2} ground) is determined from planting density *PD* and the pinnae's dry weight W_{pinnae} and specific leaf area *SLA* (m^2 leaf kg^{-1} DM):

$$L = W_{pinnae} \times SLA \times PD / 10000 \tag{79}$$

Any balance of assimilates after vegetative growth respiration is available for generative growth G_{gen} (kg CH_2O palm^{-1} day^{-1}) as:

$$G_{gen} = MAX\left[0,\ G_{growth} - A_{growth}\right] \tag{80}$$

7.3 Generative organ growth

The so-called 'boxcar train' technique described by van Kraalingen et al. (1989) was used to track the growth progress of generative organs, namely the male flowers, immature bunches (female flowers) and mature bunches (Fig. 10). Each train (one for each generative organ) comprises several boxcars whereby each boxcar represents an age class. The life-span of both male flowers and immature bunches are taken as 210 days (beginning from the time when sex spikelet is first visible), and 150 days before harvest for the mature bunches (Corley et al. 1995). Consequently, the boxcar train for male flowers and immature bunches consist of 210 one-day age classes and for mature bunches 150 one-day age classes.

The algorithm of this boxcar train technique essentially consists of two steps: 1) the weight of a generative organ is shifted to successively older age classes and 2) the weight is incremented according to the calculated growth rate. The lifespan of male flowers is set at 210 days, after which they will die. For the immature bunches, however, the day after 210 days marks the point of pollination, after which the growth of the pollinated bunch (i.e., mature bunch) will occur. To determine the growth rates of the generative organs, the following general equations are used:

$$G_{gen,(part)} = \left(f_{e,(part)} \times G_{gen} \times CVF_2\right) / n_{(part)} \quad (81a)$$

$$CVF_2 = 0.70\left(f_{e,maleflo} + f_{e,immbunch}\right) + 0.44 f_{e,matbunch} \quad (81b)$$

where $G_{gen,(part)}$ is the growth rate for a given generative organ (male flowers, immature bunches and mature bunches) (kg DM day^{-1}); G_{gen} is the total amount of assimilates available for generative growth (kg DM day^{-1}); CVF_2 is a factor (kg DM kg^{-1} CH_2O) to convert dry matter weight into CH_2O weight; $n_{(part)}$ is the number (count) of a given generative organ currently maintained; and $f_{e,immbunch}$, $f_{e,matbunch}$ and $f_{e,maleflo}$ are the scaled fraction of dry matter to immature bunches, mature bunches and male flowers, respectively (all in fraction), where they are determined by

$$f_{e,(part)} = \frac{f_{(part)}}{\sum_{generative\ organs} f_{(part)}} \quad (82a)$$

$$f_{(part)} = DM_{(part)} \times n_{(part)} / N_{(part)} \quad (82b)$$

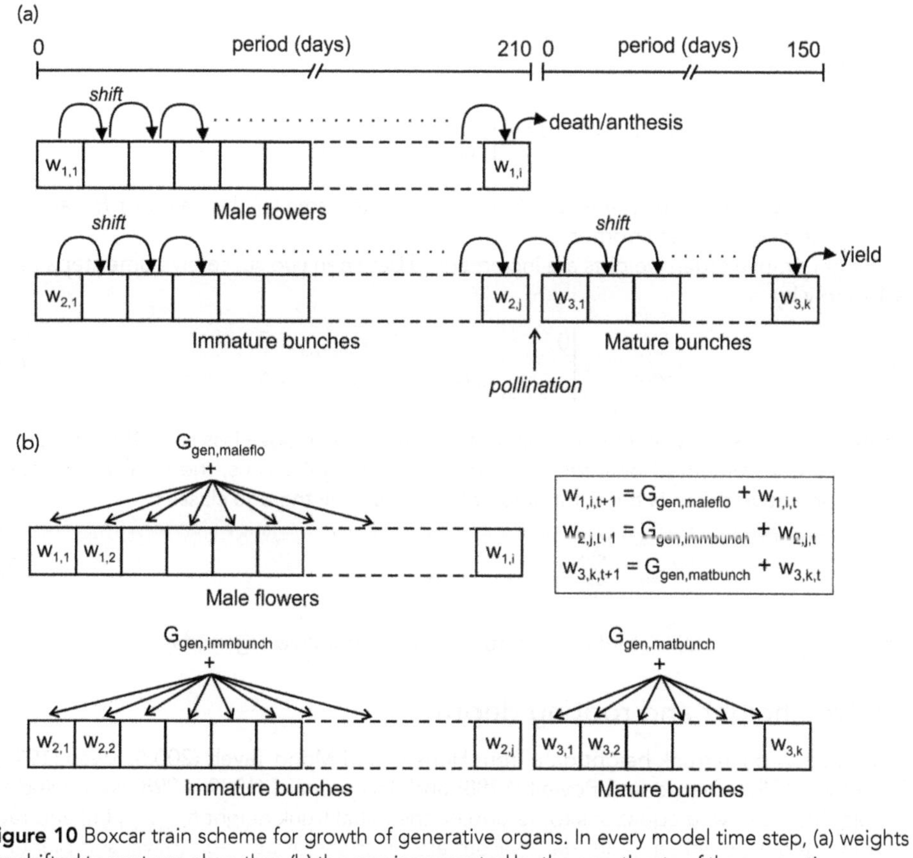

Figure 10 Boxcar train scheme for growth of generative organs. In every model time step, (a) weights are shifted to next age class, then (b) they are incremented by the growth rate of the generative organs.

where for a given generative organ, $DM_{(part)}$ is its share of assimilates available for generative growth (fraction) and $N_{(part)}$ is its total growth (maturity) length (210 days for male flowers and immature bunches and 150 days for mature bunches; Fig. 10). $DM_{(part)}$ for male flowers, immature bunches and mature bunches are 0.159, 0.159 and 0.682, respectively.

In the original model from van Kraalingen et al. (1989), both male and female flowers occur in equal proportions (50:50%) and without any flower abortion. PySawit model introduces an algorithm for flower sex determination whereby if a generated random value is smaller than or equal to a preset female probability p_{female}, then the new flower is female, else male:

$$sex\ of\ new\ flower = \begin{cases} female & r \leq p_{female} \\ male & r > p_{female} \end{cases} \tag{83}$$

where p_{female} is the probability of having female flowers (0 to 1; 0 = no chance of female and 1 = always female) and r is a random variable uniformly distributed on [0 to 1). If p_{female} = 0.5, the tendency is to have equal proportions of male and female flowers. Consequently, the p_{female} parameter can be regarded as the genetic tendency of an oil palm planting material to produce female flowers.

Flower abortion occurs at 9 months before harvest (Corley et al. 1995), and abortion occurs if a generated random value is larger than the crop water stress level R_{Dc} (from Eq. 64b):

$$flower\ aborted? = \begin{cases} no & r \leq R_{Dc} \\ yes & r > R_{Dc} \end{cases} \tag{84}$$

Consequently, the higher the level of water stress, the smaller the value of R_{Dc} and the greater the risk of flower abortion.

The generative organ weights are incremented between two successive time steps t and t+1 simply by

$$w_{(part),i,t+1} = \begin{cases} 0 & w_{(part),i,t} = 0 \\ w_{(part),i,t} + G_{gen,(part)} & w_{(part),i,t} > 0 \end{cases} \tag{85}$$

where $w_{(part),i}$ is the dry weight of a given generative organ in age class i (kg DM). The organ weights are incremented only for non-zero current weights. Thus, the total weight for a given generative organ is the summation of weights in all the age classes:

$$W_{(part)} = \sum_{i=1}^{N_{(part)}} w_{(part),i} \tag{86}$$

where $N_{(part)}$ is the total maturity period for a given generative organ (days).

7.4 Tree height and rooting depth

Measured oil palm trunk height data from Henson and Mohd Tayeb (2003), Kwan (1994), Rao et al. (1992), Breure and Powell (1988) and Jacquemard (1979, 1998) were used to develop the following equations to determine the initial trunk height $h_{trunk(initial)}$ (m) and rate of trunk height growth h'_{trunk} (m day^{-1}). Both are functions of tree age (days) and planting density PD (palms ha^{-1}):

$$h_{trunk(initial)} = \exp\left(2.846 - 1980.888 / PD^2 - 5166.366 / age\right) \tag{87}$$

$$h'_{trunk} = \frac{5166.366 h_{trunk(initial)}}{age^2} \times \frac{(0.210 R_{Dc} + 0.553)}{0.7} \tag{88}$$

where R_{Dc} is the crop water stress level (Eq. 64b). Less is known on how water stress exactly affects trunk height. Nonetheless, trunk height measurements by Foong (1999) revealed that irrigated mature palms in 148 palms ha^{-1} were taller by an average 9% than non-irrigated palms with the same planting density and age. Eq. 88 was developed to reflect this effect.

Measurements of the canopy or crown height for oil palm trees aged between 1 and 20 years (N = 12) gave the following linear regression:

$$h_{canopy} = 1.5091 + 0.001382 age. \tag{89}$$

Canopy height was measured from the base of frond 41 to the tip of frond 1, so, although the rachis length will finally plateau with tree age, our measurements revealed that the length or vertical distance between the base of frond 41 and frond 1 increased in a linear manner. Consequently, tree height h (m) is the summation of trunk height h_{trunk} and canopy height h_{canopy}.

Lastly, the increase in rooting depth d_{root} between two successive time steps t and t+1 is

$$d_{root,t+1} = d_{root,t} + \left(dg_{root} \times R_{Dc}\right) \tag{90}$$

where dg_{root} is the daily increase in the rooting depth, taken as 0.002 m day^{-1}, the mean value of root growth for oil palm (Henson and Chai 1997; Jourdan and Rey 1997), and root growth is detrimentally affected by water stress R_{Dc} (Kropff 1993). Note that rooting depth will not exceed the maximum soil depth set in the model.

8 Model testing

The accuracy of the PySawit model was evaluated by comparing its simulations with several measured growth and yield parameters of oil palm in Merlimau estate (2.253213°N, 102.451753°E), Melaka. The Merlimau data set is independent and not part of the other data sets used during the development of the PySawit model.

At Merlimau, commercial dura x pisifera (Deli x AVROS) palms were planted at the following densities: 120, 135, 148, 164, 181, 199, 220, 243, 268 and 296 palms ha^{-1}. Field plantings for all densities were simultaneously done on April 1987 when the palms were 1 year old. At field planting, the mean (± s.e.) initial dry weights (kg DM palm^{-1}) for the palm parts pinnae, rachis, trunk and roots were 0.40 ± 0.07, 0.70 ± 0.11, 0.10 ± 0.01 and 0.20 ± 0.03, respectively. Field measurements continued every year until the palms were 19 years old (December 2006). Non-destructive methods by Corley and Tinker (2016), Breure (2003), Corley et al. (1971) and Hardon et al. (1969) were used to estimate leaf area and dry weights of frond (inclusive of pinnae, rachis and petiole) of oil palms. These are the oil palm industry standard methods which are currently being used to estimate the crop's dry matter production. At palm maturity stage, the fertilizers applied were ammonium sulphate, muriate of potash, rock phosphate and kieserite at rates 3.5, 3.5, 2.0 and 1.25 kg palm^{-1} yr^{-1}, respectively.

The soil type at Merlimau is a sandy clay loam Rengam soil series (Typic Paleudults), and the first 0.45 m soil depth has a mean (± s.e.) clay, sand, organic C and total N contents of 27.4 ± 0.7, 67.3 ± 0.8, 1.7 ± 0.1 and 0.14 ± 0.01%, respectively. Available soil water content was, however, not measured.

PySawit required the following model inputs to be supplied: site latitude; daily weather (minimum and maximum air temperatures, wind speed, and rainfall amount); N and mineral contents in the individual plant parts of the oil palm tree, as well as the specific leaf area *SLA* (data obtained from van Kraalingen et al. 1989); soil properties (sand, clay, and organic matter contents); and the initial dry weights of the individual plant parts.

Three goodness-of-fit statistical indexes were used to summarize the degree of agreement between PySawit's predictions and observations. These indexes were Normalized Mean Bias Error (NMBE), Normalized Mean Absolute Error (NMAE) and the revised Willmott's index of agreement (d_r) (Willmott et al. 2012; Yu et al. 2006). These indexes are calculated as follows:

$$NMBE = \frac{\sum_{i=1}^{N} P_i - O_i}{\sum_{i=1}^{N} O_i} \tag{91}$$

$$NMAE = \frac{\sum_{i=1}^{N} |P_i - O_i|}{\sum_{i=1}^{N} O_i} \tag{92}$$

$$d_r = \begin{cases} 1 - p/2o & p \leq 2o \\ 2o/p - 1 & p > 2o \end{cases} \tag{93a}$$

$$p = \sum_{i=1}^{N} |P_i - O_i| \text{ and } o = \sum_{i=1}^{N} O_i - \bar{O} \tag{93b}$$

where P_i and O_i are the *i*-th pair of predicted and observed values, respectively (i = 1 to N pairs); and $\bar{O}$ is the mean of all observed values. NMBE (−1 to +∞) indicates a model's tendency to under- or overestimate relative to the mean observations. The larger the NME value, the larger the model's tendency for overestimation. NMAE (0 to +∞) indicates the mean absolute difference between predicted and observed values relative to the mean observations. Larger NMAE values indicate larger mean departures between model predictions and observations. The revised index of agreement d_r ranges between −1 and +1, where increasingly smaller positive or larger negative values indicate increasingly worse or inaccurate model predictions (particularly when $d_r < 0$). For a perfect model, NMBE = 0 (no overall model bias), NMAE = 0 and d_r = +1 (the latter two indicating perfect agreement between model predictions and observations).

9 Results and discussion

The mean (± s.e.) minimum and maximum air temperatures recorded at Merlimau from 1987 to 2006 were 23.71 ± 0.01 and 32.19 ± 0.03°C, respectively. Mean daily wind speed was 1.60 ± 0.01 m s^{1} and the mean daily total solar irradiance was 19.27 ± 0.02 MJ m^{-2}.

Mean annual rainfall was 1918 ± 12 mm, with two notable dry years which occurred during 1997–8 and 2004–5 (year 11 and 18 of field planting, respectively), both of which corresponded to the occurrence of El Niño. The amount of rain received for 1997–8 and 2004–5 were 15 and 32% less than the average annual rainfall at Merlimau, respectively.

Overall, PySawit showed good agreement, with little to no bias, in predicting the growth and yield parameters of oil palm (Fig. 11 to 17). Predictions were especially good for TDM,

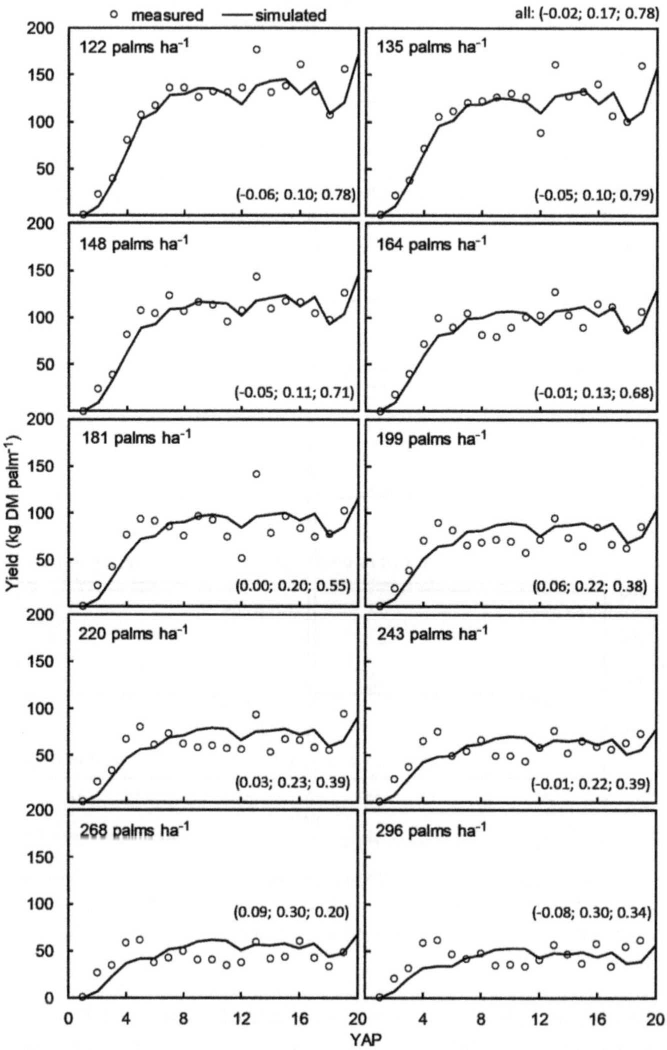

Figure 11 Comparisons between model simulations and observations for fresh fruit bunch (FFB) yield. Years 11 and 18 recorded El Niño events. Values in brackets denote NMBE, NMAE and d_r, where NMBE is the normalized mean bias error, NMAE is the normalized mean absolute error and d_r is the revised index of agreement. YAP is year after planting.

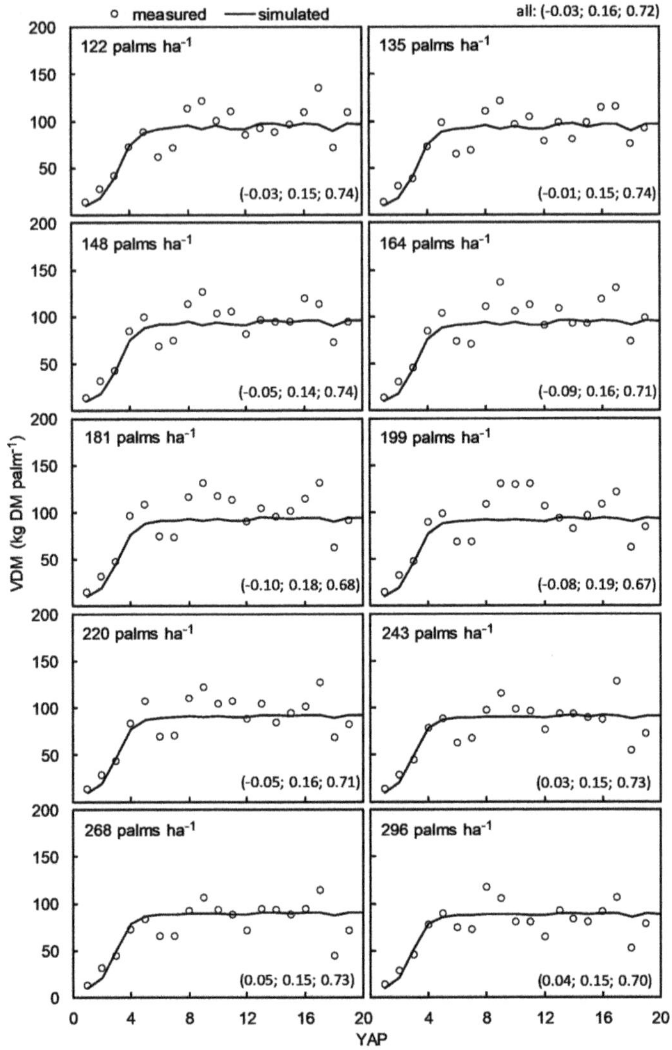

Figure 12 Comparisons between model simulations and observations for aboveground vegetative dry matter (VDM). Values in brackets denote NMBE, NMAE and d_r, where NMBE is the normalized mean bias error, NMAE is the normalized mean absolute error and d_r is the revised index of agreement. YAP is year after planting.

LAI and trunk height, as indicated by their small NMAE values (≤13%), large positive d_r values (>0.8) and near zero NMBE values. Moreover, model accuracy remained stable across all planting densities.

However, discrepancies between yield predictions and observations were found to increase with increasing planting density (Fig. 11). Yield prediction for planting density 122 palms ha^{-1} (PD122), for instance, had a high d_r value of 0.78, but this declined to

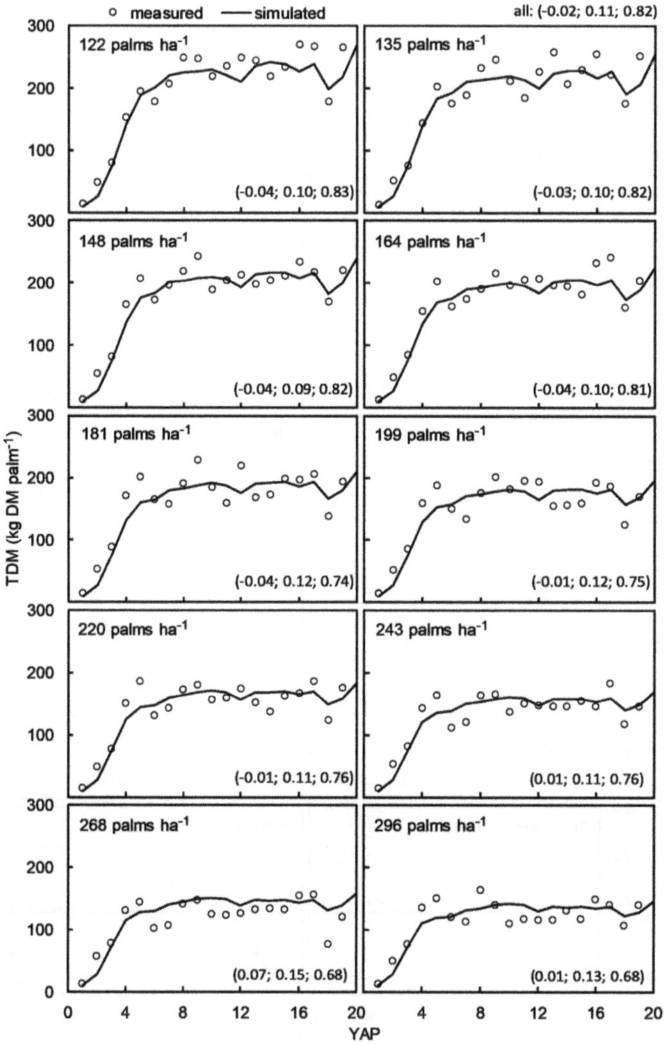

Figure 13 Comparisons between model simulations and observations for total dry matter (TDM), which is the sum of yield and aboveground vegetative dry matter (VDM). Values in brackets denote NMBE, NMAE and d_r, where NMBE is the normalized mean bias error, NMAE is the normalized mean absolute error and d_r is the revised index of agreement. YAP is year after planting.

0.34 for PD296. The degree of model bias also increased with increasing planting density. Nonetheless, paired sample t-test showed no significant differences (p>0.05) between yield predictions and observations across all planting densities, even for PD296, although the level of significance generally declined with increasing planting density. For instance, yield predictions for PD122 was significantly different from observations only at 62% level of significance, but for PD296, this level of significance was 10%.

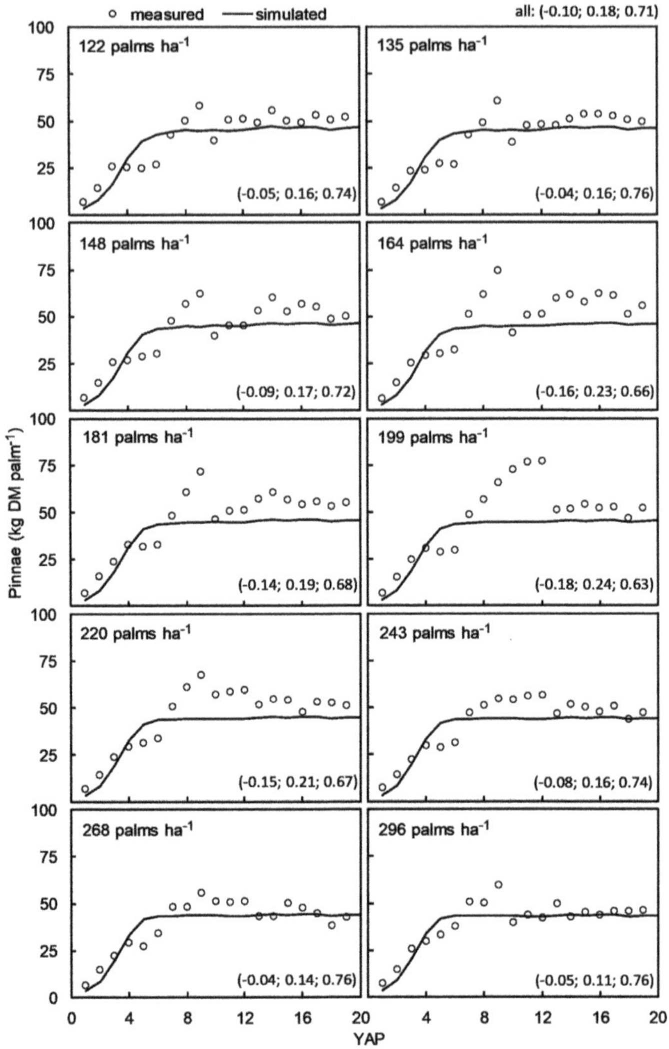

Figure 14 Comparisons between model simulations and observations for pinnae biomass. Values in brackets denote NMBE, NMAE and d_r, where NMBE is the normalized mean bias error, NMAE is the normalized mean absolute error and d_r is the revised index of agreement. YAP is year after planting.

Both years 11 and 18 recorded El Niño events which caused oil palm yields to decline in year 12 (one year after the first El Niño event) and year 18 (the same year as the second El Niño event). Relative to the mean yields from year 10 to 19, yields from all planting densities were reduced by an average (± s.e.) of 12.7 ± 4.3 and 11.6 ± 4.4% in years 12 and 18, respectively. Model simulations also showed a likewise decline in yields for both these years: by an average (± s.e.) of 8.6 ± 0.3% in year 12 and 18.2 ± 0.3% in year 18 (Fig. 11). Yield simulations for mature palms (years 10 to 19) remained rather stable, where the

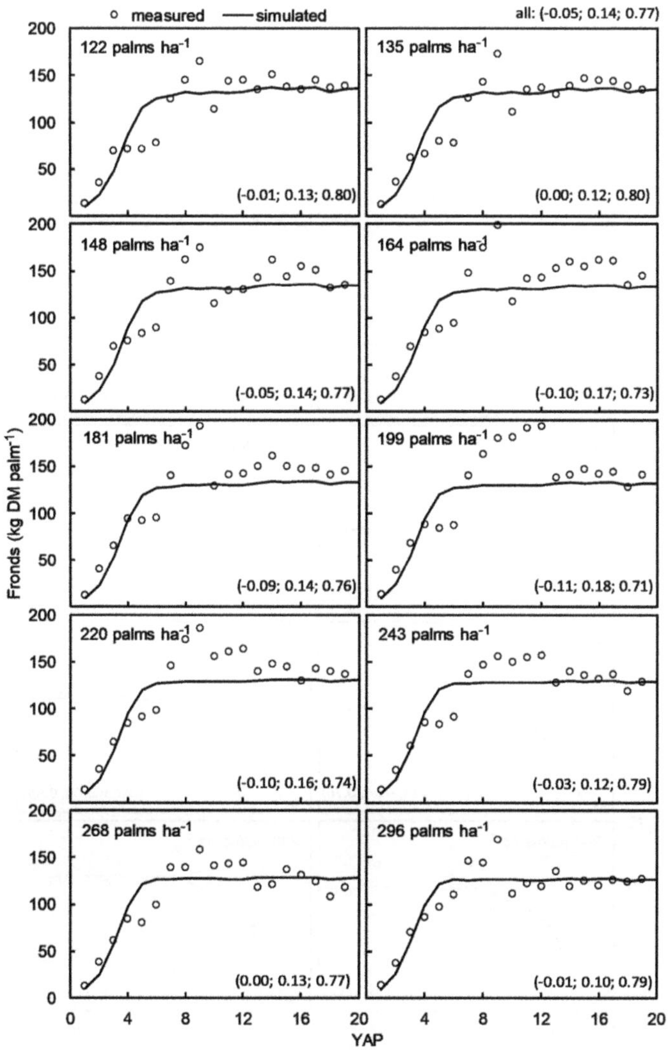

Figure 15 Comparisons between model simulations and observations for fronds (pinnae and rachis) biomass. Values in brackets denote NMBE, NMAE and d_r, where NMBE is the normalized mean bias error, NMAE is the normalized mean absolute error and d_r is the revised index of agreement. YAP is year after planting.

goodness-of-fit indexes for mature oil palm yield simulations were: 0.03 for NMBE, 0.17 for NMAE and 0.77 for d_r, only a slight decrease compared to yield simulations for all palm ages (year 2 to 19).

The large variations in the measured aboveground VDM for all planting densities especially year 5 onwards (Fig. 12) are unlikely to be real, and they suggest large measurement errors in the field. Nevertheless, model simulations showed a much more

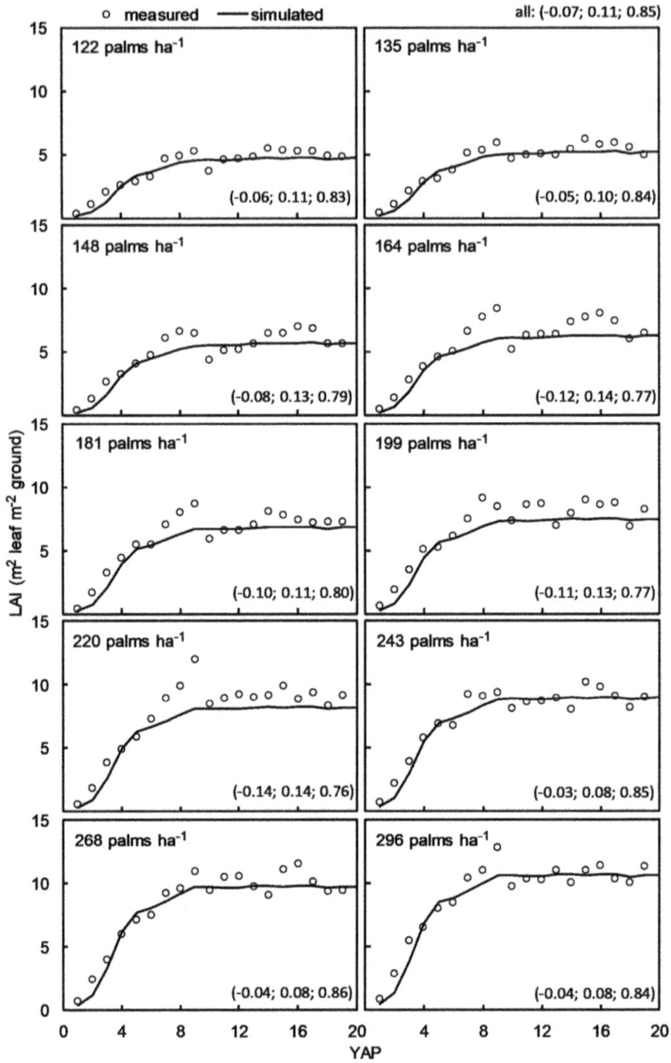

Figure 16 Comparisons between model simulations and observations for total leaf area index (LAI). Values in brackets denote NMBE, NMAE and d_r, where NMBE is the normalized mean bias error, NMAE is the normalized mean absolute error and d_r is the revised index of agreement. YAP stands for the year after planting.

steady temporal trend, and the *mean* of measured VDM from years 5 to 19 agreed closely with the model simulations for the same period.

Simulations also revealed that at Merlimau, plant water deficit increased with increasing planting density. In PD122, the mean annual plant water deficit was 218 mm but in PD296, the deficit was 350 mm. Mean annual plant water deficit increased by average of 7 mm for every 10 palms ha^{-1} increase in planting density.

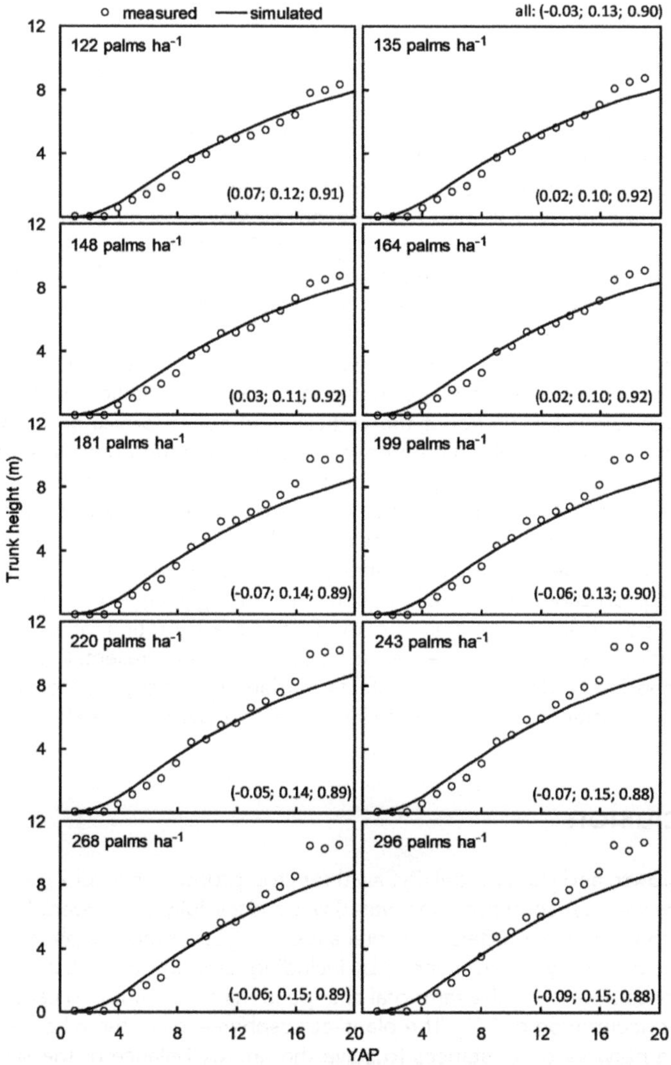

Figure 17 Comparisons between model simulations and observations for trunk height. Values in brackets denote NMBE, NMAE and d_r, where NMBE is the normalized mean bias error, NMAE is the normalized mean absolute error and d_r is the revised index of agreement. YAP stands for the year after planting.

It is well established that crops in higher planting densities will produce lower yields and biomass on a per plant basis than in lower planting densities. Specifically for oil palm, PySawit revealed that mean gross photosynthesis in PD122 was 878 kg CH_2O palm^{-1}, of which nearly two-thirds (65%) of the assimilates were used for total (maintenance and growth) respiration. The balance of the assimilates (35% or about 303 kg CH_2O palm^{-1}) was then available for flower and bunch development. But the mean gross photosynthesis rate for the highest planting density of PD296 was at a lower level of 624 kg CH_2O palm^{-1} (29% lower than in PD122), but total respiration consumed 83% of the assimilates, leaving only 17% (105 kg CH_2O palm^{-1}, which is about a third less than that used in PD122) for flower and bunch development.

There are several reasons for the lower gross photosynthesis at higher than lower planting densities, one of which is the increase in the proportion of shaded to sunlit canopies at higher planting densities compared with that at the lower planting densities. Increasing the proportion of shaded to sunlit canopies by 30%, for instance, lowers gross photosynthesis by nearly the same percentage (see Eq. 11). Increased leaf area at higher planting densities would also increase the wind speed extinction coefficient, thus reducing air flow within and below the canopies. In turn, this results in lower gas exchanges and lower photosynthetic rates.

Although many studies have been carried out on oil palm, there remain considerable gaps in knowledge, one of which relates to the microclimate conditions under the oil palm canopies. Much less is known about the vertical wind speed profile and how soil water content and plant water uptake differ between oil palm planting densities. The accurate method of scaling up leaf to canopy conductance also needs to be studied in greater detail. Knowledge of these properties would allow oil palm models to better portray the daily variability of microclimate conditions under the oil palm canopies and to produce yield predictions that better match the variability or fluctuations in observed yields. That PySawit's yield prediction errors would become increasingly large for increasingly higher planting densities is indicative that the model's characterization of the microclimate conditions particularly under very dense canopies is still not representative enough. More work is also needed to determine the effects of planting density on the dry matter and nutrient content partitioning between the individual tree parts of oil palm.

10 Conclusion

An oil palm growth and yield model (PySawit) for crop production level 2 (where growth is only limited by weather conditions and water) was successfully developed. PySawit mainly differs from other oil palm models in several aspects. PySawit models photosynthesis in a more rigorous, physically based manner, by including, among others, actual oil palm leaf measurements to determine the temporal change in maximum Rubisco capacity rate and intercellular concentration of CO_2. The plant–atmosphere–soil of the oil palm system was modelled as a network of resistances to solve the energy balance of the system. Actual measurements to determine the relationship between oil palm leaf stomatal conductance with PAR and VPD were additionally used. This allowed us to determine the simultaneous loss of water via oil palm transpiration and soil evaporation.

An independent data set from Merlimau oil palm trials (with various planting densities) was used to evaluate PySawit's accuracy. Model predictions generally agreed with observed values for several growth and yield parameters, and model accuracy was stable across all planting densities. Model predictions were especially good (small discrepancies between predictions and observations and very little to none model bias) for TDM, LAI and trunk height parameters. Yield decreases due to two El Niño events (which caused dry periods) were also reflected well in the model simulations. However, discrepancies between yield predictions and observations increased with increasing planting density. This is possibly because the microclimate conditions particularly under very dense canopies is not well understood and insufficiently characterized by the model. More work is also needed to determine the effects of planting density on soil water content, plant water use and dry matter and nutrient content partitioning between the individual tree parts. Knowledge of these properties would further improve the model's accuracy and extend its range of applications.

PySawit was implemented as a computer program using Python computer language. The source code for PySawit can be downloaded from www.christopherteh.com/pysawit.

11 Where to look for further information

The following is a list of several possible directions for future research: 1) to measure the daily soil water content, soil evaporation and oil palm transpiration under different planting densities, 2) to determine the vertical wind speed profile above and below oil palm canopies and 3) to model the uptake of nutrients by oil palm and their effect on the tree growth and yield.

Universiti Putra Malaysia (UPM), one of 20 public universities in Malaysia, has 16 faculties, one of which is the oldest faculty, the Faculty of Agriculture. This faculty comprises seven departments: Crop Science, Plant Protection, Land Management, Animal Science, Agribusiness and Bioresource Economics, Agriculture Technology and Aquaculture. These departments were established based on the concept of unification or integration of the key elements in the agricultural system: plants, animals, soil, technology and management.

Sime Darby Plantation has been involved in agriculture research and development since the early 1900s. Areas of research include best management practices, zero burning replanting techniques, oil palm nutrition, water conservation and irrigation, Integrated Pest Management, and the conversion of Empty Fruit Bunches and Palm Oil Mill Effluent into compost. Sime Darby Plantation is also involved in the sequencing, assembling and annotating the oil palm genome.

12 Acknowledgement

The authors would like to thank Sime Darby Research Sdn. Bhd. for permission to use the oil palm trial and weather data for this modelling study.

13 List of main symbols

Symbol	Description	Units
γ	psychometric constant (0.658)	mbar K^{-1}
Γ^*	CO_2 compensation point	μmol CO_2 mol^{-1} air
Δ	slope of the saturated vapour pressure curve	mbar K^{-1}
θ	solar inclination (from vertical)	radians

θ	volumetric soil water content	$m^3\ m^{-3}$
$\theta_{0/33/1500}$	volumetric soil water content at saturation (0), field capacity (33), or permanent wilting point (1500)	$m^3\ m^{-3}$
θ_{root}	volumetric soil water content in the root zone	$m^3\ m^{-3}$
θ_{cr}	critical volumetric soil water content	$m^3\ m^{-3}$
Θ_i	soil water content	m
λ	pore-size distribution index	–
λET	total latent heat flux density	$W\ m^{-2}$
$\lambda ET_{c/s}$	latent heat flux density of crop (c) or soil (s)	$W\ m^{-2}$
Λ_{canopy}	instantaneous gross canopy assimilation	$\mu mol\ CO_2\ m^{-2}$ ground s^{-1}
$\Lambda_{canopy,d}$	daily gross canopy assimilation	kg CH_2O $palm^{-1}$ day^{-1}
$\Lambda_{sl/sh}$	gross leaf assimilation rate of CO_2 for sunlit (sl) or shaded (sh) leaves	$\mu mol\ CO_2\ m^{-2}$ leaf s^{-1}
$\rho_{df/dr}$	reflection coefficient for diffuse (df) or direct photosynthetically active radiation (PAR)	–
ρc_p	volumetric heat capacity of air (1221.09)	$J\ m^{-3}\ K^{-1}$
τ	CO_2 / O_2 specificity factor	$\mu mol\ O_2\ \mu mol^{-1}\ CO_2$
τ	atmospheric transmittance (sky clearness)	–
τ	soil tortuosity	–
τ_b	canopy gap fraction	–
ϕ_p	soil porosity	$m^3\ m^{-3}$
ω, ω_0	canopy clustering coefficient at current sun position and at when the sun is at zenith, respectively	–
age	tree age	days
A	total energy supply to crop and soil	$W\ m^{-2}$
$A_{c/s}$	energy available to the crop (c) or soil (s)	$W\ m^{-2}$
$C_{a/i}$	CO_2 concentration in ambient air (a) and within stomata (i)	$\mu mol\ CO_2\ mol^{-1}$ air
CVF, CVF_2	conversion from CH_2O to dry (DM) weight	kg DM kg^{-1} CH_2O
d	zero plane displacement	M
d_{root}	rooting depth	m
dg_{root}	rooting depth growth rate	m day^{-1}
D	air vapour pressure deficit (VPD)	mbar
$D_{0/leaf}$	air VPD at mean canopy flow (0) or leaf VPD (leaf)	mbar
DL	day length	hour

$DM_{(part)}$	fraction of dry matter to a given plant part	–
e_a	air vapour pressure	mbar
e_m	quantum efficiency or yield (0.051)	μmol CO_2 μmol^{-1} photons
$e_s[T]$	saturated vapour pressure at temperature T	mbar
$ET_{c/s}$	potential transpiration (c) or evaporation (s)	m day^{-1}
$f_{D/PAR/water}$	reduction of stomatal conductance due to VPD (D), low PAR, or water stress (water)	–
gst, gst_{max}	current and maximum stomatal conductance, respectively	m s^{-1}
G	soil (ground) heat flux density	W m^{-2}
$G_{death,(part)}$	death rate of a given plant part	kg DM palm^{-1} day^{-1}
G_{gen}	total assimilates available for generative organs growth	kg CH_2O palm^{-1} day^{-1}
$G_{gen,\ (part)}$	growth rate of a given generative organ	kg DM palm^{-1} day^{-1}
G_{growth}	total assimilates available for vegetative growth respiration	kg CH_2O palm^{-1} day^{-1}
$G_{growth,\ (part)}$	growth rate of a given plant part	kg DM day^{-1}
h	tree height (trunk + canopy)	M
h_{canopy}, h_{trunk}	canopy and trunk height, respectively	m
h'_{trunk}	rate of growth for trunk height	m day^{-1}
H	total head (matric and gravity)	M
H	total sensible heat density	W m^{-2}
$H_{c/s}$	sensible heat flux density of crop (c) or soil (s)	W m^{-2}
$I_{df/dr}$	diffuse (df) or direct (dr) solar irradiance	W m^{-2}
I_{et}, I_t	extraterrestrial and total solar irradiance, respectively	W m^{-2}
k	von Karman's constant (0.4)	–
$k_{df/dr}$	canopy extinction coefficient for diffuse (df) or direct (dr) solar radiation	–
$k_{e/w}$	extinction coefficient for eddy diffusivity (e) or wind speed (w)	–
K_θ, $\overline{K_\theta}$	unsaturated and mean hydraulic conductivity, respectively	m day^{-1}
$K_{c/o}$	Michaelis–Menten constant for CO_2 (c) or O_2 (o)	μmol CO_2/O_2 mol^{-1} air
L, L_{eff}	total and effective leaf area index, respectively	m^2 leaf m^{-2} ground
$L_{sh/sl}$	shaded (sh) or sunlit (sl) leaf area index	m^2 leaf m^{-2} ground

$L_{max,PD}$	maximum leaf area index for a given planting density	m^2 leaf m^2 ground
m	optical mass number	–
$M_{(part)}$	maintenance respiration for a given plant part	kg CH_2O palm^{-1} day^{-1}
$N_{(part)}$	fraction by weight of nitrogen content in given plant part	–
O_a	air ambient concentration of O_2	μmol O_2 mol^{-1} air
PD	planting density	palms ha^{-1}
P_g, P_{net}	gross and net daily rainfall, respectively	M
q	water flux	m day^{-1}
$\widehat{q}_i$	net water flux	m day^{-1}
$Q_{df/dr}$	diffuse (df) or direct (dr) PAR component	μmol m^{-2} leaf s^{-1}
$Q_{sh/sl}$	total PAR absorbed by shaded (sh) or sunlit (sl) leaves	μmol m^{-2} leaf s^{-1}
Q_{10}	relative change for every 10 °C change	–
$\overline{Q_{p,df}}$	mean diffuse PAR component (within canopies)	μmol m^{-2} ground s^{-1}
$Q_{p,dr}$	unintercepted direct PAR component (with scattering component) (within canopies)	μmol m^{-2} leaf s^{-1}
$Q_{p,dr,\alpha}$	scattered PAR component only (within canopies)	μmol m^{-2} ground s^{-1}
$Q_{p,dr,dr}$	unintercepted direct PAR component (without scattering component) (within canopies)	μmol m^{-2} leaf s^{-1}
	aerodynamic resistance between the mean canopy flow and reference height	s m^{-1}
	bulk boundary layer resistance	s m^{-1}
	canopy resistance	s m^{-1}
	aerodynamic resistance between the soil surface and mean canopy flow	s m^{-1}
	soil surface resistance	s m^{-1}
R_n, R_{nL}	net solar radiation and net long-wave radiation, respectively	W m^{-2}
R_{Ds}, R_{Dc}	reduction factor for evaporation and transpiration, respectively	–
RH	relative humidity	%
SLA	specific leaf area	m^2 leaf kg^{-1} DM
t_h	local solar hour	Hour
t_{sr}, t_{ss}	local solar hour for sunrise and sunset, respectively	Hour
T_a, T_{dew}, T_f	air, dew point and foliage temperature, respectively	°C
$T_{min/max}$	minimum (min) or maximum (max) air temperature	°C

u_*	friction velocity	m s^{-1}
u, u_d, u_h	wind speed, mean daily wind speed and wind speed at canopy height, respectively	m s^{-1}
$u_{min/max}$	minimum (min) or maximum (max) daily wind speed	m s^{-1}
$V_{c/q/s}$	CO_2 assimilation limited by Rubisco (c), light (q), or sink (s)	μmol CO_2 m^{-2} leaf s^{-1}
V_{cmax}	Rubisco capacity rate	μmol CO_2 m^{-2} leaf s^{-1}
VDM, $VDM_{max,PD}$	current and maximum vegetative dry matter requirement for a given planting density, respectively	kg DM $palm^{-1}$ yr^{-1}
w	mean pinnae width	m
$W_{(part)}$	dry weight of a given plant part	kg DM $palm^{-1}$
$X_{(part)}$	fraction by weight of mineral content in a given plant part	–
$z_{0/s0}$	crop (0) or soil surface (s0) roughness length	m
z_r	reference height	m

14 References

Awal, M. A. (2008). Assessment of gap fraction by hemispherical photograph in oil palm plantation. *Suranaree Journal of Science and Technology*, 15, 233–41.

Bentley, A. (2007). *Interception loss in Sedenak oil palm plantation*. B.Sc. final year project report. Skudai, Malaysia: Universiti Teknologi Malaysia.

Bernacchi, C. J., Portis, A. R., Nakano, H., von Caemmerer, S. and Long, S. P. (2002). Temperature response of mesophyll conductance. Implication for the determination of Rubisco enzyme kinetics and for limitations to photosynthesis *in vivo*. *Plant Physiology*, 130, 1992–8.

Bernacchi, C. J., Singsaas, E. L., Pimentel, C., Portis Jr., A. R. and Long, S. P. (2001). Improved temperature response functions for models of Rubisco-limited photosynthesis. *Plant, Cell and Environment*, 24, 253–9.

Bittelli, M., Campbell, G. S. and Tomeu, F. (2015). *Soil Physics with Python. Transport in the Soil-Plant-Atmosphere System*. Oxford, UK: Oxford University Press.

Breure, C. J. (2003). The search for yield in oil palm: Basic principles. In T. H. Fairhurst and R. Härdter (Eds.), *Oil Palm: Management for Large and Sustainable Yields* (pp. 59–98). Singapore: Potash & Phosphate Institute of Canada (ESEAP).

Breure, C. J. and Powell, M. S. (1988). The one-shot method of establishing growth parameters in oil palm. In H. A. Halim, P. S. Chew and B. J. Wood (Eds.), *Proceedings of the 1987 International Oil Palm Conference. Progress and prospects* (pp. 203–9). Kuala Lumpur, Malaysia: Palm Oil Research Institute of Malaysia.

Brutsaert, W. (1982). *Evaporation into the Atmosphere. Theory, History and Applications*. Dordrecht, The Netherlands: D. Reidel Publishing Company.

Campbell, G. S. and Norman, J. M. (1998). *An Introduction to Environmental Biophysics* (2nd ed.). New York, NY: Springer-Verlag.

Chong, S. Y. (2012). *Development of Simple Equations to Estimate the Net Rainfall Under Closed Tree Canopies*. M. Agr. Sc. dissertation. Serdang, Malaysia: Universiti Putra Malaysia.

Choudhury, B. J. and Monteith, J. L. (1988). A four-layer model for the heat budget of homogeneous land surfaces. *Quarterly Journal of the Royal Meteorological Society*, 114, 373–98.

Collatz, G. T., Ball, J. T., Grivet, C. and Berry, J. A. (1991). Physiological and environmental–regulation of stomatal conductance, photosynthesis and transpiration: A model that includes a laminar boundary layer. *Agricultural and Forest Meteorology*, 54, 107–36.

Combres, J. -C., Pallas, B., Rouan, L., Mialet-Serra, I., Caliman, J. -P., Braconnier, S., Soulié, J. -C. and Dingkuhn, M. (2012). Simulation of inflorescence dynamics in oil palm and estimation of environment-sensitive phenological phases: A model based analysis. *Functional Plant Biology*, 40, 263–79.

Corley, R. H. V. and Tinker, P. B. (2016). *The Oil Palm* (5th ed.). Chichester, UK: Wiley-Blackwell.

Corley, R. H. V., Hardon, J. J. and Tan, G. Y. (1971). Analysis of growth of the oil palm (*Elaeis guineensis* Jacq.) I. Estimation of growth parameters and application in breeding. *Euphytica*, 20, 307–15.

Corley, R. H. V., Ng, M. and Donough, C. R. (1995). Effects of defoliation on sex differentiation in oil palm clones. *Experimental Agriculture*, 31, 177–89.

Damih, A. 1995. *Keberkesanan pemintasan air hujan oleh pokok kelapa sawit di dalam mengurangkan air larian permukaan* [Effectiveness of rainfall interception by oil palm trees in reducing surface runoff]. B.Sc. final year project report. Skudai, Malaysia: Universiti Teknologi Malaysia.

de Wit, C. T. and Penning de Vries, F. W. T. (1982). L'Analyse des systemes de production primaire [Analysis of primary production systems]. In F. W. T. Penning de Vries and M. A. Djiteye (Eds.), *Une etude des sols, des vegetations et de l'exploitation de cette ressource naturelle* [A study of the soils, vegetations and exploitation of this natural resource] (pp. 20–27). Agricultural Research Reports 918. Wageningen, The Netherlands: Pudoc.

Dimas, F. A., Gilani, S. I. H. and Aris, M. S. (2011). Hourly solar radiation estimation using ambient temperature and relative humidity data. *International Journal of Environmental Science and Development*, 2, 188–93.

Dufrêne, E. and Saugier, B. (1993). Gas exchange of oil palm in relation to light, vapour pressure deficit, temperature and leaf age. *Functional Ecology*, 7, 97–104.

Dufrêne, E., Ochs, R. and Saugier, B. (1990). Oil palm photosynthesis and productivity linked to climatic factors. *Oléagineux*, 45, 345–55.

Ephrath, J. E., Goudriaan, J. and Marani, A. (1996). Modelling diurnal patterns of air temperature, radiation, wind speed and relative humidity by equations from daily characteristics. *Agricultural Systems*, 51, 377–93.

Fan, Y., Roupsard, O., Bernoux, M., Le Maire, G., Panferov, O., Kotowska, M. M. and Knohl, A. (2015). A sub-canopy structure for simulating oil palm in the Community Land Model (CLM-Palm): phenology, allocation and yield. *Geoscientific Model Development*, 8, 3785–3800.

Farahani, H. J. and Ahuja, L. R. (1996). Evapotranspiration modeling of partial canopy/residue-covered fields. *Transactions of the ASAE*, 39, 2051–64.

Farquhar, G. D., von Caemmerer, S. and Berry, J. A. (1980). A biochemical model of photosynthetic CO_2 assimilation in leaves of C3 species. *Planta*, 149, 78–90.

Foong, F. S. (1999). Impact of moisture on potential evapotranspiration, growth and yield of oil palm. In D. Arifin, K. W. Chan and S. R. S. A. Sharifah (Eds.), *Proceedings of the 1999 PORIM International Palm Oil Congress (Agriculture)* (pp. 265–87). Bangi, Malaysia: Palm Oil Research Institute of Malaysia.

Gerritsma, W. (1988). *Simulation of oil palm yield*. Report. Department of Theoretical Production Ecology. Wageningen, The Netherlands: Wageningen Agricultural University.

Goudriaan, J. (1977). *Crop micrometeorology: A simulation study*. Simulation Monograph. Pudoc, Wageningen: Centre for Agricultural Publishing and Documentation.

Goudriaan, J. and van Laar, H. H. (1994). *Modeling Potential Crop Growth Processes. A Textbook with Exercise*. Current issues in production ecology. Dordrecht, The Netherlands: Kluwer Academic Publishers.

Hansen, F. V. (1993). *Surface Roughness Lengths*. ARL Technical Report. White Sands Missile Range, NM 88002-5501: U.S. Army.

Hardon J. J., Williams C. N. and Watson I. (1969). Leaf area and yield in the oil palm in Malaya. *Experimental Agriculture*, 5, 25–32.

Hasanuzzaman, M., Nahar, K., Alam, M. M., Roychowdhury, R. and Fujita, M. (2013). Extreme temperature responses, oxidative stress and antioxidant defense in plants. In K. Vahdat and C. Leslie (Eds.), *Abiotic Stress – Plant Responses and Applications in Agriculture* (pp. 169–205). Vienna, Austria: InTech.

Henry, P. (1955). Leaf growth morphology in Elaeis. *Revue Generale de Botanique*, 32, 66–7.

Henson, I. E. (1989). *Modelling Gas Exchange, Yield and Conversion Efficiency*. Workshop on Productivity of Oil Palm. Bangi, Malaysia: Palm Oil Research Institute of Malaysia.

Henson, I. E. (1995). Carbon assimilation, water use and energy balance of an oil palm plantation assessed using micrometeorological techniques. In S. Jalani, D. Ariffin, N. Rajanaidu, D. Mohd Tayeb, K. Paranjothy, W. Mohd Basri, I. E. Henson and C. K. Chan (Eds.), *Proceedings of the 1993 PORIM International Palm Oil Congress – Update and Vision (Agriculture)* (pp. 137–58). Bangi, Malaysia: Palm Oil Research Institute of Malaysia.

Henson, I. E. (2000). Modelling the effects of 'haze' on oil palm productivity and yield. *Journal of Oil Palm Research*, 12, 123–34.

Henson, I. E. (2006). OPFLSIM3 – An improved oil palm seasonal flowering and yield simulation model. *Oil Palm Bulletin*, 53, 1–24.

Henson, I. E. (2009). Modelling dry matter production, partitioning and yield of oil palm. OPRODSIM: A mechanistic simulation model for teaching and research. Technical manual and users' guide. Kuala Lumpur, Malaysia: Malaysian Palm Oil Board.

Henson, I. E. and Chai, S. (1997). Analysis of oil palm productivity. II. Biomass, distribution, productivity and turnover of the root system. *Elaeis*, 9, 78–92.

Henson, I. E. and Mohd Tayeb, D. (2003). Physiological analysis of an oil palm density trial on a peat soil. *Journal of Oil Palm Research*, 15, 1–27.

Hillel, D. (1977). *Computer Simulation of Soil-water Dynamics: A Compendium of Recent Work*. Ottawa, Canada: International Development Research Centre.

Hoffmann, M. P., Castaneda Vera, A., van Wijk, M. T., Giller, K. E., Oberthür, T., Donough, C. and Whitbread, A. M. (2014). Simulating potential growth and yield of oil palm (*Elaeis guineensis*) with PALMSIM: Model description, evaluation and application. *Agricultural Systems*, 131, 1–10.

Huth, N. I., Banabas, M., Nelson, P. N. and Webb, M. (2014). Development of an oil palm cropping systems model: lessons learned and future directions. *Environmental Modelling & Software*, 62, 411–19.

Jacquemard, J. C. (1979). Contribution to the study of the height growth of the stems of Elaeis guineensis Jacq. Study of the L2T x D10D cross. *Oléagineux*, 34, 492–7.

Jacquemard, J. C. (1998). *Oil palm*. London, UK: Macmillan Education Ltd.

Jourdan, C. and Rey, H. (1997). Architecture and development of the oil-palm (*Elaeis guineensis* Jacq.) root system. *Plant and Soil*, 189, 33–48.

Kallarackal, J., Jeyakumar, P. and George, S. J. (2004). Water use of irrigated oil palm at three different arid locations in Peninsular India. *Journal of Oil Palm Research*, 16, 45–53.

Kropff, M. J. (1993). Mechanisms of competition for light. In M. J. Kropff and H. H. van Laar (Eds.), *Modelling crop weed interactions* (pp. 33–61). Wallingford, UK: CAB International (in association with International Rice Research Institute).

Kustas, W. P. and Daughtry, C. S. T. (1990). Estimation of the soil heat flux/net radiation ratio from spectral data. *Agriculture and Forest Meteorology*, 49, 205–23.

Kustas, W. P. and Norman, J. M. (1999). Evaluation of soil and vegetation heat flux predictions using a simple two-source model with radiometric temperatures for partial canopy cover. *Agricultural and Forest Meteorology*, 94, 13–29.

Kwan, B. K. W. (1994). *The Effect of Planting Density on the First Fifteen Years of Growth and Yield of Oil Palm in Sabah*. Technical Bulletin No. 11. Kota Kinabalu, Malaysia: Department of Agriculture, Sabah.

Liu, B. Y. and Jordan, R. C. (1960). The interrelationship and characteristic distribution of direct, diffuse, and total solar radiation. *Solar Energy*, 4, 1–19.

Long, S. P. and Bernacchi, C. J. (2003). Gas exchange measurements, what can they tell us about the underlying limitations to photosynthesis? Procedures and sources of error. *Journal of Experimental Botany*, 54, 2393–2401.

Lubis, M. E. S. (2016). *Water Dynamics and Groundwater Quality Assessment in an Oil Palm Ecosystem.* Ph.D. dissertation. Serdang, Malaysia: Universiti Putra Malaysia.

Massman, W. J. (1987). A comparative study of some mathematical models of the mean wind structure and aerodynamic drag of plant canopies. *Boundary-Layer Meteorology*, 40, 179–97.

Massman, W. J. (1997). An analytical one-dimensional model of momentum transfer by vegetation of arbitrary structure. *Boundary-Layer Meteorology*, 83, 407–21.

Miyazaki, T. (2005). *Water Flow in Soils.* Boca Raton, FL: CRC Press.

Mohd. Desa, M. N. and Rakhecha, P. R. (2006). Deriving the highest persisting monthly 24-hour dew points in Malaysia for the estimation of PMP. In S. Demuth, A. Gustard, E. Planos, F. Scatena and E. Servat (Eds.), *Climate Variability and Change-Hydrological Impacts. Proceedings of the Fifth FRIEND World Conference held at Havana, Cuba, November 2006. IAHS Publ. 308* (pp. 287–93). Wallingford, UK: International Association of Hydrological Sciences (IAHS).

Monteith, J. L. (1965). Evaporation and the environment. *Symposium of the Society of Experimental Biology*, 19, 205–34.

Monteith, J. L. (1973). *Principles of Environmental Physics.* London, UK: Edward Arnold.

Nelson, P. N., Banabas, M., Scotter, D.R. and Webb, M. J. (2006). Using soil water depletion to measure spatial distribution of root activity in oil palm (*Elaeis guineensis* Jacq.) plantations. *Plant Soil*, 286, 109–21.

Nikolov, N. and Zeller, K .F. (2003). Modeling coupled interactions of carbon, water, and ozone exchange between terrestrial ecosystems and the atmosphere. I: Model description. *Environmental Pollution*, 124, 231–46.

Paterson, R. R. M., Kumar, L., Taylor, S. and Lima, N. (2015). Future climate effects on suitability for growth of oil palms in Malaysia and Indonesia. *Scientific Reports*, 5, 1–11.

Rao, V., Rajanaidu, N., Kushairi, A. and Jalani, S. (1992). Density effect in the oil palm. In *Proceedings in the ISOP International Workshop of Yield Potential in the Oil Palm* (pp. 71–9). Bangi, Malaysia: International Society for Oil Palm Breeders and PORIM.

Rey, H., Quencez, P., Dufrêne, E. and Dubos, B. (1998). Oil palm water profiles and water supplies in Côte d'Ivoire. *Plantations, Recherche, Développement*, 5, 47–57.

Saxton, K. E. and Rawls, W. J. (2006). Soil water characteristic estimates by texture and organic matter for hydrologic solutions. *Soil Science Society of America Journal*, 70, 1569–78.

Sharkey, T. D. (2015). What gas exchange data can tell us. *Plant, Cell & Environment*, 39, 1161–3.

Sharkey, T. D., Bernacchi, C. J., Farquhar, G. D. and Singsaas, E. L. (2007). Fitting photosynthetic carbon dioxide response curves for C3 leaves. *Plant, Cell and Environment*, 30, 1035–40.

Shen, L. and Chen, Z. (2007). Critical review of the impact of tortuosity on diffusion. *Chemical Engineering Science*, 62, 3748–55.

Shuttleworth, W. J. and Wallace J. S. (1985). Evaporation from sparse crops - an energy combination theory. *Quarterly Journal of the Royal Meteorological Society*, 111, 839–55.

Skillman, J. B. (2008). Quantum yield variation across the three pathways of photosynthesis: Not yet out of the dark. *Journal of Experimental Botany*, 59, 1647–61.

Sopian, K., Hj. Othman, M. Y. and Wirsat, A. (1995). Data bank. The wind energy potential of Malaysia. *Renewable Energy*, 6, 1005–16.

Su, Z., Schmugge, T., Kustas, W. and Massman, W. (2001). An evaluation of two models for estimation of the roughness height for heat transfer between the land surface and the atmosphere. *Journal of Applied Meteorology*, 40, 1933–51.

Szeicz, G. and Long, I. F. (1969). Surface resistance of crop canopies. *Water Resources Research*, 5, 622–33.

Tan, Y. P. and Ng, S. K. (1977). Spacing for oil palms on coastal clays in Peninsular Malaysia. In D. A. Earp and W. Newall (Eds.), *International Developments in Oil Palm. Proceedings of the*

Malaysian International Agricultural Oil Palm Conference (pp. 183–91). Kuala Lumpur, Malaysia: Incorporated Society of Planters.

Teh, C. B. S. (2006). *Introduction to Mathematical Modeling of Crop Growth: How the Equations are Derived and Assembled into a Computer Program.* Boca Raton, FL: Brown Walker Press.

Thornley, J. H. M. (1970). Respiration, growth and maintenance in plants. *Nature*, 227, 304–5.

van Keulen, H. and Seligman, N. G. (1987). *Simulation of Water use, Nitrogen and Growth of a Spring wheat Crop.* Simulation Monographs. Wageningen, The Netherlands: Pudoc.

van Kraalingen, D. W. G. (1985). *Simulation of Oil Palm Growth and Yield.* Ph.D. thesis. Wageningen, The Netherlands: Wageningen Agricultural University.

van Kraalingen, D. W. G., Breure, C. J. and Spitters, C. J. T. (1989). Simulation of oil palm growth and yield. *Agricultural and Forest Meteorology*, 46, 227–44.

van Noordwijk, M., Lusiana, B., Khasanah, N. and Mulia, R. (2011). WaNuLCAS 4.0: Background on a model of water, nutrient, and light capture in agroforestry systems. Bogor, Indonesia: World Agroforestry Centre (ICRAF).

Wilkerson, G. G., Jones, J. W., Boote, K. J., Ingram, K. T. and Mishoe, J. W. (1983). Modeling soybean growth for crop management. *Transactions of the ASAE*, 26, 63–73.

Willmott C. J., Robeson S. M. and Matsuura, K. (2012). A refined index of model performance. *International Journal of Climatology*, 32, 2088–94.

Yin, X. and van Laar, H. H. (2005). *Crop Systems Dynamics. An Ecophysiological Simulation Model for Genotype-by-environment Interactions.* Wageningen, The Netherlands: Wageningen Academic Publishers.

Yu, S., Eder, B., Dennis, R., Chu, S. -H. and Schwartz, S. E. (2006). New unbiased symmetric metrics for evaluation of air quality models. *Atmospheric Science Letters*, 7, 26–34.

Zulkifli, Y., Geoffery, J., Saw, A. L. and Norul, S. T. (2006). Preliminary study on throughfall spatial variability and stemflow characteristics under oil palm catchment. In *Proceedings of the National Water Conference (in CD)*. Kuala Lumpur, Malaysia: Malaysian Hydrological Society.

Chapter 11

Improving soil health and crop nutrition in oil palm cultivation

Jean-Pierre Caliman, Suhardi and Pujianto, Smart Research Institute, Indonesia

1 Introduction

Thousands of years ago, before the development of agriculture, soil was considered merely a physical support for crops. With the development of modern agriculture, a wider role for soil has emerged. A greater interest in soil has been associated with the discovery of the role nutrients and water play in plant growth and development, and subsequently their intensive use as part of the development of intensive agriculture in the middle of the 20th century, required to meet exponential population growth. This transformation, later called 'the Green Revolution' in agriculture after the Second World War, was based on extensive use of ploughing of soil, mineral fertilizers and irrigation, combined with new high-yielding cultivars. Concerns to reduce a perceived over-reliance on fertilizers lead to a new interest in soil health.

Over the following decades, farmers, those providing extension services and researchers have paid more attention to the physical, chemical and biological fertility of soil. Most attention has been paid to the physical and chemical components of soil fertility, probably because they are easiest to study. The majority of journal publications and guidelines to good agricultural practice have associated crop performance with the chemical and physical components of soil fertility.

The importance of soil characteristics was first recognized with attempts to classify 'life zones' in the early years of the nineteenth century, when Russian pedologists found evidence for a correlation between climate and soil (Holdridge, 1966). It was increasingly realized that soil was an important component of the ecosystem (Holdridge, 1947; Allee,

http://dx.doi.org/10.19103/AS.2017.0018.11

1949; Kendeigh, 1961; Wiltaker, 1962; Holdridge, 1966; Wiltaker, 1970, 1975). New emphasis has been placed on the importance of soils in the assessment of ecosystem services undertaken by the Millennium Ecosystem Assessment institution (2005) (www.millenniumassessment.org) and the Economics of Ecosystems and Biodiversity Initiative led by the United Nations (www.teebweb.org). These have highlighted the importance of soils since they play a significant part in all four categories of services that ecosystems provide to society. These services are:

- *Provisioning services:* Soils provide the medium for growing food as well as raw materials for biofuels. They also ensure water purification, control the hydrological cycle and play a key role in the storage and availability of water and nutrients to plants.
- *Regulating services:* Soils contribute to disease and pest control, as well as carbon sequestration and storage. They are home to microorganisms that contribute to plant health and regulate nutrient availability and to the decomposition and mineralization of organic residues and waste. Soil fertility is recognized as a key factor for plant growth, while soil degradation, such as erosion, undermines ecosystem services and biomes (Roose et al., 2006; Roose, 2017).
- *Habitats and supporting services:* Soils support biodiversity and has been an integral part of habitats, supplying food and water to both plants and animals.
- *Cultural services:* Soils contribute to landscape diversity and thus activities such as tourism and recreation.

This new understanding of the role and importance of soil has led to the development of new soil management systems required to achieve sustainable agriculture. It is not surprising that the Food and Agriculture Organization of the United Nations (FAO) declared 2016 the year of soils, and 2015–24 the decade of soils.

The range of ecosystem services provided by or involving soils means the multiple functions that soils perform need to be considered, rather than just focusing on 'simple' fertility components. The interactions between the chemical, physical and biological properties of soils should not be seen as a 'simple' list of functions but as part of the ecosystem services soil supports. This has required researchers, agronomists and farmers to pay greater attention to monitoring soil condition and properties. It has also required them to develop new tools that take into account both spatial and temporal scales. These tools can then provide the basis for site-specific precision agricultural practices. This recent emphasis on ecosystem services has encouraged researchers to look at soils in a more comprehensive and holistic way, acknowledging the multiple functions of soils, their interconnections, interdependence and capacity for transformation (Bispo et al., 2016). As a result, the number of scientific papers on these topics has increased significantly in recent years (there have been almost 20 000 articles related to soil ecosystem services since 2010, according to Google Scholar). It is now common to find articles focusing on the environmental functions of soils, although this is often not linked back to crop growth or yield.

This change in attitudes to the role of soil is particularly true for research into a new crop such as oil palm (*Elaeis guineensis* Jacq.). While palm oil has been used in Africa for centuries, its cultivation only started to develop on a major scale during the twentieth century. The first large-scale plantations were developed in 1911 in Indonesia and then in Malaysia in 1917 (Chandran, 2017). Large national programmes aimed at developing commercial plantations started in the 1960s in Africa, then in South East Asia in the 1970s, first in Malaysia, then in Thailand, Indonesia and Papua New Guinea. These developments led to growing research into areas such as breeding, nutrient management and crop protection with the aim of achieving higher crop productivity in terms of oil yield per cultivated hectare.

This chapter discusses the current state of knowledge about soil and nutrient management in optimizing oil palm cultivation. It highlights the challenges that still need to be addressed in order to make the crop more sustainable in the long term. It considers emerging technologies such as sensors and drones and the tremendous possibilities brought up in molecular biology, supported by dramatic improvements in computing power and information processing. The first part of the chapter discusses soil functions as they relate to more sustainable oil palm cultivation. The second part focuses on nutrient management for both higher productivity and improved sustainability, through better mineral nutrition that takes account of environmental needs. As noted earlier, the chapter acknowledges the multiple functions of soils, their interconnections and interdependence.

2 Soil fertility

The first classification of soils suitable for oil palm cultivation was based on the physical and chemical characteristics that supported the growth of palms and led to the production of large amounts of fruit and oil (Olivin, 1968). In 1976, the FAO proposed a framework for land evaluation based largely on soil characteristics, building on systems developed independently by several countries in line with their respective programmes for agricultural development. The shared vision was to guarantee food production for an increasing world population. Ten out of 17 criteria in the framework related to soil; the remaining seven referred to climatic conditions.

This document formed the basis of the land evaluation system framework proposed by Poeloengan (1989) for oil palm development, which included seven criteria for soil characteristics (five related to the physical condition of soils and two to their chemical conditions) and two criteria related to climate (rainfall and temperature). The system established the criteria used to identify suitable land for the development of oil palm in Africa, as described by IRHO (Institut de Recherche pour les Huiles et Oléagineux) in the 1960s (Olivin, 1968; Nguyen et al., 1984a,b). Paramananthan (1982, 2000) used a similar framework with more detailed parameters in order to classify soils in Malaysia and developed it as a tool for management classification to be implemented by plantation managers. It is worth noting that, while the descriptions of land and soil quality attributes were focused on how they contributed to crop productivity, the supporting services provided by soils were implicitly recognized. Indeed, several authors have recognized the particular sensitivity of these functions and the risk of their degradation (Olivin, 1980; Webb et al., 2015).

2.1 Soil fertility: the chemical component

The chemical components of soil have been studied intensively, paralleling the development of oil palm cultivation in the 1960s through to the 1980s. Such interest is due to the important part played by soil nutrient management in maximizing the potential production of oil palm. It is also due to the fact that fertilizers account for the largest share of maintenance costs in plantations. Therefore, intensive research has been dedicated to the relationship between soil chemical fertility and oil palm productivity for most types of soils suitable to the crop. A large number of studies have been published, mostly in journals dedicated to the oil palm industry such as *Oléagineux*, *Planters*, *Palmas* or the *Journal of Oil Palm Research*, and through the proceedings of international oil palm conferences (PIPOC in Malaysia, Cenipalma in Columbia and IOPRI in Indonesia).

These studies detail both the relationship between soil chemical conditions (nutrients, acidity, etc.) and their management by fertilizers and manuring systems, as well as scientific research into their kinetics. As an example, potassium dynamics in soils in both Africa (Ivory Coast) and South East Asia (Indonesia) have been studied using techniques such as isotopic dilution measurements of exchangeable potassium (Ouvry, 1985; Fallavier et al., 1989). Research has focused not only on the nutrients in soil that are essential for palm oil production (i.e. nitrogen, phosphorus, potassium, magnesium) but also on other chemical components in soil, such as aluminium, that have a direct or indirect impact (Cristancho, 2009) and acidity (Delval, 1981; Caliman et al., 1987; Nelson et al., 2010). A simple classification of soil chemical fertility, used for direct evaluation, has been developed and updated by Fairhurst (Fairhurst, 1997, 2005).

2.2 The move to organic fertilization

The impact of organic manuring on the chemical characteristics of soil, mainly through recycling solid waste such as empty fruit bunches (EFB) and liquid palm oil mill effluents (POME), has been relatively well studied (Caliman et al., 2001; Chiew and Rahman, 2002; Carron et al., 2015a, 2016). The kinetics of soil chemical changes, such as exchangeable potassium (Fig. 1) and other nutrients and soil pH (Fig. 2), have inspired management practices that can fully integrate waste recycling into fertilizer management plans, including rates and frequencies of field applications. Information in terms of spatial distribution of impacts for both POME and EFB applications is also available (Carron et al., 2015b).

The positive impact on oil palm yield is also well documented. Repeated application of EFB as a substitute for mineral fertilizers resulted in FFB yields similar to or even higher than in fields where mineral fertilizers were the only source of nutriments (Fig. 3).

Similar results have also been found when EFB are applied after composting. An appropriate rate of compost application results in FFB yields that are comparable to a control treatment with mineral fertilizer (Fig. 4).

Several fields of research still require more attention in order to fully understand soil nutrient functions and dynamics in oil palm agronomics. These fields include the biogeochemical cycles of nutrients in soils, although interesting research has been conducted by teams at Papua New Guinea Oil Palm Research Association, Applied Agricultural Resources S. B. in Malaysia, SMART Research Institute in Indonesia and Cirad, France. Such studies are increasingly needed in order to develop statistical (and preferably dynamic) models aimed at quantifying nutrient cycles in the oil palm agronomic system (Banabas, 2007; Huth et al., 2014; Pardon et al., 2016a, b) and their impact on the environment (in terms of risks to water and atmospheric pollution) as well as on soil fertility. Such biogeochemical studies must also include the carbon cycle, which directly impacts the environment and soil fertility.

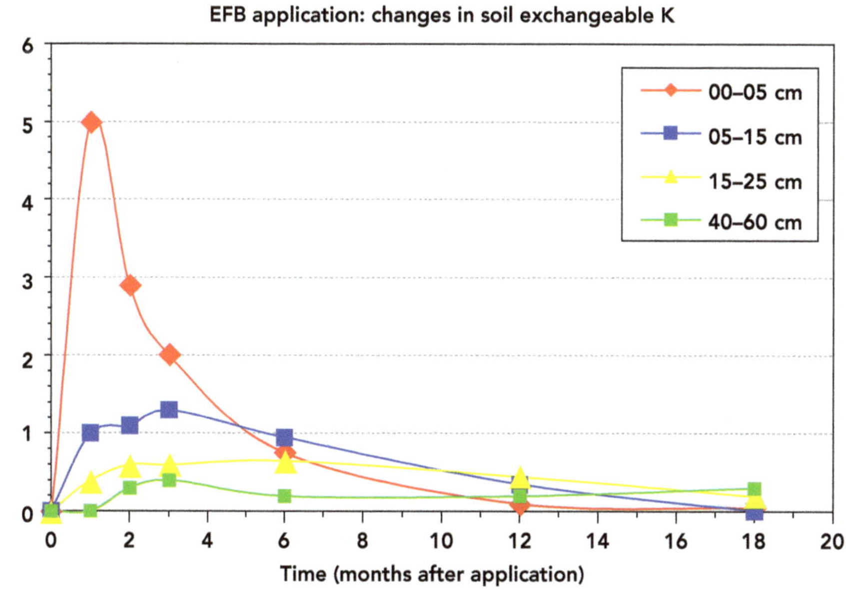

Figure 1 The impact of EFB application (60 t fresh matter/ha) on the variation of exchangeable potassium with time for several layers of soils. The legend shows the depths of the layers observed.

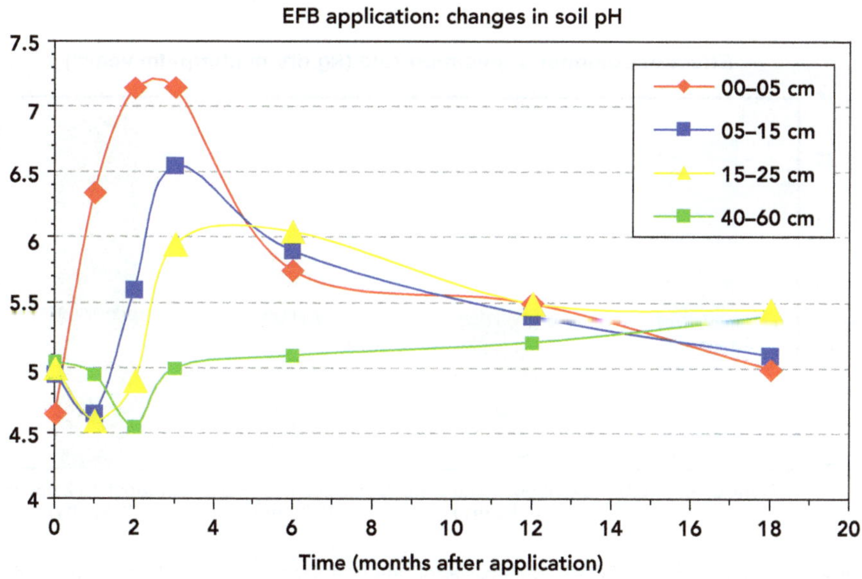

Figure 2 The impact of EFB application (60 t fresh matter/ha) on changes in pH with time in different soil layers. The legend shows the depths of the layers observed.

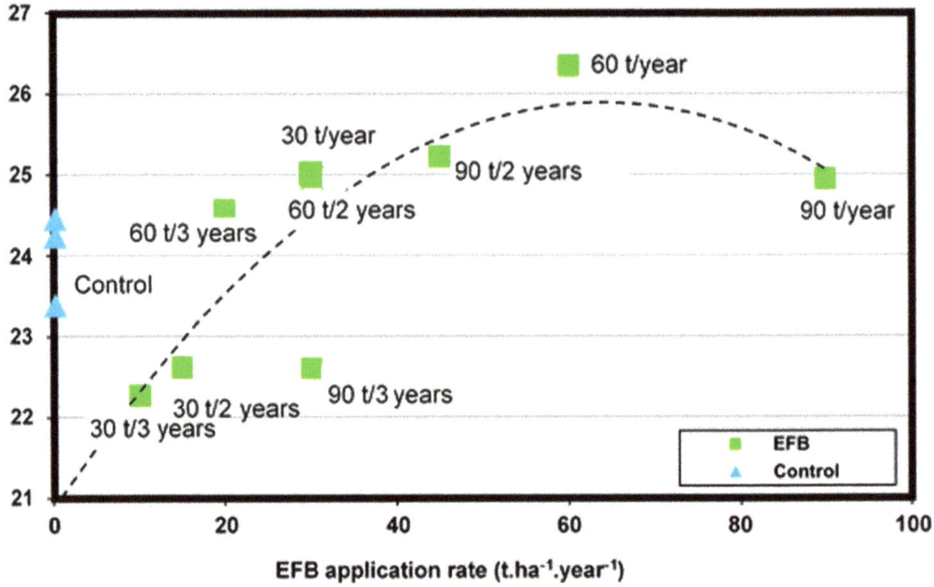

Figure 3 The impact of EFB applications performed at various rates (0, 30, 60 and 90 t/ha) and frequency (every year, every two years and every three years) on the yield of the palms. Blue triangles show the performance of palms treated with mineral fertilizer applications only. Five years' average yields are shown.

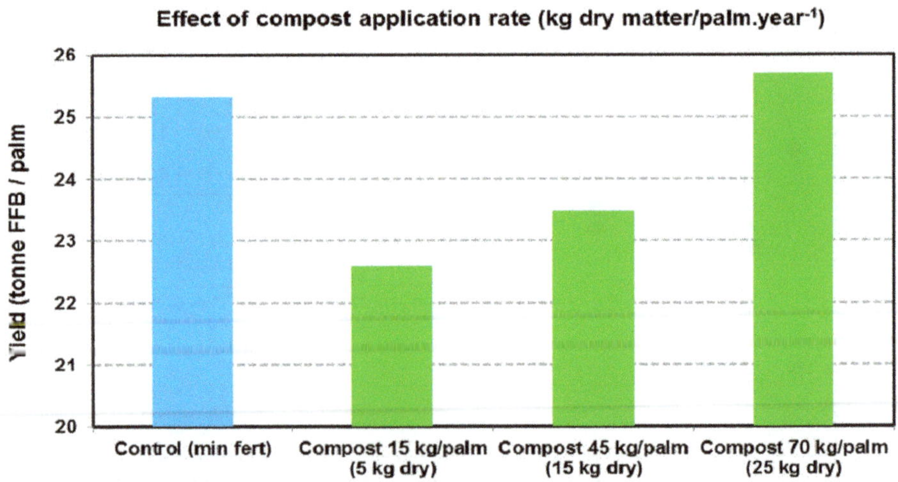

Figure 4 The impact of EFB+POME compost application at various rates (5, 15 and 25 kg of dry compost per palm per year). The blue column shows the performance of palms treated with mineral fertilizer applications only. Five years' average yields are shown.

There are two other challenges currently faced by the oil palm industry that require in-depth research on soil chemical functions:

- The increasing use of poorer soils on marginal land
- The need to improve soil management by smallholders

Principally for conservation reasons, as well as land availability, new oil palm developments are progressively being pushed towards marginal land or poor soils, including sandy soils, or those, for example, with high levels of nickel in ultrabasic soils. Cultivating oil palm in sandy soils requires developing best practices in terms of water and nutrient management, in which the monitoring of organic matter is key. Crop development in toxic soils requires a precise understanding of soil micronutrients, including accurate measurements of toxicity in both soil and plant (Fig. 5). In addition, research to alleviate toxicity through biological means and/or through a dedicated breeding programme to adaptatively select cultivars should be initiated.

The second trend observed in the recent development of oil palm plantations relates to the importance of smallholders, who currently manage 40–50% of the area devoted to oil palm cultivation. As oil palm is cultivated in many developing countries, a common characteristic of smallholders' cropping systems is poor access to national extension services, leading to inadequate soil management. Poor understanding and management of soil fertility can result in a declining trend in soil chemical quality. Literature on this subject is scarce and does not reflect the fact that proper soil management would ensure higher yields with subsequent less pressure for new land use change, and consequently lower rates of deforestation. Additionally, an adequate soil management system is able to

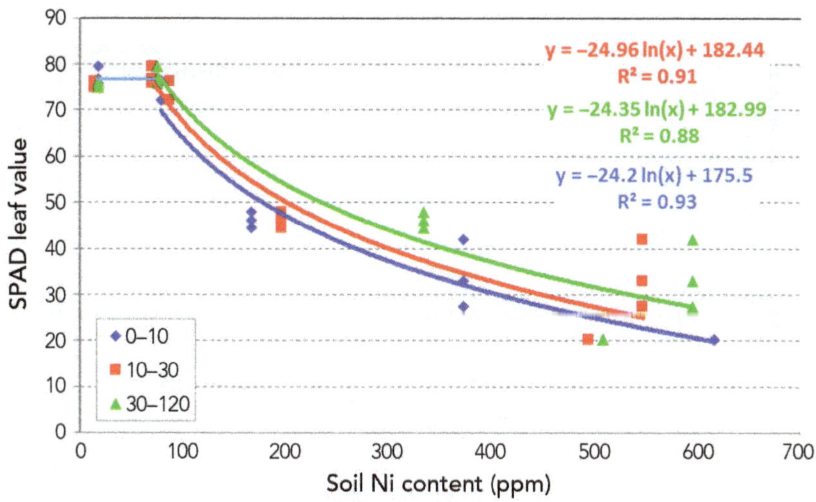

Figure 5 SPAD estimation of chlorophyll content in palms planted in soils with various nickel content. Colour shows the depth of the soil observed. High Ni level in soil starts at around 100 ppm, with negative impact on chlorophyll content.

improve soil's capacity for carbon storage, allowing the potential for soils to contribute to climate change mitigation.

While this chapter is not intended to provide an exhaustive list of fields of research requiring more development, it is important to keep in mind the new role of biomass as a recyclable source of energy. The huge biomass produced by the oil palm, especially that accumulated in the trunk, has started to attract energy-consuming sectors such as the steel industry. If biomass is exported from fields for energy use, this would likely result in diverting large amounts of organic matter away from the fields, with potentially negative effects on soil fertility in the short-, medium- and long-term. Research needs to be implemented in order to identify best practices adapted to each set of circumstances, as the systematic exportation of biomass might jeopardize the positive trend in organic matter content often observed in well-managed plantation soils (Fig. 6 and 7).

2.3 Soil fertility: the physical component

It is quite surprising that there is so little literature on soil physics in oil palm cultivation. The impact of soil compaction on the performance of palms was studied in West Africa during the 1980s (Caliman et al., 1987; Caliman, 1990). In this case, mechanized land preparation of a savannah concession caused the upper soil layers to compact, leading to a reduction in root development and a decrease in the water retention capacity of the soil. As a direct consequence, a significant 30% yield decline was recorded. Compaction intensity and spatial extent were measured using penetrometry (Fig. 8), which measures the resistance of soils to the development of the root system of the palms. A corrective cultivation system was subsequently proposed and tested, based on deep subsoiling, in

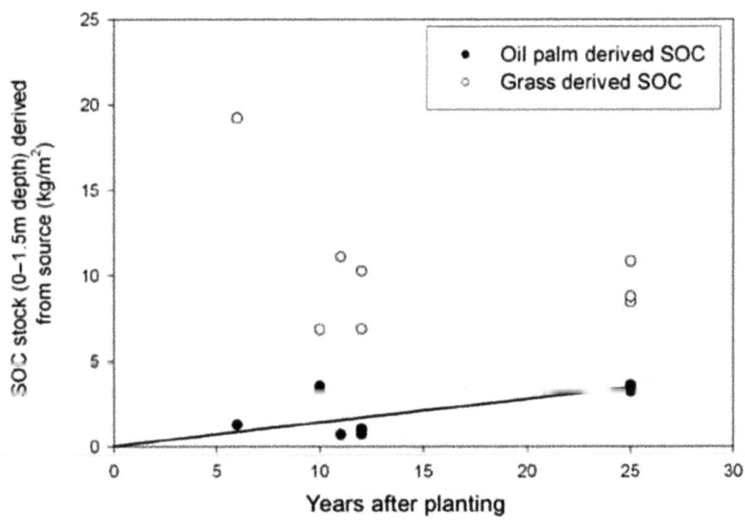

Figure 6 Soil organic carbon trend in oil palm plantation (Goodrich et al., 2012).

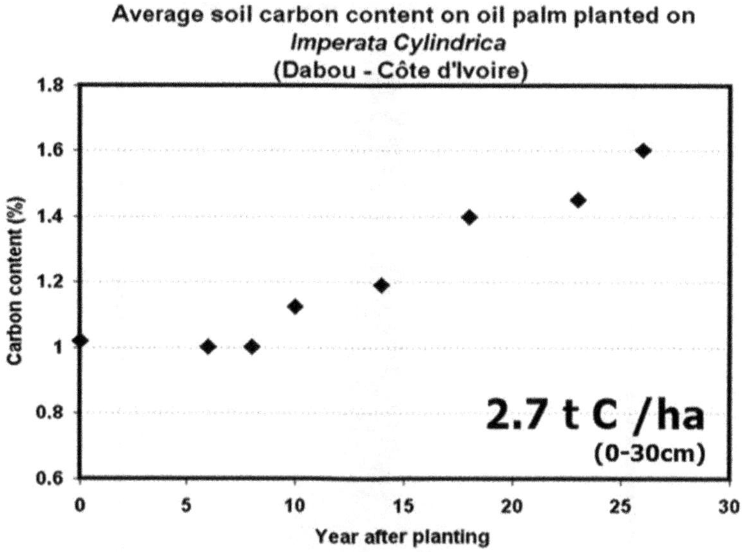

Figure 7 Soil organic carbon trend in oil palm plantation planted on savannah in Côte d'Ivoire (the increase in organic carbon during one full cycle of oil palm cultivation reaches 2.7 t/ha).

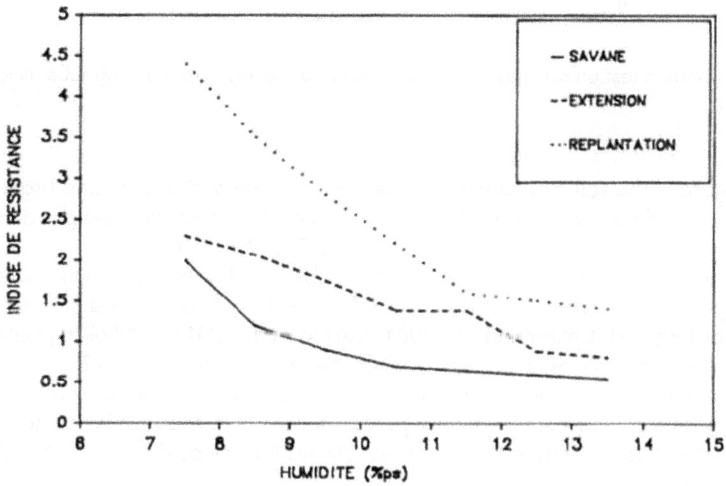

Figure 8 Comparative measurement of soil penetration resistance in several types of oil palm cultivations (namely original savannah, extension and replanting). The second oil palm generation shows a higher level of soil compaction (Caliman, 1990).

Figure 9 A vibration test on soil aggregates developed to identify soils requiring subsoiling (Caliman, 1990).

order to improve the soil structure along the whole profile, without diluting organic matter from the soil surface into deeper layers. A dedicated system was designed to identify types of soil compatible with subsoiling practices (Fig. 9 and 10).

The impact of POME on soils when used as a substitute for mineral fertilizers has yielded contrasting results. While applications of EFB were found to improve soil permeability to rainwater (Fig. 11) (Caliman et al., 2001; Carron et al., 2015a, 2016), repeated POME applications resulted in greater instability of the soil structure (Fig. 12) and have been linked to a dramatic increase in the amount of exchangeable potassium in the soil. It is therefore important to support new practices with robust scientific evidence and to remember that organic fertilisation does not always have a positive impact and must be reassessed for every new situation.

Erosion, generally due to run-off, is a major source of soil degradation around the world (Roose et al., 2006; Roose, 2017). In oil palm cultivation, several studies have been carried out to study factors affecting erosion and to test cultivation practices to control erosion (Kee and Chew, 1996; Banabas et al., 2008; Comte et al., 2012; GAR, 2013, 2015; Fuady et al, 2014). However, there is still a huge amount of research work to be done, especially in synthesizing results from the many studies already implemented. The objective should

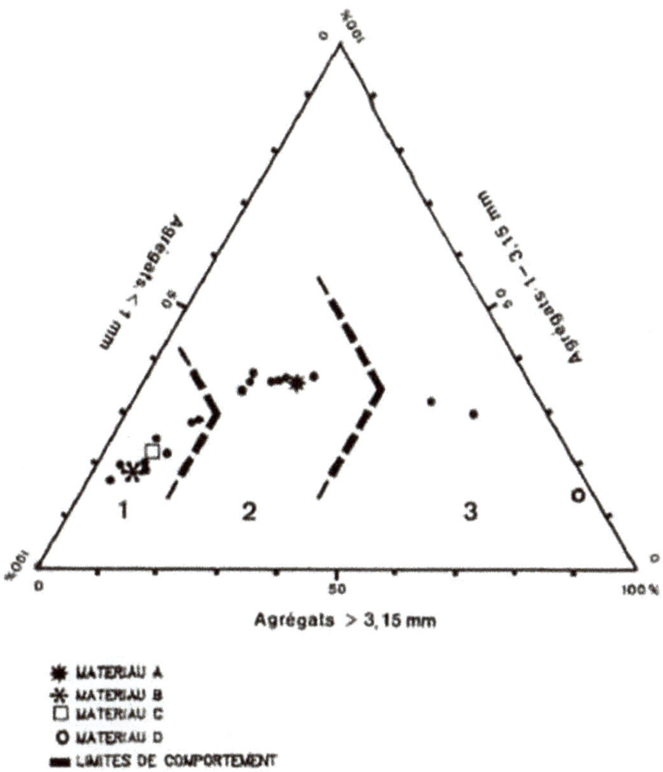

Figure 10 Results of the vibration test on aggregate to identify soils appropriate for subsoiling, classified as category 2 and category 3 soils (Caliman, 1990).

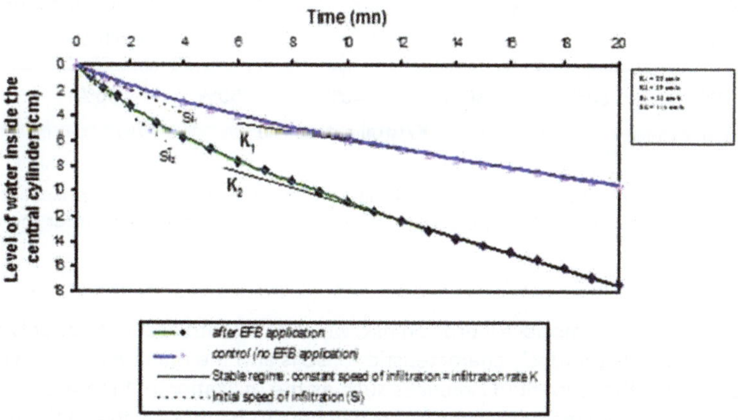

Figure 11 Changes in soil infiltrability after application of empty fruit bunches (Caliman et al., 2001).

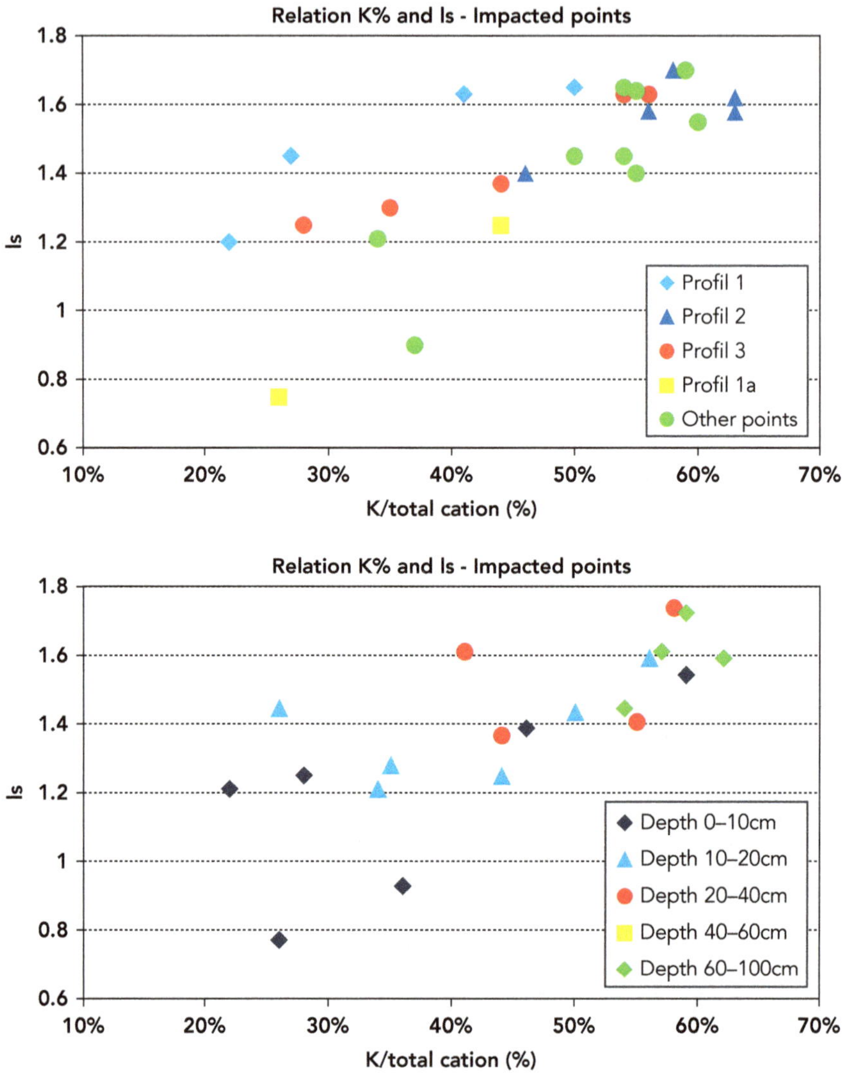

Figure 12 Soil aggregate instability in areas submitted to palm oil mill effluent application. Palm oil mill effluent is characterized by high potassium content. The impact can be seen for the whole soil profile depth.

be to develop a general model for erosion risk, which takes into account not only the slope of the land but also the intrinsic characteristics of soils and the management of vegetation ground cover, together with field practices such as the utilization of biomass.

There are at least two major reasons for researchers to be more active in the field of soil physics:

1 Oil palm productivity is very dependent on water availability. As an example, a 100 mm deficit of water due to drought resulted in a 10–15% yield decrease over two years (Caliman and Southworth, 1998). The capacity of soil to store water and to promote root growth is therefore crucial for the productivity of oil palms.
2 There is a significant trend towards mechanization in most plantation operations. This is already the case for fertilizer application, in-field fruit collection, organic solid waste recycling and several other field activities. Current research for further mechanization is in progress. Such a shift in terms of field practice will certainly have an impact on soil physical characteristics, including compaction.

The oil palm industry can seek solutions from other agricultural communities that have experienced and quantified the impact of mechanization on the physical properties of soil. Nevertheless, it is of paramount importance to open a 'new' (at least for the oil palm industry) field of research, with two main objectives to i) monitor the condition ('health') of soils and ii) identify, then test, appropriate solutions that will be both preventive and corrective. As mentioned earlier, literature in this area is scarce. Zuradai et al. (2015a,b, 2017) are among the few authors to have carried out some research on the effect of mechanization of routine operations on the physical properties of soil and oil palm performance. They found that the above-ground biomass (frond size, trunk diameter and biomass) and the root biomass of the palms were negatively impacted.

2.4 Soil fertility: the biological component

Of the three traditional components of soil fertility, the biological aspect has attracted the most recent attention from researchers. Undoubtedly such interest relates to growing concern over the last two decades about the environmental and ecological impact of oil palm cultivation, together with the global sustainability objectives of agriculture taken as a whole (Breure et al., 2004).

Soil biodiversity is associated with important ecosystem functions. For example, many members of soil mesofauna, such as earthworms or termites, are considered soil 'engineers', decomposing organic material (Lavelle et al., 2006). They also play a critical role in greenhouse gas emissions (Lubbers et al., 2015). The diversity and number of micro fauna are also responsible for regulating emissions such as nitrous oxide (Soares et al., 2016) as well as affecting plant nutrition through the role of mycorrhizae (Garbaye, 2013).

The biological characteristics of soils are studied at macro, meso and microbiological levels. New developments in molecular biology have led to a dramatic increase in DNA and RNA sequencing capacity and a new range of metagenomics and proteomics tools (Nesme et al., 2016). These developments have led to renewed interest in studying microorganism populations in soils and their relationship (in terms of both abundance and diversity) with field practices, on the one hand, and soil fertility, on the other. However, recent results show that this field of research is still at an early stage of development. There have been difficulties in relating studies of soil bacterial communities (Lee-Cruz et al., 2013) to studies looking at soil functions (Soares et al., 2016). Consequently, studies of the impact of agriculture in general and of oil palm cultivation in particular on soil bacterial communities provide contrasting results, depending on the community's scale, spatial observations and the functions analysed (Lee-Cruz et al., 2013). Biological fertility studies are often based on comparative approaches, with standard practices used as a control treatment, aimed at assessing practices simulating natural ecological situations.

When considering population abundance and diversity, a higher diversity in species is often related to a higher soil quality.

Studies on biological soil fertility seek to address a number of questions:

- Comparing mineral fertilizer versus organic manuring. Studies have demonstrated the potential use of organic manure as a substitute for mineral fertilizers in the short-, medium- and long-term, to maintain an optimal level of soil biological activity (estimated through the feeding activity of meso-organisms on lamina baits placed across the upper soil layers) (Tao et al., 2014, 2015, 2017).
- Measuring spatial heterogeneity in soil biological activity related to plantation structure and management, for example, the impact of weeding, harvesting or residues recycling on soil biological activity (Carron et al., 2015a).
- Exploring the possible relationship between microbial characteristics of soil and the development of diseases such as Ganoderma fungus (Goh et al., 2015).

Growing interest in the biological properties of soil has also been mirrored by an increasing number of soil conditioning products available in the market, all of which claim some favourable impact on soil health and are often used in substitution of, or addition to, standard fertilizing practices.

3 Nutrient management

Together with breeding, nutrient management has been a key factor in improving oil palm productivity. Oil palm usually responds well to nutrient inputs. Fertilizer management is one of the most challenging activities for the plantation manager, since fertilizers account for the highest proportion (60–65%) of the running costs of a plantation.

In the early days of plantation development in West Africa in the late 1940s, it was reported that, in sandy ferralitic soils, potassium was the major limiting factor for oil palm growth and development, and repeated applications of appropriate quantities of potassium fertilizer resulted in a seven-fold increase in yields (Caliman and Aubry, 1998). More recently, a global field trial network involving more than 50 fertilizer trials established by SMART Research Institute (Pt SMART Tbk R&D center) in Indonesia on various types of soil located in several provinces of Sumatra, Kalimantan and Papua revealed that, on new developments, most fertilizer applications were resulting in a significant increase in yield (Table 1).

Indeed, 90%, 95% and 100% of the trials responded favourably to applications of phosphorus, potassium and nitrogen fertilizers, respectively. In comparison, only 65% of the trials showed a yield increase following applications of magnesium fertilizer. The highest yield increase recorded reached 200% of the control treatment. This scenario is not wholly surprising, considering that tropical soils are often washed by heavy rains. While this results in a high weathering process of the soils, and therefore a rapid release of nutrients from the soil material itself and consequently a fast availability to plant nutrition, it subsequently puts such released nutrients at high risk of losses through leaching and run-off.

Given the importance of soil characteristics, one would have thought that any smart management tool for nutrient and fertilizer application should be based on soil diagnosis.

Table 1 Statistics of yield response to nutrients applied calculated from a network of fertilizer trials in Indonesia

Nutrient		Number of field fertilizer trials showing a yield increase	Number of field fertilizer trials showing a yield increase >25%	Maximum yield increase recorded in at least one field fertilizer trial
Nitrogen	N	100%	50%	130%
Phosphorus	P	90%	50%	75%
Potassium	K	95%	33%	200%
Magnesium	Mg	60%	6%	40%

This is indeed the case for most annual crops, such as corn or wheat. However, while information regarding the soil's chemical status is key (especially the balance and availability of nutrients), several considerations have encouraged oil palm agronomists to develop diagnosis tools based instead on plant status through the mineral analysis of vegetative tissue. There are a number of reasons for this:

- The spatial heterogeneity of soil characteristics (in chemical, physical and biological characteristics) created by oil palm cultivation techniques. These techniques include relatively low planting density per hectare (generally fewer than 150 palms) as well as the planting design, and subsequently the rapid creation of heterogeneous zones (zones such as the weeded circle, harvesting path, frond heaps, etc.).
- The differences in the management of different zones. When mineral fertilizers are applied manually, this is carried out on a specific area, located mostly in the circle of the palms or around the palms and sometimes on the frond heaps. This application of fertilizer further increases soil spatial heterogeneity.
- Biomass recycling. This includes the regular disposal of fronds, which are pruned from palm trees and arranged in frond heap zones. Recycling of high quantities of EFB also occurs between rows and individual palms. This causes even more spatial heterogeneity between blocks of trees with or without EFB applications.
- The management of vegetation ground cover in these zones. When combined with regular weeding of the circles around the palms and on harvesting paths, this induces selective development of weeds in each zone. As an example, soft ferns become the dominant weed species in the frond heap zone, adding to soil heterogeneity .
- Subsurface fertilization. Some planters prefer to apply fertilizer in pockets below ground level, created in the first soil layers.
- The greater potential of leaf analysis to account for soil variability. The preference for leaf analysis is based on the assumption that palms are able to buffer the effects of soil variability, due to their highly extended root system. It is assumed that the levels of nutrients in palm leaves balance out the variability between various spatial zones.

In addition to these considerations, the correct interpretation of soil analyses is not an easy task, as the results are always influenced by a given method of analysis. Last but not least, soil analysis is also significantly more expensive than plant tissue analysis.

Consequently, most oil palm plantations manage the mineral nutrition of palms based on leaf diagnosis, coupled with field trial networks to fine-tune recommendations (Ng, 1986). The aim is to determine the optimum leaf nutrient content and match it to fertilizer

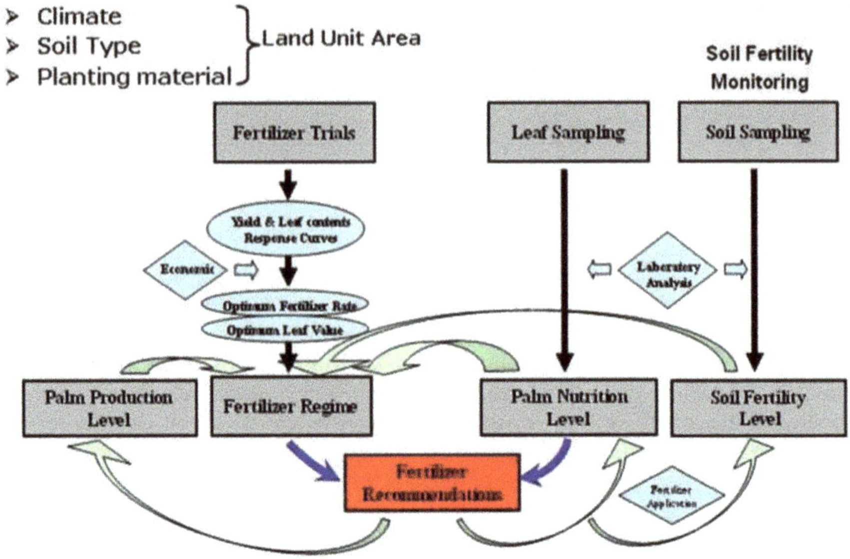

Figure 13 Simplified diagram for mineral fertilizer management in oil palm plantations (Caliman et al., 1994, 2003).

types and rates, in order to reach optimum nutrition levels. Corley and Tinker (2016) have reviewed several approaches to mineral nutrition management developed by oil palm agronomists around the world.

Figure 13 shows the general diagram used by Caliman et al. (1994, 2004) for fertilizer management.

1. Soil characteristics (together with climate and planting material origin) are used to determine the land unit area in which a uniform fertilizer management system can be implemented.
2. Field fertilizer trials are set up within these units in order to determine the optimum leaf nutrient value and the amount of fertilizer to be applied.
3. Leaf nutrient contents are analysed yearly in blocks of oil palms in order to estimate the nutritional status of the palms.
4. Soil analysis is carried out every three to five years to ensure that the fertilizer regime applied maintains (and preferably improves) the fertility of the soil.
5. The system also relies on the analysis of the impact of the fertilizer regime on the productivity of palms, on their mineral status and on changes in the chemical status of soils.

While the results of experiments on mineral nutrition management have been extensively interpreted and analysed in many articles and conferences (see review from Corley and Tinker, 2016), current and future research is or should be focusing on changes in fertilizer types (compound fertilizers versus single formulations), controlled release fertilizers (CRFs), organic fertilizers and soil conditioners.

The last decade has seen a shift in the types of fertilizer used in plantations. While for decades single formula fertilizers were used (as they were considered the cheapest source of nutrients), an increasing number of plantation managers are starting to request compound fertilizers as a way of reducing the number of applications in the field, thereby decreasing management costs. They also prefer a more flexible schedule for applications as a smart way to adapt to unexpected changes in local rainfall patterns.

While such a shift might tackle the issue of labour shortage in some places, it remains the case that the cost of fertilizer is still higher than the cost of its application. Purchasing more expensive fertilizer such as compound formulas to reduce the cost of application does not therefore seem logical. However, when fertilizer applications are mechanized, there is a lower resulting application cost and labour requirement. Fertilizer suppliers see this trend positively because of the higher profit when selling compound fertilizers compared to single formulas, and are therefore pushing plantation managers in this direction.

More recently, many fertilizer suppliers are going a step further by proposing CRFs, a coated formulation commonly used in horticulture. The utilization of such products should significantly reduce application rates compared to other fertilizers, due to the fact that CRFs are less sensitive to nutrient losses through leaching. This potential benefit has gained the attention of both planters and environmentalists, as CRFs could reduce both production costs and risks to the environment. However, there is only a limited number of experiments underway using CRFs, and therefore their comparative advantage is yet to be confirmed. In addition, the decrease in application rates is often not enough to cover the price difference to make the use of CRF economically viable for a crop such as oil palm. The exception to this is during the first year(s) after planting, when fertilizer applications are usually applied manually. In this case, the application of a smaller quantity of CRF in the planting hole contributes to a reduction in fertilizer application rounds during the first eight to 12 months of cultivation after planting.

As far as research is concerned, there is a need to study and quantify the current and potential efficiency of all these products, from single forms to CRF, in order to quantify the benefit of each type of fertilizer and to propose informed and documented field practices that could be adapted to each field situation (soil type, terrain slope, rainfall pattern, ground cover management, etc.). This should be done using a scientific approach, applying such techniques as lysimetry studies, and/or isotopic approaches, which are relatively complex but provide good results.

A large number of new suppliers have emerged, offering organic products to either replace mineral fertilizers or to enhance the natural nutrient availability in soils. These soil conditioners often include a combination of microbes (bacteria and fungi) to promote nutrient availability. Several products also include specific molecules, such as organic acids, which claim to have a positive impact on the performance of palm trees. Other fertilizing products (such as humic acids and other soil conditioners) are also available. The major difficulty for a manager looking for the perfect formula is the poor existing regulation regarding the specification and quality of such products. Quality control can only be carried out in specialized laboratories, and it is often impossible to undertake a reliable comparative analysis through laboratory results alone. Consequently, field tests are currently the only way to evaluate such products, and it usually takes several years before any conclusions can be drawn. Collaborative research work is clearly needed to validate the agronomical, environmental and economic claims of such products.

3.1 Fertilizer efficiency: reduction of nutrient losses

In order to reduce production costs and risks to the environment without affecting yields, it is important to maximize the efficiency of fertilizer use. A limited number of studies have focused on nutrient losses in plantations, primarily on losses through run-off, leaching and erosion (Kee and Chew, 1996; Banabas et al., 2008; Comte et al., 2012; GAR, 2013, 2015; Fuady et al., 2014). All these studies report significant nutrient losses. It is clear there is room for improvement through better agricultural field practices. A recent study carried out by SMART Research Institute, in collaboration with CIRAD, confirms that simple field practices are able to significantly reduce the quantity of nutrients lost through run-off and erosion (GAR, 2014, 2015).

Fertilizer management is increasingly important, both in terms of economics (60–65% of production cost) and environmental risks linked to the pollution of surface and/or ground water. Fertilizers are the third biggest source of greenhouse gas emissions, after land use change and methane emissions from liquid wastes. There is an urgent need for more research in this field (Pardon et al., 2016a, b). The challenge is the wide range of different conditions in terms of factors such as topography, soil types, climate, vegetation ground cover and the palms. A possible solution is the use of modelling.

3.2 Supporting site-specific fertilization: towards precision agriculture

For many years, there has been consensus among agronomists and managers about the potential for improving nutrient management in oil palm plantations through the implementation of precision agriculture principles. However, very little has been achieved on any scale. Smartly managed plantations consider that they implement site-specific management when they work at a scale of a commercial block, that is, 25–30 ha. Other plantations adjust their fertilizer rates at a much larger scale, sometimes a few hundred hectares. Today's technology is improving at a rapid pace, in order to more accurately manage palm nutrition on a much smaller scale.

There are at least three major goals to be achieved, should the industry wish to successfully apply precision agriculture in fertilizer management:

1. Monitoring the condition of palms at the tree scale;
2. Adjusting fertilizer rates at any spatial scale and
3. Applying fertilizer based on rainfall forecast.

3.2.1 Changing the scale for the monitoring of the nutritional status of palms

The current practice for determining the condition of palms involves tissue analysis via leaf sampling. Samples are generally taken on around 1% of cultivated palms then combined in order to get one sample per commercial block (i.e. one sample for 25–60 ha). The availability of satellite imaging since the early 1980s opened up the possible use of image analysis. However, reliable algorithms to translate the radiometric values measured by satellites into recommendations for fertilizer management have not yet been developed for oil palm (Nguyen et al., 1993). In addition, difficulties in achieving cloud-free images in tropical zones have made this method less reliable.

A key improvement around ten years ago was the development of spectral signatures extending from the visible to the near infrared (up to 1000 μm or even 2400 μm) at close range, sometimes down to 1 or 2 μm (Lelong et al., 2006, 2009; Jorgensen et al., 2007; Liew et al., 2008). This technology was more accurate than traditional RGB or other limited-range wavelength cameras.

Recent innovations in sensors, together with the development of drones specially designed for agricultural applications, are heralding a new era for digital agriculture in general, and mineral nutrition management in particular. Success will depend on developing dedicated algorithms, but it seems quite certain that new tools will be developed in a relatively short timescale. The change of scale allowed with such tools will be tremendous, shifting from a single sample per commercial block, to a value for each individual palm. Computing capacity has followed a similar trend, now enabling the treatment and interpretation of 'big data', generated by a series of high throughput technologies applied to agricultural production. Similar tools might be available in the near future to monitor soil health. Despite difficulties related to the spatial variability of recorded soil fertility characters, such sensors will certainly help in improving decision-making in nutrition management.

3.2.2 Accurate fertilizer applications at the palm scale

Fertilizer application, both manual and mechanical, can be adjusted to fit the needs of individual palms if required. However, this requires a significant investment of management time and resources. Current developments in agricultural devices equipped with geographical positioning systems, together with the ability to upload programmed maps with variables fertilizer rates, flow metres and other sensors, now allow for a relatively high level of automation and precision in fertilizer application. Mechanical fertilizer spreaders with this technology are already available, opening the way to truly site-specific nutrient management.

3.2.3 Applying fertilizer at the right time

Standard operational procedures prepared by companies for fertilizer applications generally include actual rainfall records as one criterion to determine the timing of application. However, this is often considered as a second priority by plantation managers, who unfortunately often forget that the first action of a farmer when getting up in the morning is to have a 'look at the sky' and the weather before deciding on what to do on the farm. Both the farmers and plantation managers know the importance of rainfall and soil humidity for maximum fertilizer efficiency. However, the latter might have less flexibility in their daily work programme, or require a more accurate decision support system than just 'looking at the sky'. There is no doubt that, in the very near future, weather forecasts for the short- (hours) (Tollefson, 2017), medium- (days) and longer-term (months) will be developed and be made available for plantation managers and planters, allowing them to make more informed and relevant decisions. This will undoubtedly result in a lower risk of fertilizer losses, while increasing the efficiency of fertilizer application.

4 Mineral nutrition and planting material

Figure 13 highlights the importance of planting material as a key parameter in defining a land unit area for a particular fertilizer regime. Currently, most planting material is supplied in the form of seeds and planted in a commercial block, including various different crossings. Although coming from genetically related parents, each crossing can present differences in nutrient use efficiency and therefore should be managed differently. Evidence for such variability in fertilizer rate requirement is noted in several publications (Caliman et al., 2003, 2004; Ollivier et al., 2017). Future best practices will call for management of plantations (planting, manuring etc.) based on individual planting material spatial distribution instead of mixing.

5 The specific case of smallholders

Defining a fertilizer system suitable for smallholders is a particular challenge. This is especially true in regions where a significant number of smallholders have developed their fields independently of any plantation programme. These independent farmers typically have little or no access to extension services.

Smallholders have the capacity for precise application of fertilizers in terms of space and time, since they work their own land with a limited area to maintain. However, they often have little information regarding the status of their palms beyond their visual appearance. There is a need to develop new decision-making tools to improve nutrition management in such situations. Technology is currently expensive, but might be affordable in the future, through appropriate cooperative organization, or/and through digital applications.

In the shorter term, it might be possible to develop biological indicators that could help smallholders assess the condition of their land and adjust their fertilizer regime accordingly. One simple tool could be based on the census of weed species growing naturally in their fields, as some of these species are more adapted to a particular level of nutrient abundance or deficiency (Fig. 14).

Smallholders who live close to large plantation companies (either through governmental development programmes, or because they have signed a fruit supply agreement with a private company), can generally benefit from some level of extension service in terms of fertilizer management. As a result, unless their financial situation does not allow it, these farmers can manage fertilization programmes in a similar way to those implemented by large-scale plantations. They can even achieve a better application strategy in terms of quality (scheduling and amount of fertilizers used).

6 Future trends and conclusion

After 75 years of intensive field research, a satisfactory level of understanding of soil and mineral nutrition management in oil palm plantations has been achieved. Such progress has significantly contributed to the high efficiency of the crop in terms of oil productivity.

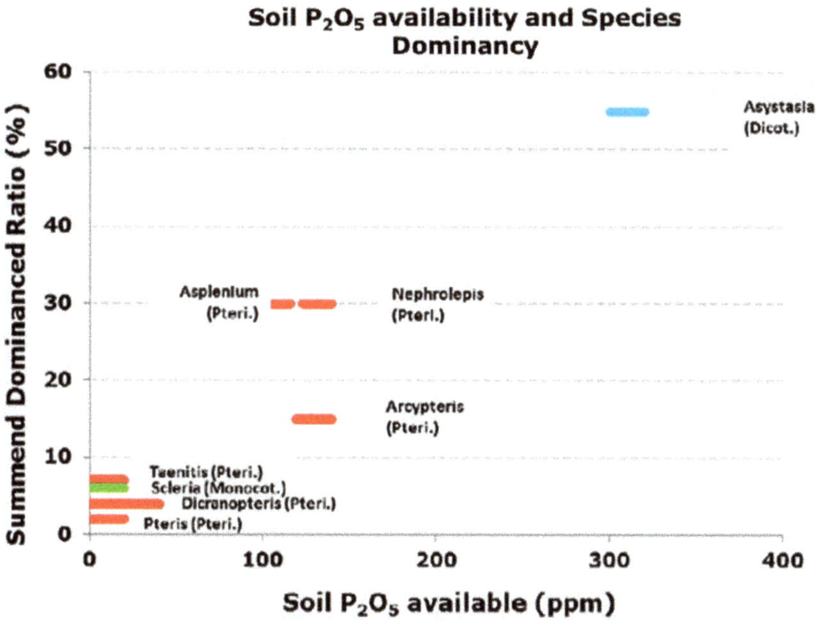

Figure 14 Weed species as a biological indicator of soil phosphorus level (Caliman et al., 2015). This biological indicator serves as the base for a simple visual tool able to assess soil nutrient content, helping to adjust fertilizer types and rates accordingly.

During the last decade, research work on the biological component of soil fertility has increased, in line with a general concern to improve sustainability in agriculture.

Traditional concepts of soil fertility have started to shift towards the more holistic concept of soil health, taking into account the multiple ecological functions of soil, as well as their interactions and changes with time and space. A new phase of scientific research around the concept of soil health has begun. It needs to develop quickly if the oil palm industry is to maintain its current productivity advantage over competing oil crops, while improving its environmental impact (Byerlee et al., 2016). In particular, there is a need to identify indicators which can translate the assessment of the multiple functions of soils into a practical decision support system (Velasquez et al., 2007). However, given the diversity of functions in such a complex habitat as soil, the difficulty might be in developing new tools that are affordable (intellectually, practically and economically) for extension services staff or even for farmers and/or their cooperatives.

Mineral nutrition management will also undergo significant changes in the future, beginning with the integration of new technologies in several areas, including accurate monitoring of the condition of palms and the soil, a more informed decision-making process and more advanced equipment for field operations. The adequate use of current or emerging fertilizer products also requires a more holistic understanding of agroecology.

7 Where to look for further information

7.1 Websites:

www.cenipalma.org
www.cirad.fr
www.fao.org
www.goldenagri.com.sg
www.icope-series.com
www.iopri.org
www.millenniumassessment.org
www.mpob.gov.my
www.oilpalmbiodiversity.com
www.teebweb.org

8 References

Allee, W. C. (1949). *Principles of Animal Ecology*. Sounders Co., Philadelphia. www.biodiversitylibrary.org/bibliography/7325#.

Banabas, M. (2007). Study of nitrogen loss pathways in oil palm (*Elaeis Guineensis* Jacq.) growing agro-ecosystems on volcanic ash soils in Papua New Guinea. PhD thesis, Massey University, Palmerston North, New Zealand.

Banabas, M., Turner, M. A., Scotter, D. R. and Nelson, P. N. (2008). Losses of nitrogen fertilizer under oil palm in Papua New Guinea: 1. Water balance, and nitrogen in soil solution and runoff. *Australian Journal of Soil Research*, 46, 4, 332–9.

Byerlee, D., Falcon, W. P. and Naylor, R. L. (2016). *The Tropical Oil Crop Revolution: Food, Feed, Fuel, and Forests*. Oxford University Press, Oxford, UK, pp. 288

Bispo, A., Guellier, C., Martin, E., Sapijanskas, J., Soubelet, H. and Chenu, C. (2016). Les sols: Intégrer leur multifonctionnalité pour une gestion durable. Editions Quae, 2016, 379p. ISBN: 978-2-7592-2392-3, ISSN: 1952–1251.

Breure, A. M., Mulder, C., Rutgers, M., Schouten, T., de Zwart, D. and Bloem, J. (2004). A biological indicator for soil quality. In: *Proceedings from an OECD Expert Meeting Rome, Italy, March 2004: Agricultural Impacts on Soil Erosion and Soil Biodiversity: Developing Indicators for Policy Analysis*, pp. 485–94.

Caliman, J. P. (1990). Dégradation de propriétés physiques conditionnant la fertilité des sols sous culture du palmier à huile en Côte d'Ivoire. Essai de correction. Thèse de Doctorat ès- Sciences, Université de Bourgogne, Dijon, France, p. 225

Caliman, J. P. and Aubry, M. (1998). Passé et Avenir du palmier à huile dans la savane de Dabou (Côte d'Ivoire). *Oléagineux Corps gras Lipides*, 5, 113–15.

Caliman, J. P. and Southworth, A. (1998). Effect of drought and haze on the performance of oil palm. In: *International Oil Palm Conference*, 23–25 September 1998, Bali – Indonesia.

Caliman, J. P., Budi, M. and Saletes, S. (2001). Dynamics of nutrient release from empty fruit bunches in field conditions and soil characteristics changes. In: *MPOB International Palm Oil Congress (PIPOC)*, 20–23 August 2001, Kuala Lumpur, Malaysia.

Caliman, J. P., Daniel, C. and Tailliez, B. (1994). Oil palm mineral nutrition. *Plantations, Recherche, Développement*, 1(3/93), 36–54

Caliman, J. P., Dubos, B., Tailliez, B., Robin, P., Bonneau, X. and Barros, I. (2003). Oil Palm mineral nutrition management: Current situation and prospects. In: *XIV International Oil Palm Conference*, 23–26 September 2003. Cartagena de Indias, Colombia.

Caliman, J. P., Dubos, B., Tailliez, B., Robin, P., Bonneau, X. and Barros, I. (2004). Manejo de nutricion mineral en palma de aceite: Situation actual y perspectivas. *PALMAS*, 25(Tomo1), 42–60.

Caliman, J. P. Olivin, J. and Dufour, O. (1987). Degradation of sandy femalitic soils in oil palm cultivation through acidification and compaction. Correction methods. *Oléagineux*, 42(11), 393–401.

Caliman, J. P., Pujianto, Advento, A. D., Ribka Sionita, T., Marichal, R., Wohlfahrt, J., Purnomo, D., Snaddon, J. L., Turner, E. C. and Foster, W. A. (2015). Agri-Environmental indicators for a sustainable palm oil production. In: *18th International Oil Palm Conference*, 22–25 September 2015. Cartagena de Indias, Colombia.

Carron, M. P., Auriac, Q., Snoeck, D., Villenave, C., Blanchart, E., Ribeyre, F., Marichal, R., Darminto, M. and Caliman, J. P. (2015a). Spatial heterogeneity of soil quality around mature oil palms receiving mineral fertilization. *European Journal of Soil Biology*, 66, 24–31

Carron, M. P., Pierrat, M., Snoeck, D., Villenave, C., Ribeyre, F., Suhardi, Marichal, R. and Caliman, J. P. (2015b). Temporal variability in soil quality after organic residue application in mature oil palms plantations. *Soil Research*, 53(2), 205–15. http://dx.doi.org/10.1071/SR14249.

Carron, M. P., Auriac, Q., Snoeck, D., Villenave, C., Blanchart, E., Ribeyre, F., Marichal, R., Darminto, M. and Caliman, J. P. (2016). Do the impact of organic residues on soil quality extend beyond the deposition area under oil palm? *European Journal of Soil Biology*, 75, 54–61.

Chandran, M. R. (2017). The French connection. Celebrating oil palm's Malaysian centennial. *The Edge*, 8 March 2017. Forum 50–51. http://www.theedgemarkets.com/my/article/my-say-French-connection.

Chiew, L. K. and Rahman, Z. A. (2002). The effect of oil palm empty fruit bunches on oil palm nutrition and yield, and soil chemical properties. *Journal of Oil Palm Research*, 14, 1–9.

Comte, I., Colin, F., Whalen, J. K., Grünberger, O. and Caliman, J. P. (2012). Agricultural practices in oil palm plantation and their impact on hydrological changes, nutrient fluxes and water quality in Indonesia. A review. *Advances in Agronomy*, 116, 71–124.

Corley, R. H. V. and Tinker, P. B. (2016). *The Oil Palm*. Wiley Blackwell, Oxford, 639p.

Cristancho, A., Hanafi, M., Omar, S. and Yusof, R. (2009). Chemical characteristics of representative high aluminum saturation soil as affected by addition of soil amendments in a closed incubation system. *Malaysian Journal of Soil Science*, 13, 13–28.

Delval, D. (1981). Evaluation des sols sous palmeraie. Master thesis, Ecole nationale supérieure d'agronomie, Nancy.

Fairhurst, T. (1997). *Nutrient Deficiency Symptoms and Disorders in Oil Palm (Eleais guineensis* Jacq.*)*. Potash and Phosphate Institute, , Atlanta, GA, 32pp.

Fairhurst, T., Caliman, J. P., Hardter, R. and Witt, C. (2005). *Oil Palm: Nutrient Deficiencies and Nutrient Management*. Oil Palm Series Volume 7 Potash and Phosphate Institute, Atlanta, GA, 122pp. USBBI: 981-05-2452-8, ISBN: 2-87614-292-8, EAN: 9782876145924.

Fallavier, P., Breysse, M. and Olivin, J. (1989). Experimental study of potassium dynamics in tow tropical soils used for oil palm growing. *Oléagineux*, 44(9), 197–207.

Fuady, Z., Sastriawan, H. and Mayani, N. (2014). Runoff, erosion and nutrient sediment due the vegetative soil conservation on oil palm plantation. *Journal Soil Science and Agroclimatology*, 11(2), 95–103.

GAR (2013). Sustainability report. www.goldenagri.com.sg.

GAR (2014). Sustainability report 2013, 72p. www.goldenagri.com.sg.

GAR (2015). Sustainability report, www.goldenagri.com.sg.

Garbaye, J. (2013). *La symbiose mycorhizienne: Une association entre les plantes et les champignons*. Editions Quae, 2013. ISBN: 978-2-7592-1963-6, ISSN: 1777–4624.

Goh, Y. K., Goh, Y. K., Wong, W. C., Tung, H. J., Chen, Z. Y., Cheah, L. W., Tarren, N. M. and Goh, K. J. (2015). Ganoderma research updates. AAR-MEOA Seminar, 21 January 2015.

Goodrich, I., Nelson, P. N., Banabas, M., Warster, C. and Bird, M. I. (2012). Soil carbon balance following conversion of grassland to oil palm. In: *Soil Solutions for Diverse Landscapes*, p. 1. From: Joint ASSI and NZSSS Soil Science Conference: Soil solutions for diverse landscapes, 2–7 December 2012, Hobart, Australia.

Holdridge, L. R. (1947). Determination of world plant formation from simple climatic data. *Science*, 105, 367–8.

Holdridge, L. R. (1966). The life zone system. *Adansonia*, VI, 199–203.

Huth, N. I., Banabas, M., Nelson, P. N. and Webb, M. (2014). Development of an oil palm cropping systems model: Lessons learned and future directions. *Environment Modelling Software*, 62, 411–19.

Jorgensen, R. N., Christensen, L. K. and Bros, R. (2007). Spectral reflectance at sub-leaf blade including the spatial distribution discriminating NPK stress characteristics in barley using multiway partial least square equation. *International Journal of Remote Sensing*, 28, 943–62.

Kee, K. K. and Chew, P. S. (1996). A13: Nutrient losses through surface runoff and erosion-implications for improved fertilizer efficiency in mature oil palm. Applied Agricultural Research Sdn. Bhd. Locked Bag no. 212.

Kendeigh, S. C. (1961). *Animal Ecology*.Prentice-Hell, Englewood cliffs, NJ.

Lavelle, P., Decaens, T., Auvert, M., Barrot, S., Blouin, M., Bureau, F., Margerie, F., Mora, P. and Rossi, J. P. (2006) Soil invertebrates and ecosystem services. *European Journal of Soil Biology*, 42(1), 3–15.

Lee-Cruz, L., Edwards, D. P., Tripathi, B. M. and Adams, J. M. (2013). Impact of logging and forest conversion to oil palm plantations on soil bacterial communities in Borneo. *Applied and Environmental Microbiology*, 79–23, 7290–7.

Lelong, C., Lanore, M. and Caliman, J. P. (2006). Evaluation of Hyperspectral relevance to estimate oil palm trees nutrition status. In: Sobrino, J. A. (Ed.), *Second Recent Advances in Quantitative Remote Sensing*. Auditori de Torrent, Spain 25–29 September 2006. publicacions de la Universidat de Valencia. ISBN: 84-370-6533-X; 978-84-370-6533-5.

Lelong, C., Sitorus, N. A., Raharja, D. A., Dubertret, F., Bregand, S., Lanore, M., Sulistyanto, A. W., Prasetya, Roger, J. M. and Caliman, J. P. (2009). Developing non-destructive tools for the diagnostic of Ganoderma attack-level on oil palm trees: Potentialities of reflectance spectroscopy. In: *MPOB International Palm Oil Congress (PIPOC)*, 09–12 November 2009, Kuala Lumpur, Malaysia.

Liew, O. W., Chong, P. C. J., Li, G. and Asundi, A. K. (2008). Signature optical clues: Emerging technologies for monitoring plant health. *Sensors*, 8, 3205–39.

Lubbers, I. M., Van Groenigen, K. J., Brussaard, L. and Van Groenigen, W. J. (2015). Reduced greenhouse gas mitigation potential of no-tillage soild through earthworm activity. *Nature Scientific Reports*, 5, 13787. DOI:10.1038/srep13787.

MEA (2005). *Ecosystems and Human Well-being: General Synthesis*. Millenium ecosystem assessment series. Island press, Washington DC, 160p.

Nelson, P. N., Berthelsen, S., Webb, M. J. and Banabas, M. (2010). Acidification of volcanic ask soils under oil palm in Papua New Guinea: Effects of fertiliser type and placement. In: Gilkes, R. J. and Prakongkep, N. (Eds), *Proceedings of the 19th World Congress of Soil Sciences*, 1–6 August 2010. International Union of Soil Science, Brisbane, Australia, pp. 8–11.

Nesme, J., Achouak, W., Agathos, S. N., Bailey, M., Baldrian, P., Brunel, D. and Kruus, K. L. (2016). Back to the future of soil metagenomics. *Frontiers in Microbiology*, 7, 73.

Ng, S. K. (1986). Phosphorus nutrition and fertilization of oil palms. *Oléagineux*, 41(7), 307–13.

Nguyen, H. V., Olivin, J. and Ochs, R. (1984a). Oil palm and coconut soils in West Africa: Part 1. *Oléagineux*, 39(3), 117–19.

Nguyen, H. V., Olivin, J. and Ochs, R. (1984b). Oil palm and coconut soils in West Africa: Part 2. *Oléagineux*, 39(4), 198–200.

Nguyen, H. V., Caliman, J. P., Flori, A. and Lukman, F. (1993). Spot image to assess sanitary, nutrient and general status of oil palm plantations. In: *PORIM International Palm Oil Congress*, 20–25 September 1993, Kuala Lumpur, Malaysia.

Olivin, J. (1968). Etude pour la localisation d'un bloc industriel de palmiers: Huile (I). *Oléagineux*, 23(8–9), 499–501.

Olivin, J. (1980). Relation between ecology and plantation agriculture. *Oléagineux*, 35(2), 65–78.

Ollivier, J., Flori, A., Cochard, B., Amblard, P., Turnbull, N., Syahputra, I. and Gasselin, D. T. (2017). Genetic variation in nutrient uptake and nutrient use efficiency of oil palm. *Journal of Plant Nutrition*, 40(4), 558–73.

Ouvry, J. F. (1985). Etude expérimentale de la dynamique du potassium sur deux sols tropicaux. Thése de Docteur-Ingénicur, ENSA, Montpellier, 254p.

Paramananthan, S. (1982). Towards a common system of classification in Malaysia with special reference to soils on sedimentary rocks. In: Paramanathan, S. (Ed.), *Trans. Second Meeting of the National Soil Correlation Subcommittee*, Kota Kinabalu, Sabah, April 1981.

Paramananthan, S. (2000). *Soils of Malaysia: Their Characteristics and Identification*, Volume 1. Academy of Sciences Malaysia, Kuala Lumpur, Malaysia. ISBN: 983-9445-06-5.

Pardon, L., Bessou, C., Nelson, P., Dubos, B., Ollivier, J., Marchal, R., Caliman, J. P. and Gabrielle, B. (2016a). Key unknowns in nitrogen budget for oil palm plantations: A review. *Agronomy Sustainable Development*, 36, 21.

Pardon, L., Bessou, C., Soint-Geours, N., Gabrielle, B., Khasanah Ni'matul, Caliman, J. P. and Nelson, P. (2016b). Quantifying nitrogen losses in oil palm plantations: Models and challenges. *Biogeosciences*, 13, 5433–52.

Poeloengan, Z. (1989). Physical land evaluation using a parametric method application to oil palm plantation in North Sumatra, Indonesia. In: *Proceeding of 1989 PORIM International Palm Oil Development Conference*.

Roose, E. J. (2017). *Restauration de la productivité des sols tropicaux et méditerranéens*. IRD Editions. Collection Synthèses. ISBN: 978-2-7099-2277-7.

Roose, E. J., Lal, R., Feller, C., Barthes, B. and Stewart, D. A. (2006). Soil erosion and carbon dynamics. In: *Advances in Soil Science*. CRC Press, Boca Raton, FL, 352pp.

Soares, J. R., Cassman, N. A., Kielak, A. M., Pijl, A., Carmo, J. B., Lourenco, K. S., Laanbroek, H. J., Cantarella, H. and Kuramae, E. E. (2016). Nitrous oxide emission related to ammonia- oxidizing bacteria and mitigation options from N fertilization in a tropical soils. *Nature Scientific Reports*, 6, 30349. DOI:10.1038.srep30349.

Tao, H. H., Snaddon, J. L., Wahyu, R., Caliman, J. P. and Willis, K. J. (2014). Effects of empty fruit bunch (EFB) application on soil fauna feeding activity in oil palm plantations. In: *International Conference on Oil Palm and the Environment ICOPE 2012*, 12–14 February 2014, Bali, Indonesia.

Tao, H. H.; Slade, E. M., Willis, K. J., Caliman, J. P. and Snaddon, J. L. 2015. Effects of soil management practices on soil fauna feeding activity in an Indonesian oil palm plantation. *Agriculture, Ecosystems and Environment*, 218, 133–40.

Tao, H. H., Snaddon, J. L., Slade, E. M., Caliman, J. P., Widodo, R. H., Suhardi and Willis, K. J. (2017). Long-term crop residue application maintains oil palm yield and temporal stability of production. *Agronomy Sustainable Development*, 37, 33.

Tollefson, J. (2017). Rainforecast go mobile. *Nature*, 544, 146–7.

Velasquez, E., Lavelle, P. and Andrade, M. (2007). GISQ, a multifunctional indicator of soil quality. *Soil Biology and Biodiversity*, 39, 3066–80

Webb, M. J., Nelson, P. N., Bessou, C. and Caliman J. P. (2015). Sustainable management of soil in oil palm plantings. In: *Proceedings of a Workshop Held in Medan, Indonesia*, 7–8 November 2013.

Whittaker, R. H. (1962). Botanical review. *Classification of Natural Communities*, 28(1), 1–239.

Whittaker, R. H. (1970). *Communities and Ecosystems*. MacMillan Publishing Company, Inc., Toronto and New York, pp. 51–64. https//books.google.com/books?id=T6m ØAAAAIAAS.

Whittaker, R. H. (1975). *Communities and Ecosystems*. MacMillan Publishing Company, Inc, New York.

Zuraidah, Y., Zulkifli, H., Haniff, M. H., Nur Zuhaili, H. A. Z. A., Nordiana, A. A., Shuib, A. R., Afifah, A. R. and Nur Maisarah, J. (2015a). Alterations of soil physical properties due to mechanisation activities under oil palm on Bernam Series soil. *International Journal of Agriculture Innovations and Research*, 3(5), 1435–46. ISBN (on line): 2319–1473.

Zuraidah, Y., Haniff, M. H. and Zulkifli, H. (2015b). Oil palm roots adaptation under soil compacted by mechanisation. *International Journal of Agriculture Innovations and Research*, 5(4), 331–42. ISSN (P): 2250–0057; ISSN(E): 2321–0087.

Zuraidah, Y., Haniff, M. H. and Zulkifli, H. (2017). Does compaction affect oil palm standing biomass. *Journal of Oil Palm Research*, 29(3), 352–7.

Chapter 12

Maintaining soil health in oil palm cultivation

Bernard Dubos and Didier Snoeck, CIRAD, France

1 Introduction

In West and Central Africa, initiatives to set up organized oil palm plantations were reported as early as the nineteenth century (Berger and Martin, 2000). However, the first modern plantations to contribute significantly to oil and fat production only appeared after World War II in the Belgian Congo, Côte d'Ivoire, Nigeria, Ghana and Cameroon (Corley and Tinker, 2016). The first plantations in Southeast Asia were created between 1910 and 1920. The areas increased rapidly and, after World War II, Malaysia and Indonesia were definitively the leading region in the world in terms of areas planted (Rival and Levang, 2014). In Latin America, plantations started appearing in the 1950s in Colombia and Ecuador, and more recently in Brazil and Central America.

The climatic requirements of the crop mean that the cultivated areas are mainly located in an equatorial band between the latitudes ten degrees North and South. This area is also the home of equatorial and tropical rainforests. Oil palm plantations set up after cutting primary or secondary forests were therefore very frequent at the beginning of the twentieth century in Africa and Asia. After an initial cycle of around 30 years, replanting with oil palm started while the rapid spread of cultivated areas in Southeast Asia continued to take place with the conversion of natural forests or forests managed for timber. In Indonesia and Malaysia, depending on regions, it is estimated that 55–90% of the areas planted have been established on forests (Pauli et al., 2014).

In tropical soils under forest, the topsoil is where the fertility lies due to its physico-chemical properties, developed through an accumulation of organic matter (OM) and intense biological activity. That layer is protected by the plant cover and the formation

http://dx.doi.org/10.19103/AS.2017.0018.32

of a humus-bearing layer. Erosion and sediment transportation processes are minimal under tropical forests, despite being subjected to high rainfall (Hartemink, 2006a). The quality of soils inherited from forests converted to oil palm plantations led to high palm oil yields, which contributed to the crop expansion. In Côte d'Ivoire, Ollagnier et al. (1978) investigated the changes arising from cultivation by comparing soil properties at different ages (up to 14 years) with those still under forest, down to a depth of 90 cm. They concluded that after a rapid change in the first four years after felling of the associated trees, the topsoil evolved towards a new chemical state corresponding to 60–90% of the initial state depending on the variables examined (C, N, CEC, pH and exchangeable cations). According to these authors, the observed changes towards a new stable state did not constitute a degradation likely to have an adverse effect on yield potential over the long term if the best management practices at the time were respected.

The main challenges for soil health were thus established to preserve, or even improve, physico-chemical properties that enable high yields. Along with preserving the chemical reserves of soils, protecting them from the risks of physical degradation through erosion also needs to be added. Finally, cultivated soil under oil palm should not be a threat to the environment, which can occur when leached mineral nutrients contaminate water tables and courses.

2 Key issues and challenges

2.1 Soil: an essential factor for crop sustainability

Although yields depend on several factors, especially of genetic and climatic order, the physico-chemical properties of the topsoil down to a depth of 30 cm play an essential role in influencing the crop's supply in water and nutrients. Oil palm is grown on a wide range of soils whose profiles differ depending on their origin (nature of the parent material, climate and topography) and how they evolve after planting. So, in each situation, the properties of the different layers of the soil profile and the limitations for the crop will vary.

Several authors have proposed keys for understanding to define classes in which potential yield is reduced by certain limitations. Indeed, Olivin (1968) defined a framework of five agronomic classes based on texture, the existence of coarse elements, the adequacy of drainage and chemical fertility. Such classification gave good results in Africa where the criteria had been defined. Paramanthan (2000) developed a similar approach involving five ability classes based on several precise variables, taking into account topography and other factors that seemed more specific to Asia, such as soil salinity or the existence of an acid sulphate horizon. Setting aside these two aspects, which are not particularly widespread in the oil palm world, the variables identified by Olivin and Paramanthan converge to describe the main functions of soil for the crop, which are: (i) to provide a sufficient water supply for the compensation of evaporative demand, especially during periods of low rainfall. Besides, soil water saturation must not be excessive and drainage must be adequate enough to avoid the establishment of an anaerobic environment over long periods during the rainy season; (ii) to constitute a reservoir of nutrients which must be both available for the plant and minimally exposed to losses through leaching; and (iii) to protect the soil from erosion and run-off, which contribute to losses of both sediment and OM.

The ecosystem functions of soils linked to their biological activities also need to be added to the above functions.

2.2 Soil and water supply

The oil palm is highly susceptible to water deficits that occur in climates with a marked dry season, especially in Africa. As soil is the reservoir that helps to mitigate drought effects, its depth is one of the most important physical properties which govern that function. The topsoil is most colonized by roots, though primary and secondary roots have been found down to a depth of more than 6 meters (Jourdan, 1995). The absence of an indurated horizon in the first 2 meters is considered as favouring adequate rooting.

The soil texture also plays an essential role in the occurrence of structures which are able to promote an adequate water reserve. The quality of such structure (dimensions of aggregates and voids, porosity, stability of aggregates) determines the volume of water available over a given depth and is highly dependent on several other parameters, notably the biology of the soil, plant cover and cultural practices. Both the depth and structure properties of the soil determine the available water reserve at a given stage of the crop under given climatic conditions.

2.3 Soil and nutrient supply

The chemical properties of the topsoil, which are used by most authors as an indicator of soil fertility, are the amounts of OM and N, the cation exchange capacity (CEC), the pH and the amounts of exchangeable potassium (K), calcium (Ca) and magnesium (Mg), along with base saturation.

Some authors also mention soil phosphorus (P) reserves as an indicator of fertility, although this aspect of soil chemical fertility is mainly related to the management of crop fertilization. Soil salinity problems are rare and the methods to be implemented for keeping a saline water table at a suitable depth do not call for a specific approach for the oil palm. Moreover, sodization of the soil profile is controlled by usually abundant rainfall which naturally keeps sodium at a depth if soil drainage is effective.

Tropical soils undergo the process of weathering and leaching of cations and anions due to heavy rainfall and high temperatures which are typical of the oil palm cultivation zone. These phenomena, combined with the frequent presence of 1:1 layer clay minerals, mean that the CEC is generally weaker than in soils developed under a temperate climate. OM therefore plays a more important role when the texture is sandy and the pH is acidic.

Finally, erosion processes directly reduce the ability of the soil to provide the crop with water and nutrients by altering the topsoil horizon, which is by far the most useful. This also leads to the decay of fine absorbing roots once they are exposed to the open air. The factors that increase erosion are described, and rainfall intensity, slopes and soil cover were found to be the most important (Hartemink, 2006b). It is practically impossible to control the intensity of incident rainfall at the plantation scale; the installation of terraces and suchlike along contour lines can help in some cases to overcome slopes, although such techniques are expensive and usually limited in area. This is why one should be focusing more on managing biomass under oil palm in order to protect the soil from erosion. These techniques are all the more promising in that they can be applied to even the gentlest slopes.

3 Management practices and soil health in oil palm plantations

Operations in oil palm plantations modify soil properties throughout the life span of the crop by creating distinct zones with various different activities (Fig. 1A). Although many configurations exist, in most cases a distinction is made between at least three management zones (or management units, Nelson et al., 2015): the chemically weeded circle (WC); the harvesting path (HP), which is mostly used for harvesting and dispersion of fertilizers; and the area in which pruned fronds are deposited (FP-Frond pile), which is where herbaceous or woody regrowth is generally more abundant. In some plantations, in order to avoid accident due to the spiny petiole, this part of the leaf is separated to form a petiole stack (PS). Some authors also distinguish an intermediate zone (BZ – Between Zone) which is the area located outside the transit zones (HP and WC) but which is not covered by pruned fronds. The relative sizes of each of the sectors do vary with the practices adopted by each plantation manager and with the planting density of oil palms: with a radius of between 2.0 and 2.5 metres, WCs account for around 20% of the area planted; with a width of 3 metres, HPs account for around 20%; and the FP areas vary from 20% to 30% depending on the spreading adopted. In theory, this zone may be expanded to cover the entire area, apart from the circles, that is 80% of the planted area. BZs cover the additional area.

In addition to these different zones individualized by management practices, it is also necessary to consider the effect of fertilizers which are applied to specific areas and the

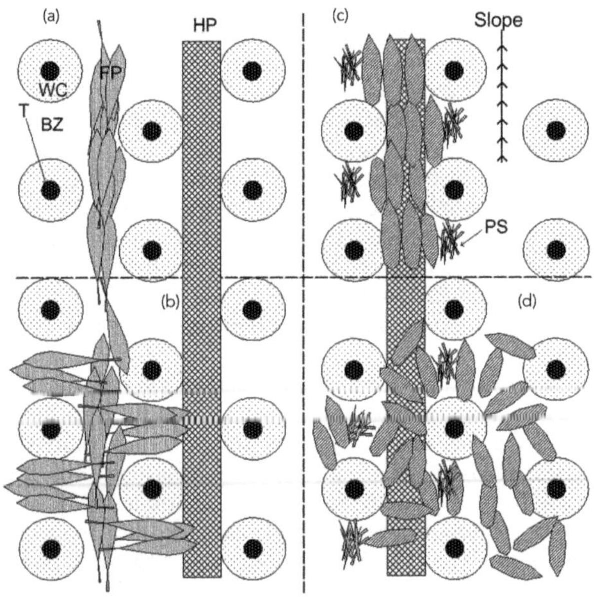

Figure 1 Various types of frond disposal. (a) traditional with alternate frond piles (FP) and harvesting path (HP), (b) U-shaped frond disposal, (c) frond disposal in the centre of the interrow exposed to erosion processes and petiole stacks (PS), (d) random disposal and petiole stacks.

periodic recycling of mill by-products in the form of fresh empty fruit bunches, compost, and in some cases even the dispersion of liquid effluents or sludge. Depending on the choices adopted for all these practices, the physico-chemical properties of soils will change either negatively or positively. Interaction with other characteristics which are site specific (soil texture, topography, rainfall pattern) may also increase the resulting differentiation from the properties of the original soil.

3.1 Effect of pruned frond recycling on some soil properties

In a survey conducted on smallholdings in Benin, Aholoukpé et al. (2016) compared soil properties under frond piles with those of soils where all fronds were removed from the plot. The purpose of the study was to assess the long-term effects of this practice, which consists in using pruned fronds as domestic fuel by local populations. The effects of fronds recycling on the total C, total N, water pH and exchangeable K variables depending on soil depth are recapped in Fig. 2. A very clear improvement in pH, total organic C, total N, CEC and exchangeable cations can be observed even at a depth of 25 cm, beyond which the effect of frond OM becomes negligible.

Figure 3 describes a lower soil density under frond piles (0–20 cm) than without OM recycling. Frond recycling therefore helps in increasing soil porosity and available water reserves in the horizon immediately below the frond layer. The same layer is also the one colonized by fine roots, and it is essential for the supply of water to oil palm.

Pruned fronds amount to an average ten tonnes of dry matter per hectare in mature oil palm plantations (Redshaw, 2003). They provide a source of OM which is cheaply available on site. This biomass has long been spread in every other inter-row (Fig. 1A), the idea being to limit the area covered by fronds so as not to hinder operations (upkeeping, fertilization, harvesting).

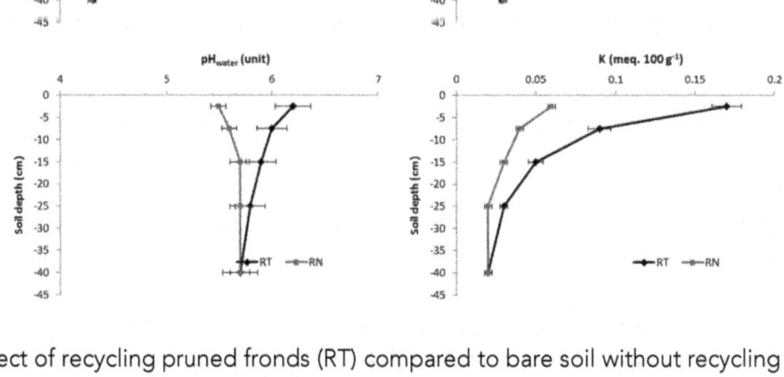

Figure 2 Effect of recycling pruned fronds (RT) compared to bare soil without recycling (RN) on 4 soil properties in Benin (according to Aholoukpè et al. 2016). Each mean is presented with the confidence interval (at $p < 0.05$).

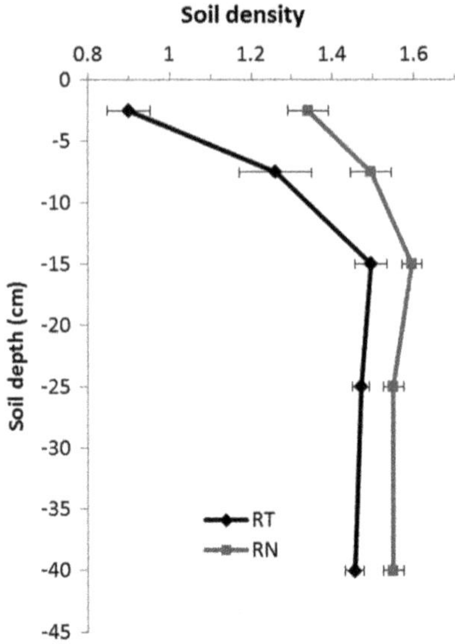

Figure 3 Changes in soil bulk density under total recycling (RT) and no recycling (RN) for the 0–50 cm soil layer under mature palms (13–24 years) (according to Aholoukpè et al. 2016). Each mean is presented with the confidence interval (at $p < 0.05$).

The results obtained for OM effects on soil properties have led growers to adapt how they position the fronds, seeking to amplify their beneficial effects. Depending on the local situation and constraints, several effects may be sought after:

- *Better topsoil porosity and colonization by roots:* The increase in pH and in CEC, and the appearance of organic humus compounds promote the generation of a stable structure enabling adequate aeration and drainage conditions. To a certain degree, this area will be protected from evaporation by a mulch effect due to the plant debris, which is useful under a climate with a marked dry season. Among smallholders, who do not have the constraints of mechanized operations, unlike agro-industrial estates, virtually the entire area, apart from the weeded circles, can be covered by fronds (Fig. 4 and Fig. 1D).
- *Greater ability of the topsoil to retain fertilizing inputs:* This function is all the more efficient as the amended area promotes the development of absorbing roots. Although fertilizers are sometimes spread over the surface of the weeded circle, it is recommended that they be applied on the areas covered by OM as an improvement in the pH and CEC values will favour cation fixation on the surface and thus will reduce the risk of leaching.
- *Erosion and run-off control* (Fig. 5)*:* Pruned fronds can be laid on areas subject to run-off, in order to create obstacles which will reduce the flow rate and curb the erosion process (Quencez, 1986). This is generally the first step that encourages further colonization by plants whose roots will fix the soil. In addition to pruned fronds, which sometimes need to be fixed to the soil, other sources of OM can be used, notably EFB.

Figure 4 Frond disposal in Ecuador.

Figure 5 Run-off and erosion processes in Ivory Coast.

- Finally, and this is definitely the most essential aspect over the long term to provide a habitat for soil organisms that contribute to several soil-based goods and services (Wall, 2012). Among the most important are soil OM synthesis linked to the conversion of organic C, the nutrient cycle and soil structure maintenance (Kibblewhite and Ritz, 2008).

These services involve the arthropods, nematodes, fungi and bacteria making up the soil biota (Carron et al., 2016; Tao et al., 2016).

3.2 Effects of some mineral fertilizers on the chemical properties of soils

Fertilization trials are conducted to assess the requirements of the plant in order to reach or approach the maximum yield. These are determined for each situation by several variables, notably the climate and the planting material. The experimental designs selected will serve in determining the leaf contents that will be considered as optimum for reaching yield targets. By using an indicator of the mineral composition of the foliage in each plantation block (25–50 ha), annual fertilizer requirements will be determined. This scheme totally ignores the status of the soil's reserves, and no accurate decision-making tool has been proposed based on soil analyses for the industrial management of fertilization (Goh, 2004). However, although managing mineral nutrition of plantations using leaf analyses provides satisfactory results in terms of both vegetative growth and oil yields, the fact that a stable composition for the aboveground organs can be maintained does not guarantee that an equivalent balance exists in soils. It is therefore reasonable to explore changes in the soil's mineral reserves on the long term. Apart from changes in the physical properties of soils that are linked to cultural practices (e.g. compaction due to the passage of agricultural machinery) and are not covered in the present chapter, two questions are frequently recurring about the chemical fertility of soils under oil palms: (i) Does mineral fertilization modify soil chemical properties such as pH or cation balances? (ii) On the long term, are the so-called mining effect processes occurring that alter the soil nutrient reserves that feed the soil solution?

In Ecuador, Dubos et al. (2016) used two long-term fertilization trials in order to investigate these issues. After ten years of urea and KCl applications using different protocols (Tables 1 and 2), the topsoil (0–20 cm) was analysed while distinguishing between the zones directly receiving fertilizers and those deprived of treatment.

In both trials, N application did not improve FFB production (data not shown), but the consequences for the chemical properties of the fertilized soils were substantial, since it resulted in a drop in pH following cation depletion. The effects were greater when the CEC and pH values in the original soils were high, as shown in CP08 trial. To explain the variation in pH found between the two trials, the authors proposed that leaching of excess N occurred under the form of nitrates towards the underlying horizons.

The consequences of KCl applications on the fertilized soil were also marked: the exchangeable K content increased very significantly and could exceed 1 cmol kg^{-1}, which is a very high value (Goh, 2004) for the types of tropical soils where oil palm is grown. This

Table 1 Effects of N fertilizer on pH and exchangeable cations in the fertilized zone after 10 years of application (according to Dubos et al. 2016)

		CP06				CP08		
		N0	N1	N2	*P*-value	N0	N1	*P*-value
N applied (kg palm^{-1} yr^{-1})		0	0.45	1.35		0	1.35	
pH $_{water}$		4.95	4.80	4.63	0.089	5.88	4.71	**0.000**
Ca	cmol kg^{-1}	1.563a	0.751b	0.500b	**0.002**	1.953	1.805	**0.000**
Mg	cmol kg^{-1}	0.538	0.390	0.306	0.088	1.101	0.525	**0.001**
K	cmol kg^{-1}	0.426	0.438	0.370	0.776	1.108	0.785	0.054
Al	cmol kg^{-1}	1.637b	2.374ab	2.972a	**0.016**	0.028	0.946	**0.002**
H	cmol kg^{-1}	0.089b	0.106ab	0.126a	**0.008**	0.015	0.109	**0.000**
CEC	cmol kg^{-1}	5.89	5.73	5.86	0.770	7.50	4.90	**0.000**

P-value: significant results are shown in bold. Means followed by the same letter are not significantly different, according to Tukey's test.

Table 2 Effects of K fertilizer on exchangeable cations in the fertilized zone after 10 years of application (according to Dubos et al. 2016)

		CP06				CP08		
		K0	K1	K2	*P*-value	K0	K2	*P*-value
K applied (kg palm^{-1} yr^{-1})		0	0.5	1.5		0	1.5	
Ca	cmol kg^{-1}	1.090	0.903	0.821	0.343	4.041	2.716	**0.007**
Mg	cmol kg^{-1}	0.366	0.419	0.449	0.660	0.986	0.640	**0.012**
K	cmol kg^{-1}	0.030 b	0.193 b	1.010 a	**0.000**	0.261	1.631	**0.000**
Total	cmol kg^{-1}	1.49	1.52	2.28		5.29	4.99	

P-value: significant results are shown in bold. Means followed by the same letter are not significantly different, according to Tukey's test.

imbalance affected Ca and Mg contents, whose significant drop compensated for the increase in K (CP08).

The harmful effects of N and K fertilizers on the chemical properties of soils were found to increase in line with application rates. Indeed N and K are the two nutrients most used by oil palm, be it by the aboveground vegetative organs or by the FFB. The observed changes in exchangeable cations may be reflected in the unfavourable evolution of the soil structure, as seen in Côte d'Ivoire (Caliman et al., 1987). When replanting oil palm plantations in savannah in the Dabou region, these studies showed that a drop in productivity between the first and second cycles was linked to the unfavourable evolution of the topsoil structure, which limited its colonization by roots. The measures that were successfully tested consisted in restoring the structure by subsoiling and applying gypsum to rebalance the monovalent:bivalent cation ratio.

Environmental risks are directly linked to the leaching of excess nitrates and K. Indeed the decrease in CEC observed for fertilized soil increases the risk of water table pollution, particularly as NH_4 and K adsorption is low. It is therefore definitely worth not over-fertilizing the plants, meaning that the mineral status of plants and soil in oil palm plantations needs to be accurately assessed, ensuring that fertilizer uptake efficiency is optimum and losses are limited. Foong (1993) estimated maximum N and K losses by leaching at 5.8 and 4.4%, respectively, in mature plantations in Indonesia, which seem to be low values. However, Banabas et al. (2008) estimated N losses under the root layer at 47% when incident rainfall was very intense. Further studies would therefore be worthwhile (Corley and Tinker, 2016) in the aim of determining the conditions that tend to reduce the risks of leaching. There are roughly two mechanisms that promote cation retention in the soil, and which can help in reducing NH_4, K, Ca and Mg losses: a high CEC and the presence of fine absorbing roots. OM applications do promote both mechanisms, which leads us to recommend applying fertilizers on the areas which are the richest in OM, that is, as a priority in the frond spreading zones, but also in zones where oil mill by-products are spread when practised. The favourable effect of fertilizer dispersion can also be added to the advantages of OM. In the two trials examined above, the fertilizers were applied over an area not exceeding 10 m² per palm, which is small when compared to the 70 m² potentially available with a planting density of 143 palms per ha. A theoretical calculation was made for a fertilizer area of 30 m² on a soil with a bulk density of 1.3 and a useful layer of 25 cm for absorbing roots. K application equivalent to 1 kg of KCl, that is, 1282 cmol, only amounts to 2.6% of the total CEC of the volume of fertilized soil if the CEC measured is 50 cmol/kg. This level for CEC, which is very low, is that of a very sandy soil propitious to leaching. However, the degree of disturbance caused by KCl application is intuitively acceptable when considering the risk of displacing other cations from the exchange surface.

In a study of soil properties after repeated applications of N and K fertilizer, Dubos et al. (2016) also considered soils that were not fertilized. The idea was to compare the soil reserves under unfertilized treatments with those from treatments balancing N and K requirements of the palms with fertilizer applications. After ten years without fertilizers, no mining effect could be measured, as the soil reserves were not significantly lower. This conclusion does not mean that no soil impoverishment took place, but rather that the degree of take-off over ten years was not enough to be detected given the measurement system and the accuracy of the analysis methods in the laboratory. Nevertheless, maintaining the chemical reserves of soils remains as a concern over the long term.

4 Optimizing biomass recycling: the promising way to increase yields and sustainability

Recycling available biomass is a powerful way of restoring, preserving or improving the properties of soils. OM conservation begins as soon as the crop cycle is created by adopting non-destructive practices (especially zero burning), unlike what was sometimes done in previous centuries. The recycling of oil mill by-products has become a widespread practice, be it under the form of EFB or compost. At the outset, it was a strategy to substitute mineral fertilizer with organic fertilizers, thus enabling some use of by-products while providing a technical solution for their recycling. At the plantation scale, the improvement in soil fertility was confirmed in blocks under organic fertilization, thus reflecting in an increase in pH, in organic C, in the CEC and in exchangeable cations (Comte et al., 2013). However, palm oil mill by-products, whose effects can be expressed differently depending on the nature of the soil, cannot be spread over all areas, unlike pruned fronds which are available everywhere without any need for transportation. At the palm tree scale, EFB application over a very limited zone (the harvesting path) leads to marked changes compared to application in the other zones (Rosenani et al., 2011). However, OM applications bring into play complex mechanisms that go beyond the foreseeable effects due to the chemical composition of the recycled matter. For what concerns chemical parameters, an increase in pH, Ntotal, Corg, CEC and exchangeable cations is generally measured in the application zone. However, those changes may come about through an imbalance, especially for P, K and Mg and the pH. As regards biological parameters, it was found that EFB applications impacted the communities of living organisms making up the biota of the soil, especially earthworm communities, whose movements followed variations in nutrient availability or stress associated with the application of organic by-products (Carron et al., 2015a). An analysis of the variability in soil parameters after applying organic waste brought out three distinct periods: 1) a period of strong disturbance lasting around six months, with an increase in pH, base saturation, K and the macrofauna density (especially ants), but a decrease in earthworms, centipedes and nematodes; 2) a period of resilience lasting around 12 months; and 3) a period of stimulation lasting around eight months with an increase in macrofauna (earthworms, centipedes and nematodes) activity and some improvement in the soil fertility parameters (Carron et al., 2015b). Beyond two years' time, the analysis of physical, chemical, morphological or biological indicators of soil quality revealed that EFB application helped to restore the soil under the harvesting path and thereby homogenized properties of the different zones, which were no longer different for these parameters at the palm tree scale (Carron et al., 2016).

Of the four ecosystem services identified by Brussaard (1997), OM decomposition, bioturbation and nutrient recycling are the most capable of ensuring that soil health is maintained over the long term. The soil biota, especially earthworms, break down OM, redistribute it at depth, and increase aeration and water infiltration. Some bacteria (*Rhizobia* group) facilitate nitrogen fixation by developing nodules on plant roots (regrowth, legume cover crop). Several studies have indicated that soil biodiversity is a key factor for maintaining ecosystem services (Wall, 2012). It also seems that the stability (i.e. resistance and resilience) of ecosystem services in relation to disturbance and stress generally appears to be more strongly related to species richness (Brussaard et al., 2007).

Soil biodiversity cannot be modified directly by introducing individuals and, for that reason, improving environmental conditions that are favourable to biodiversity increase seems to be the most promising strategy.

5 Future trends in research

Two avenues are opened up for going beyond the effect of OM on chemical properties and for understanding how the services provided by soil can influence crop functioning.

Studies on how EFB application rates and frequencies affect the biodiversity of soils (Carron et al., 2016) have only made it possible to describe changes in the communities of organisms at a given instant, but not to measure the impact of those changes on the plant itself. These results therefore need to be extended by linking the diversity of the soil biota to the functions that are decisive for carbohydrate production. The crop's water supply, nutrient cycle and root biomass should therefore be observed over sufficiently long time lapses for the properties of the soil and the composition of the biota to be stabilized.

Long-term monitoring of cultivated soils under oil palm is a recent concern for growers. It is often suggested that soil analyses should be carried out periodically, once every 3–5 years, in order to assess the reserves of the soil and its ability to retain nutrients (i.e. OM and CEC) (Paramananthan, 2015). It is also covered by the principles and criteria governing certification through voluntary Roundtable for Sustainable Palm Oil (RSPO) standards. Until now, the decision of undergoing or not a long-term monitoring of soil health belongs to plantation managers only and no standard has been issued for interpretation over the long term. Yet, as already mentioned, the accuracy of observations depends on the accuracy of analyses delivered by laboratories and this would probably not be enough to guarantee an objective analysis of the sustainability of soil reserves over a time lapse as short as 5 years. Nelson et al. (2015) proposed a simple method for directly taking composite samples proportionally representing the different zones of the crop cycle and of the preceding cycle in the case of replanting. This method seems to be particularly appropriate for very long time lapses (decade, culturing cycle). It can be used to calculate directly the overall reserves of the planted area (e.g. 1 ha) for a given depth. One might question the ability of this single variable (overall evaluation) to measure the sustainability of plantation management when the planted area is structured in different zones with different soil properties. Cultural practices, primarily fertilization, locally modify soil reserves (Paramananthan, 2015). It may also be a matter of transfers via the recycling of pruned fronds whose composition depends in return on the fertilization regime (Dubos et al., 2016). Ultimately, among the various zones that are gradually created while oil palm plantations are ageing, some are likely to be exposed more rapidly to two worrying phenomena: (1) an impoverishment of reserves and (2) an enrichment for certain nutrients, up to concentrations that could lead to some leaching towards deeper zones and water tables. One could therefore just as easily choose to monitor unfertilized zones that are not subjected to changes linked to agricultural activity (e.g. weeded circle and harvesting path) in order to detect the impoverishment of reserves or, conversely, to consider zones enriched by fertilizers and/or pruned fronds in order to assess potential environmental impacts. There are therefore some important choices to be made when sampling soils for a long-term monitoring.

Hartemink (2006b) gave considerable thought to the difficulties encountered when estimating the decline in soil fertility in the tropics under tree crops. In particular, he emphasized the consequences of the choice of analysis laboratory and of the methods used. Each laboratory has its own signature which is linked to its equipment and methods and that signature can change over time with the renewal or upgrading of the equipment. For each nutrient, there are several analysis methods that are supposed to estimate the fraction which is available for the plant. For instance, there are more than ten methods estimating available P. Besides, exchangeable cations are frequently extracted with ammonium acetate with a constant pH close to neutral, although cobalt hexamine extraction at a variable pH close to that of the soil solution seems to be suitable for generally acid tropical soils. For Indonesian soils, Foster and Prabowo (1996) found that K extraction with boiling HCl better reflected the K reserves available for oil palms than cation extraction with acetate. The proposed method would be more suitable for assessing K reserves in soils with a low sand content. There is therefore every reason to believe that the methodological search to optimize the most suitable analysis methods for estimating reserves available to oil palm will impact on the diagnoses to be made in the coming decades.

In this perspective, it seems necessary to set up rapidly some reference situations for plantations whose long-term fertility needs to be assessed and it might be necessary to collect samples from varied zones. Such samples describing an initial situation are particularly important when converting forest cover to oil palm plantations.

In order to compensate for a change of laboratory or the adoption of new analysis methods, it is wise to keep duplicate soil samples of the reference situations under excellent preservation conditions for possible future re-analysis.

6 Where to look for further information

General information and educational documents on soil health concepts (assessment and management) can be found in the website of the United States Department of Agriculture (http://www.nrcs.usda.gov/wps/portal/nrcs/main/soils/health/).

General information and data (soil data and soil mapping) can also be found on the ISRIC website (http://www.isric.org/). ISRIC – World Soil Information is an independent, science-based foundation set up on the initiative of the International Soil Science Society (ISSS) and of the United Nations Educational, Scientific and Cultural Organization (UNESCO).

In Malaysia, AAR (Applied Agricultural Resources Sdn. Bhd) is a research centre working on tropical tree crops, particularly oil palm. It produces summaries of R&D findings, some of which are devoted to soil management (http://www.aarsb.com.my/).

7 References

Aholoukpe, H. N. S., Amadji, G. L., Blavet, D. and et al. 2016. Effet de la gestion des feuilles d'élagage du palmier à huile sur le stock de carbone et les propriétés physico-chimiques du sol dans les palmeraies villageoises du Bénin. *Biotechnol. Agron. Soc.*, 20(2), 171–82.

Banabas, M., Scotter, D. R. and Turner, M. A. 2008. Losses of nitrogen fertiliser under oil palm in Papua New Guinea: 2. Nitrogen transformations and leaching, and a residence time model. *Aust. J. Soil Res.*, 46(4), 340.

Berger, K. G. and Martin, S. M. 2000. Palm oil. In Kiple, K. K. and Ornelas, K. C. (Eds). *The Cambridge World History of Food Vol. 1 Part II E3*, Cambridge University Press. pp. 397–411, http://www.book2look.com/vbook.aspx?id=9780521402163.

Brussaard, L., 1997. Biodiversity and Ecosystem Functioning in Soil. *Ambio*, 26(8), 563–70.

Brussaard, L., de Ruiter, P. C. and Brown, G. G. 2007. Soil biodiversity for agricultural sustainability. *Agric. Ecosyst. Environ.*, 121(3), 233–44.

Caliman, J. P., Olivin, J. and Dufour, O. 1987. Dégradation des sols ferrallitiques sableux en culture de palmiers à huile par acidification et compaction. *Oléagineux*, 42(11), 394–401.

Carron, M. P., Auriac, Q., Snoeck, D. and et al. 2015a. Spatial heterogeneity of soil quality around mature oil palms receiving mineral fertilization. *Eur. J. Soil Biol.*, 66, 24–31.

Carron, M. P., Pierrat, M., Snoeck, D. and et al. 2015b. Temporal variability in soil quality after organic residue application in mature oil palm plantations. *Soil Res.*, 53(2), 205–15.

Carron, M. P., Auriac, Q., Snoeck, D. and et al. 2016. Do the impact of organic residues on soil quality extend beyond the deposition area under oil palm? *Eur. J. Soil Biol.*, 75, 54–61.

Comte, I., Colin, F., Grünberger, O. and et al. 2013. Landscape-scale assessment of soil response to long-term organic and mineral fertilizer application in an industrial oil palm plantation, Indonesia. *Agric. Ecosyst. Environ.*, 169, 58–68.

Corley, R. H. V. and Tinker, P. B. 2016. *The Oil Palm* Fifth edition. Hoboken, NJ: John Wiley and Sons.

Dubos, B., Snoeck, D. and Flori, A. 2016. Excessive use of fertilizer can increase leaching processes and modify soil reserves in two ecuadorian oil palm plantations. *Exp. Agric.*, 53, 255–268.

Foong, S. F. Potential evapotranspiration, potential yield and leaching losses of oil palm. In Basiron, Y. et al. (Ed.), Proceedings of the PORIM International Oil Palm Conference. Progress, Prospect and Challenges Towards the 21st Century. Kuala Lumpur Malaysia, pp. 105–19.

Goh, K. J., 2004. Fertiliser recommendation systems for oil palm : estimating the fertiliser rates. In Chew, P. S. and Tan, Y. P. (Eds), MOSTA Best Practices Workshops – Agronomy & Crop Management. Kuala Lumpur Malaysia: Malaysian Oil Scientists and Technologies Association, pp. 235–68.

Hartemink, A. E., 2006a. Assessing Soil Fertility Decline in the Tropics Using Soil Chemical Data. *Adv. Agron.*, 89, 179–225. Elsevier. Available at: http://linkinghub.elsevier.com/retrieve/pii/S0065211305890042 [Accessed26 August 2016].

Hartemink, A. E., 2006b. Soil erosion: perennial crop plantations. *Encyclopedia of Soil Science*, R. Lal (Eds). Marcel Dekker, New York, NY, USA pp. 1613–17.

Jourdan, C., 1995. Modélisation de l'architecture et du développement du système racinaire du palmier à huile. Doctorat. Montpellier France: Montpellier II.

Kibblewhite, M., Ritz, K. and Swift, M. 2008. Soil health in agricultural systems. *Philos. Trans. R. Soc. Lond., B, Biol. Sci.*, 363(1492), 685–701.

Nelson, P. N., Banabas, M., Goodrick, I. et al. 2015. Soil sampling in oil palm plantations: a practical design that accounts for lateral variability at the tree scale. *Plant Soil*, 394(1–2), 421–9.

Olivin, J. 1968. Survey for the localisation of an industrial oil palm block. *Oléagineux*, 23(8), 499–504.

Ollagnier, M., Lauzeral, A., Olivin, J. and et al. 1978. Development of soils under oil palm after deforestation. *Oléagineux*, 33(11), 537–47.

Paramananthan, S. 2015. Soil properties and their influence on oil palm management and yield. In Webb, M. J., Nelson P.N., Bessou C. and et al. (Eds), *Sustainable Management of Soil in Oil Palm Plantings*. ACIAR Proceedings, 144. Canberra: ACIAR, pp. 22–6.

Paramananthan, S. 2000. Soil requirements of oil palm for high yields. In Goh, K. J. (Ed.), *Managing Oil Palm for High Yields: Agronomic Principles*. Lumut Malaysia: Malaysian Society of Soil Science, Param Agricultural Soil Survey, pp. 18–38.

Pauli, N., Donough, C., Oberthür, T. and et al. 2014. Changes in soil quality indicators under oil palm plantations following application of 'best management practices' in a four-year field trial. *Agric. Ecosyst. Environ.*, 195, 98–111.

Quencez, P. 1986. Utilisation des palmes pour lutter contre l'érosion en plantation de palmiers à huile - The use of fronds to control erosion in oil palm plantations. *Oléagineux*, 41(7), 315–20.

Redshaw, M. 2003. Utilization of Field Residues and Mill By-Products. In Fairhurst, T. and Hardter, R. (Eds), *Oil Palm: Management for Large and Sustainable Yields. PPI, PPIC and IPI.* Potashand Phosphate Institute, pp. 307–20.

Rival, A. and Levang, P. 2014. Palms of controversies: Oil palm and development challenges. Available at: http://www.cifor.org/nc/online-library/browse/view- publication/publication/4860.html.

Rosenani, A. B., Darus, S. Z., Kulaseharan, S. and et al. 2011. Effects of ten year application of empty fruit bunches in an oil palm plantation on soil chemical properties. *Nutr. Cycl. Agroecosys.*, 89(3), 341–9.

Tao, H. H., Slade, E. M., Willis, K. J., Caliman, J. P. and Snaddon, J. L. (2016). Effects of soil management practices on soil fauna feeding activity in an Indonesian oil palm plantation. *Agric. Ecosyst. Environ.*, 218, 133–40.

Wall, D. H. and Nielsen, U. N. (2012). Biodiversity and Ecosystem Services: Is It the Same Below Ground? *Nat. Educ. Knowl.*, 3(12), 8. Available at: http://www.nature.com/scitable/knowledge/library/biodiversity-and-ecosystem-services-is-it- the-96677163.

Chapter 13

Use of palm oil for biofuel

Jean-Marc Roda, CIRAD, France and Universiti Putra Malaysia, Malaysia

1 Introduction

1.1 Introducing biofuels

Fossil fuels are refined from coal, oil and gas, of which there is a limited global supply. Their consumption also releases stored CO_2 into the atmosphere. By contrast, biofuels are derived from renewable biomass and are carbon neutral since their consumption releases CO_2 previously absorbed from the atmosphere by growing the original feedstock. Humans have been using 12 to 20% of total terrestrial plant-based net primary production (Haberl et al., 2007). Terrestrial plants essentially produce sugars, starch and oils, while their main constituents are lignocellulose and water. Only 10% of them are used to provide energy, with the remainder being used for animal feed (58%), materials (20%) and human food (12%) (Krausmann and Kowalski, 2008). Consequently, any additional uptake of terrestrial biomass for bioenergy use could increase the pressure on ecosystems if it means producing and harvesting more sugar, starch or oil than is already the case. This pressure on ecosystems lies at the heart of the potential adverse effects arising from the development of biofuels.

1.2 The sustainability of biofuels

The major objections to the use of biofuels are deforestation, competition with food production and uncertainties about their ability to meet some or all of their stated

http://dx.doi.org/10.19103/AS.2017.0018.12

objectives. The global production of biofuel was more than 100 billion litres in 2010 (Wilkinson et al., 2013) raising issues about its competition with food and arable land. Over the last decade, grain and edible oil prices have increased by 70–120%. World food prices reached their peak in 2008, causing world food markets to experience their largest price shock in 30 years, a development which has been linked in part to biofuel production (Bailis and Baka, 2011; Meyer, 2009).

Biofuel production currently uses around 2–3% of arable land globally (Wilkinson et al., 2013). However, several assessments have shown how biofuels can contribute to large-scale land use change (Fargione et al., 2008). Competition occurs through direct land use change which decreases the production of food on arable lands by replacing existing crops (e.g. food or feedstock) with energy crops. Competition can also occur through indirect land use change where biofuel crops change market conditions, either resulting in crops and livestock areas being less profitable, or diverting existing production from food to energy. They may also have other unexpected negative effects: for example, some biofuel production systems emit more greenhouse gases (GHGs) than fossil fuels (Bailis and Baka, 2011). All these direct and indirect effects interact through complex feedback loops connecting agriculture, food and energy systems (Fig. 1). Such complexity is represented in Table 1 which provides a suggested (but not exhaustive) list of sustainability criteria required for responsible biofuel production.

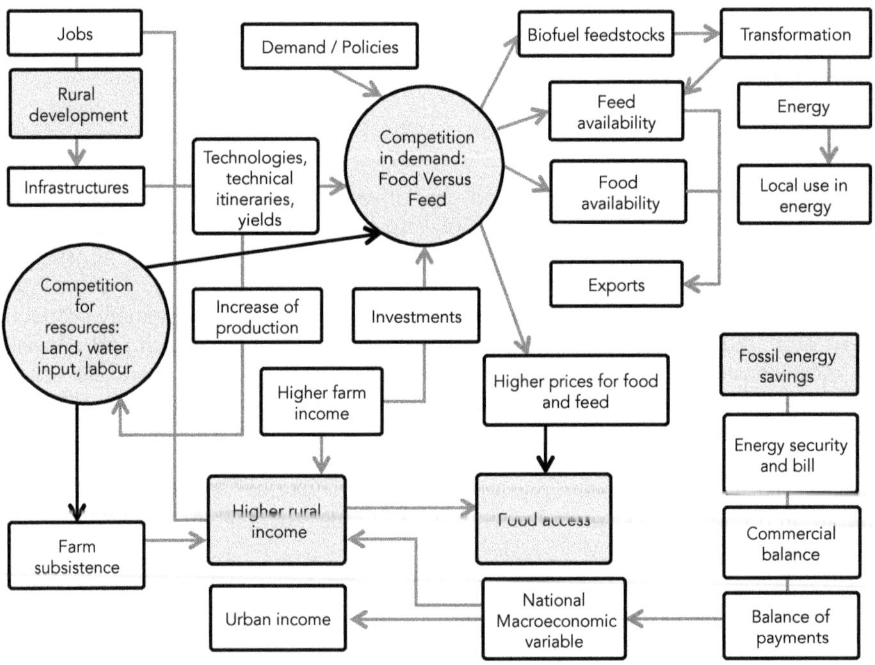

Figure 1 Main impacts and feedback in the food, agriculture and energy systems following the introduction of biofuel demand (adapted from Wilkinson et al., 2013).

Table 1 Biofuel sustainability requirements (adapted from Anon., 2008)

Sustainability criteria	Policy criteria
Do biofuels threaten food security?	Protecting the poor and food-insecure
Can biofuels help promote agricultural development?	Taking advantage of opportunities for agricultural and rural development
Do biofuels help reduce greenhouse gas emissions?	Ensuring environmental sustainability and a national system to support sustainable biofuel development
Can biofuels help achieve energy security?	Reviewing existing biofuel policies

2 Trends in biofuels

Biofuels can be first-, second- or third-generation depending on the raw material and conversion technology used. First-generation biofuels refer to fuels that are derived from crop sources such as sugars, starch and vegetable oil. Second-generation biofuels are essentially based on lignocellulosic biomass or industrial and urban waste. Lignocellulosic biomass is derived from crop residues found in fields, plantations or forestry residues. Technically all biomass can be connected to second-generation pathways and processes (Fig. 2). Third-generation biofuels essentially refer to oils obtained from algae.

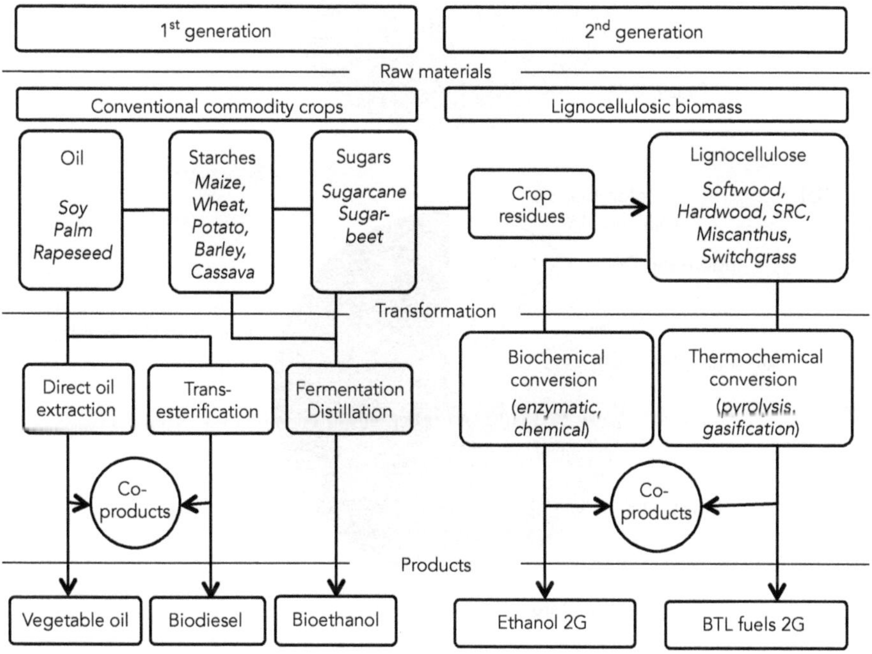

Figure 2 Pathways for producing first- and second-generation biofuels (adapted from Naik et al., 2010).

In the aftermath of the oil crises of 1973 and 1979, the United States and Brazil encouraged first-generation biofuel technology to convert corn and sugarcane into bioethanol and biodiesel. This early start gave the United States and Brazil a leading role in the biofuel sector, though other countries followed later. Between 2005 and 2008, major biofuel programmes were initiated in the European Union and the United States. By 2012, around 60 countries had targets in place to promote biofuel production for energy security, improving the balance of payments in their countries, creating new sources of income and employment, developing rural areas and reducing GHG emissions. In 2014, the United States and Brazil still dominated global biodiesel production with 78% of the world production, but the European Union has since become a major player. 72% of the world's biodiesel sources remain temperate rapeseed oil and soybean oil despite the growth in the use of other sources of biodiesel such as palm oil (Fig. 3) (Neslen,

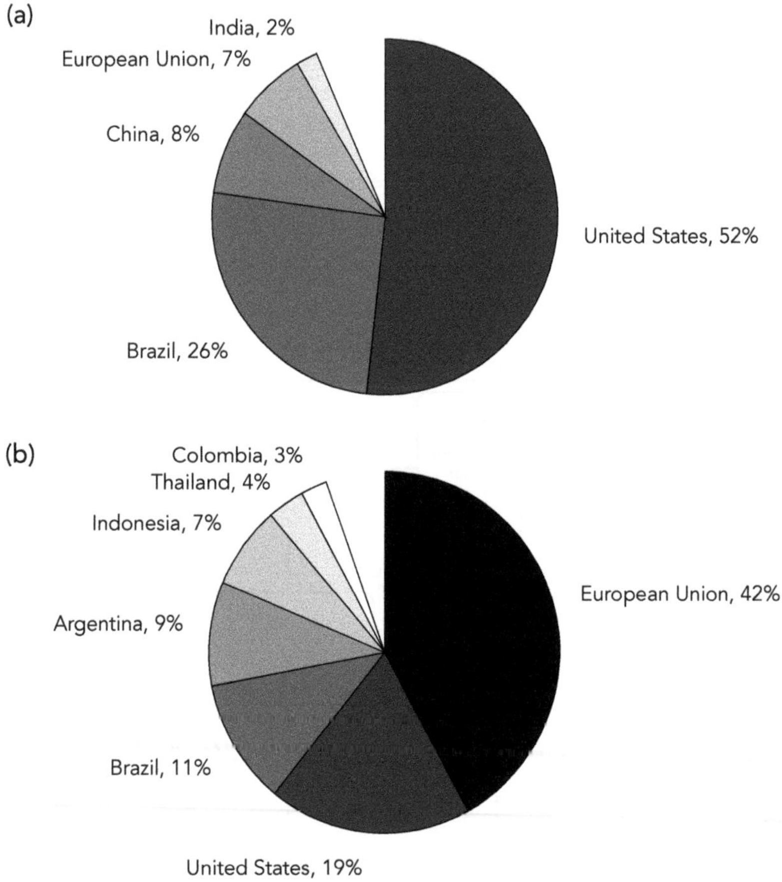

Figure 3 Major world producers of bioethanol and biodiesel in 2014 (OECD, 2015). (a) Bioethanol (113 billion L). (b) Biodiesel (30 billion L).

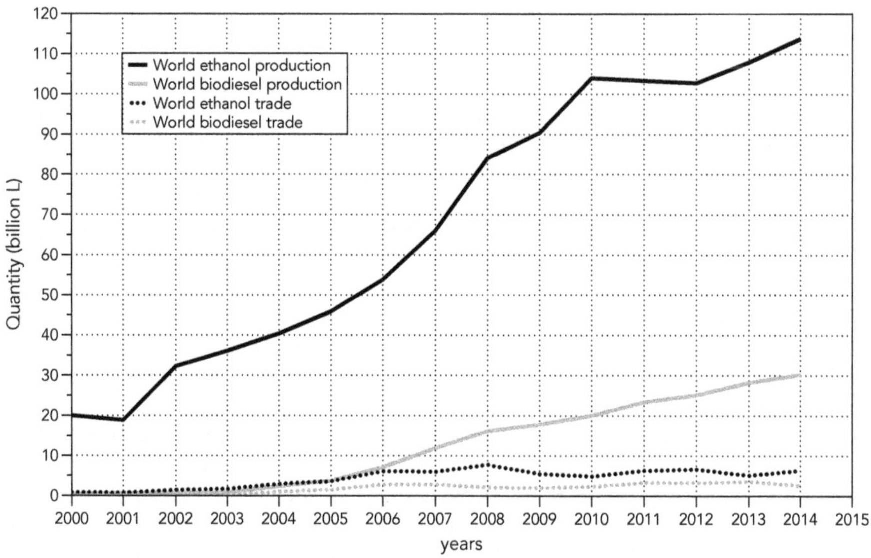

Figure 4 Growth of world production and trade of bioethanol and biodiesel (OECD, 2015).

2016; Dings, 2016). Bioethanol production really took off after 2001 and biodiesel quickly followed. In 2014, the world production of bioethanol and biodiesel reached 143 billion litres (Fig. 4) with 7–8% growth per year on average since 2001.

However, there was increasing concern that first-generation fuels could easily become unsustainable if produced in large quantities because of the potential stress that global production would place on food commodities (Gomez et al., 2008). Consequently, there has been a shift towards second-generation biofuels to mitigate the disadvantages of first-generation biofuels. Even though second-generation biofuels are theoretically attractive, a number of technical hurdles still exist and profitability needs to be improved. Besides the issue of profitability, their sustainability also needs to be more fully investigated. Among potential impacts are effects on GHG emissions, biodiversity loss, soil degradation, effects on food security, rural development issues, health hazards and social conflict (Gomez et al., 2008).

Second-generation biofuels from lignocellulosic raw materials are now produced on an industrial scale in the United States and Brazil, but still at lower volumes than first-generation biofuels. Second-generation biofuels represent less than 1% of overall biofuel production in most countries apart from Brazil (International Energy Agency, 2013; Anon., 2014). Very few tropical countries have the resources to develop second-generation biofuels given the often proprietary nature of this technology, the high capital investment required and the need for infrastructure, logistics and human capital (Wilkinson et al., 2013). In 2016 Indonesia announced its plans to invest in the production of significant volumes of lignocellulosic second-generation biofuels but simultaneously announced new plants for first-generation biodiesel production, the technology which has been in place for about a decade (Erni Ginting, 2015).

3 Production of palm-based biofuels

Oil palm fruits produce a lot of cellulosic waste and two main types of oil, palm oil from the fibrous mesocarp and kernel oil from the fruit kernel (Fig. 5). A few products have been developed on a very small scale such as solid residues for combustion, charcoal, etc. Currently, there is no large-scale second-generation biofuel produced from palm oil waste. When occasionally these wastes are used, some of the cellulosic waste (core of the trunks and part of the fronds) is sold for animal feed for a higher price than second-generation biofuels would incur. Another part of the waste is sometimes gasified in existing plants generating electricity. Some palm oil processing plants produce methane through treatment of liquid palm oil mill effluents (POMEs). Since such plants are located in remote parts of Indonesia and Malaysia, they are too distant from the grid for profitable electricity commercialisation and the excess gas is burnt (flared). Currently, the only palm oil-based biofuel developed on an industrial scale is first-generation biodiesel.

Only one tonne of kernel oil is extracted compared to 8–9 t of palm oil (Fig. 5). Kernel oil is sold at much higher prices than palm oil. Even though biodiesel production from the kernel is technically feasible, palm oil is the first choice for biodiesel production. In the conventional milling process, the fresh fruit bunches are first sterilised and then steamed in vessels pressurised up to three bars to halt natural fermentation and to have the fruits ready for the subsequent processes. The sterilised bunches are then stripped of

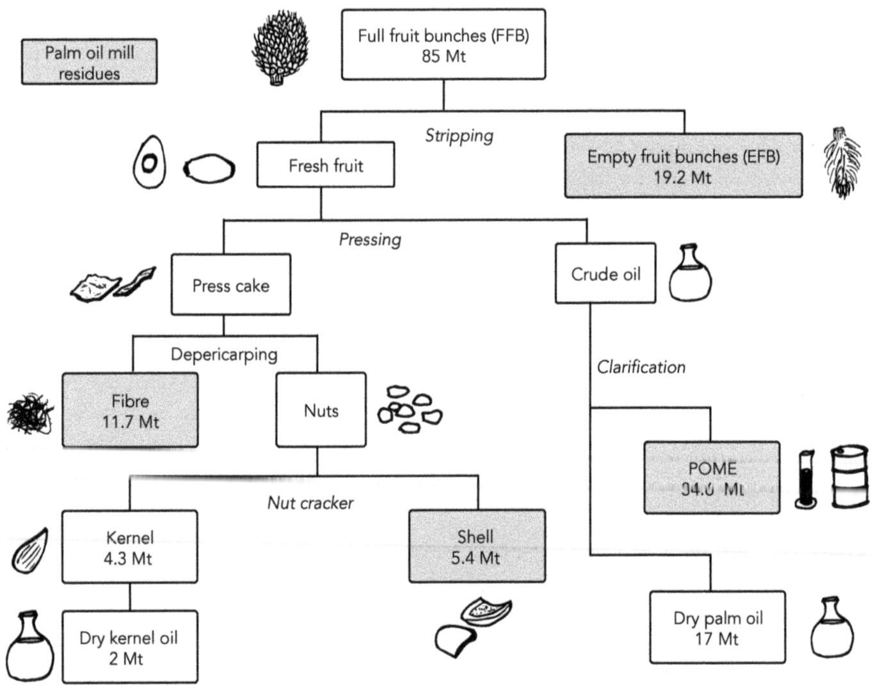

Figure 5 Oil palm mill processing flow chart in Malaysia (Roda et al., 2015).

the fruitlets in a rotating drum thresher before being conveyed to the press digesters. In the digesters, the fruits are heated using live steam and stirred continuously to loosen the oil-bearing mesocarp from the nuts and break open the oil cells present in the mesocarp. The digested mash is then pressed, extracting the oil using screw presses, and the pressed cake is conveyed to the kernel plant where the kernels are recovered. At this point empty fruit bunches can be collected. The oil from the press is diluted and pumped to vertical clarifier tanks where the clarified oil is fed to purifiers to remove dirt and moisture before being dried further in the vacuum drier. POME is unsuitable for biodiesel. Crude palm oil is a commodity that is processed directly or sold on international markets mostly for food and cosmetics as well as for biodiesel.

In theory, vegetable oils could be used in diesel engines without the need for a biorefinery. Rudolph Diesel, who wanted farmers to be able to produce their own fuel, invented the diesel engine specifically to run on vegetable oils:

> 'The diesel engine can be fed with vegetable oils and would help considerably in the development of agriculture of the countries which use it.'
> 'The use of vegetable oils for engine fuels may seem insignificant today. But such oils may become in the course of time as important as the petroleum and coal tar products of the present time.'
>
> (Diesel, 1912)

In practice, the heterogeneous nature of vegetable oils and their sensitivity to cold weather prevent efficient carburation. Biorefining allows the production of a quality-controlled and consistent product, tolerant of low temperatures, which is a perfect substitute for petrol–based diesel.

The biorefinery process for all major oils and fats is similar. The same generic technology applies to vegetable oils, animal fats and waste fats or oils. All these products are essentially composed of triglycerides in varying proportions. These triglycerides are themselves made of chains of fatty acids of varying lengths. The controlled chemical breakdown of the triglycerides into their components and transformation of the acid chains into methyl esters are the bases of the technology. This process of converting a fatty acid chain from an ester to a simple alcohol is called transesterification (Fig. 6). It is a reversible reaction, facilitated by a catalyst. Pure biodiesel consists of an array of fatty acid methyl ester chains of various lengths (Capareda and Press, 2014). The product is named 'FAME' – Fatty Acid Methyl Esters. Its properties and energy values are close to those of petrol-based diesel, allowing its use with various blends of fossil fuel. The major vegetable oils (soybean,

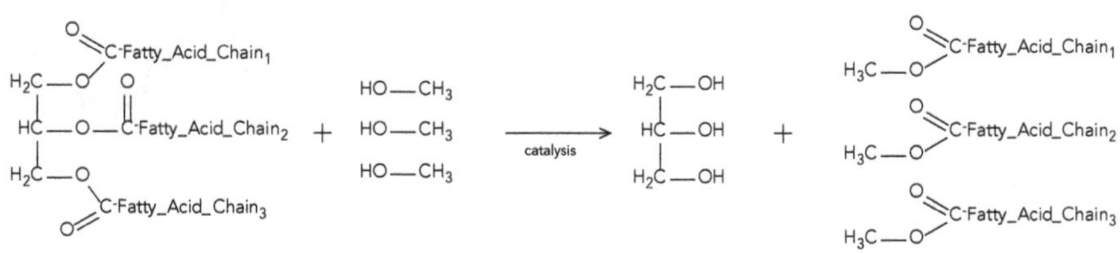

100 kg of refined vegetable oil + 10 kg of methanol ⟶ 10 kg of glycerine + 100 kg of biodiesel (methyl esters x)

Figure 6 Generic reaction of transesterification (adapted from Capareda and Press (2014)).

rapeseed, sunflower and palm) share a similar generic composition, the difference being in the proportion of chains of given lengths. These differences in proportion are important for specific food applications (for example, the property of being solid or liquid at ambient temperature), but they are completely neutral for biofuel end uses. All these oils are transesterified into FAME the generic formula of which is $CH_3(CH_2)_nCOOCH_3$. Consequently they are completely interchangeable for energy use. Most recent car engines are certified by the manufacturers for blends containing up to 30% biodiesel and in Germany a sizeable fleet of vehicles runs on pure biodiesel (B100 biodiesel).

4 Biodiesel economics

Biodiesel prices are entirely determined by vegetable oil prices. In a typical palm oil biorefinery, for example, palm oil would represent 80–90% of the overall cost of the biodiesel (Ong, 2012). Before the increase in biodiesel demand and volatility in petrol prices, supply and demand were the main drivers in agrifood markets. When the demand for fossil fuel began to rise consistently at the beginning of the millennium, it was the main cause of variations in the price of vegetable oil (Fig. 7). At the same time, motivated by both climate change and energy independence concerns, political interest in biodiesel grew and production of biodiesel quickly followed. Biodiesel is designed to be a substitute to petrol-based diesel. Fuel costs and environmental concerns are the main reasons behind this substitution. The production cost of biodiesel depends mainly on upstream plantation costs: yields, diseases, fertiliser and fossil fuel costs. Yields account for the main differences between the four major vegetable oils. Per hectare, palm oil is the most productive by far (Rival and Levang, 2014). Therefore, structurally it is the cheapest of the four (Fig. 7). Regardless of the yields, fossil fuel costs directly impact fertiliser and

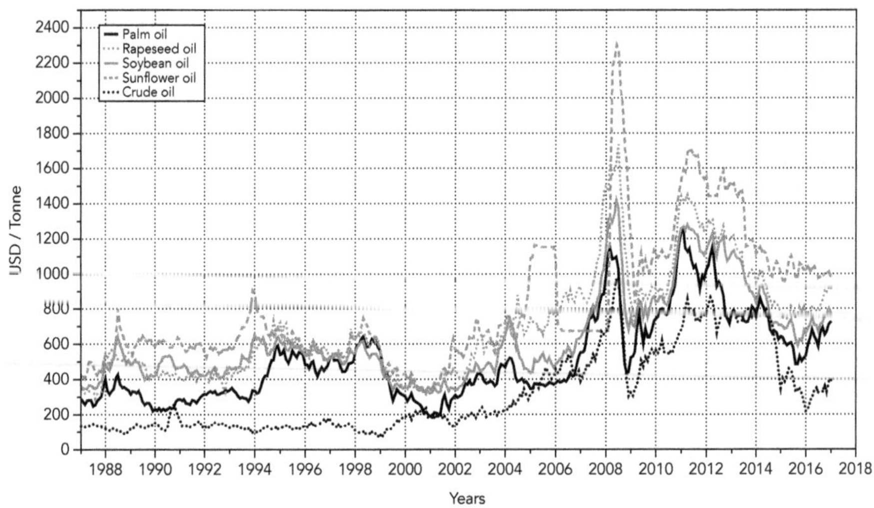

Figure 7 International prices of palm, rapeseed, soybean, sunflower and fossil crude oil.

other agricultural costs. Consequently, the international prices of all vegetable oils from industrial plantations are very similar and this has been apparent since the beginning of the millennium (Fig. 7).

In 2016, Indonesia and Malaysia produced over 85% of the total global palm oil supply. Asia accounts for more than 63% of the palm oil demand, followed by Africa (more than 13%) and Europe (more than 12%). In all the consuming countries, palm oil demand is driven by end uses as food and animal feed. Only around 27% of global palm oil production is for 'industrial domestic consumption'. Part of this industrial consumption is for biodiesel and part is for industrial non-food uses such as cosmetics. In the case of the five countries that represent 75% of the global palm oil 'industrial domestic consumption' (Indonesia, Europe [*EU-27*], Malaysia, China and Thailand), the breakdown of fuel and non-fuel uses is given in Table 2. Total palm oil-based biodiesel production uses around 12% of the palm oil grown. Interestingly, although Europe consumes more than 42% of all biodiesel globally, it only uses 2.8% of world palm oil production for biodiesel. The amount of palm oil used by Europe for biodiesel appears disproportionately small compared to the significant amount of research that has been on the adverse effects of European use of this biodiesel: indirect deforestation, climate change, loss of biodiversity and direct competition with European vegetable oil (Sihvonen, 2016a; Neslen, 2016; Dings, 2016).

The main limitation to palm oil-based biodiesel globally is the competitiveness of crude palm oil and fossil fuel prices. With an increasing demand for palm oil in food and cosmetics, it is frequently more expensive than the petrol-based diesel it is supposed to be substituting. This creates an economic barrier in countries where fuel policies impose very low or zero excise duties on gasoline and diesel, such as the United States. In countries with full or partial subsidies on gasoline and diesel, such as Malaysia, Indonesia and India, it is even more difficult to run biodiesel refineries based on palm oil feedstock. For example, in 2006, the Malaysian government introduced the use of B5 blended biodiesel – 95% petroleum diesel and 5% biofuels (Lee and Ofori-Boateng, 2013). Between 2006 and 2007, 92 biodiesel projects were approved in Malaysia but by 2012, most of them had been halted due to high feedstock prices and competition in demand for raw oil by the agro-food sector. To revive these projects, in 2013, the Malaysian government announced that B10 blended biodiesel would become mandatory but this has yet to be implemented. Altogether, only 30 biomass transformation plants are currently operating; most of them are located in the states of Johor and Selangor. Conversely, in countries applying high excise duties on gasoline and diesel such as most of the European Union

Table 2 Breakdown of palm oil feedstock for biodiesel production in 2016 for the five major palm oil industrial domestic consumers (Flach et al., 2016)

Palm oil (1000 T)	Indonesia	EU-27	Malaysia	China	Thailand
Industrial domestic consumption	3500	3040	2400	2050	1520
Use for biodiesel	2250	1790	360	500	885
Other non-food industrial uses, such as cosmetics	1250	1250	2040	1550	635
% of world palm oil production used as biodiesel	3.5%	2.8%	0.6%	0.8%	1.4%

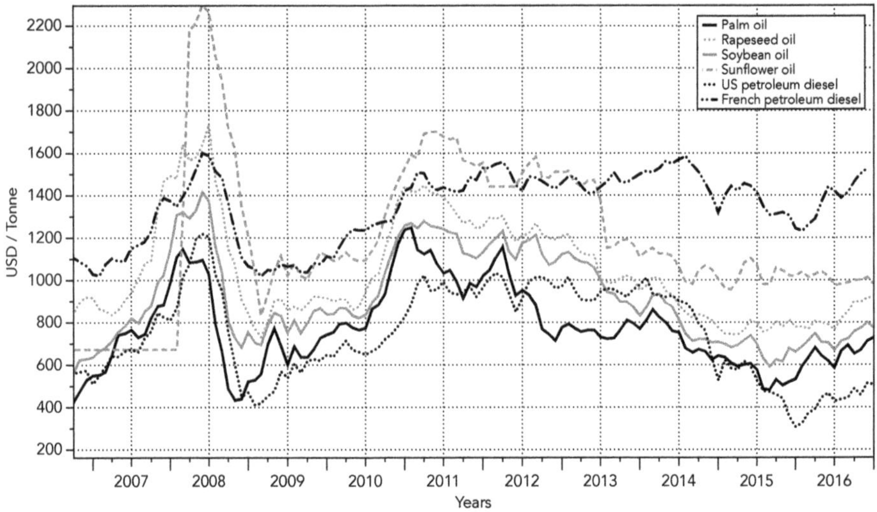

Figure 8 International prices of palm, rapeseed, soybean and sunflower oils, compared with diesel prices in France and in the United States.

members, converting vegetable oils into biodiesel is much more competitive. The result has not necessarily been that these countries have become large importers of palm oil-based biodiesel, but it is clear that such conditions are better for biorefineries using local vegetable oil.

The magnitude of possible excise duties largely determines the price range of the major vegetable oils. Regardless of the existing biofuel policies or vegetable oil trade, in every country fossil fuel policies are the first factor determining the competitiveness of biofuels. Taking as examples the United States (near the lower end) and France (near the higher end of excise duties), Fig. 8 shows how fossil fuel policies overlap with vegetable oil prices. Palm oil prices have been close to, or lower than, US diesel prices for extended periods, but higher than any subsidised petroleum diesel (as in Malaysia, for example). Soybean oil has occasionally been cheaper than the US diesel price, but prices are typically similar. This price proximity explains why the United States relies mostly on soybean as its main temperate oil feedstock. Conversely, even costly oils like rapeseed are almost always lower than the price of French petroleum diesel and the occasions when it is more expensive coincide with dips in rapeseed production in Europe.

5 Imports and defiscalisation: the odd alliance

As noted earlier, Europe has become the main global producer and consumer of biodiesel. This has been due to biodiesel 'defiscalisation' (i.e. removal of relevant taxes and use of subsidies) and high petroleum diesel prices (since all European Union members apply

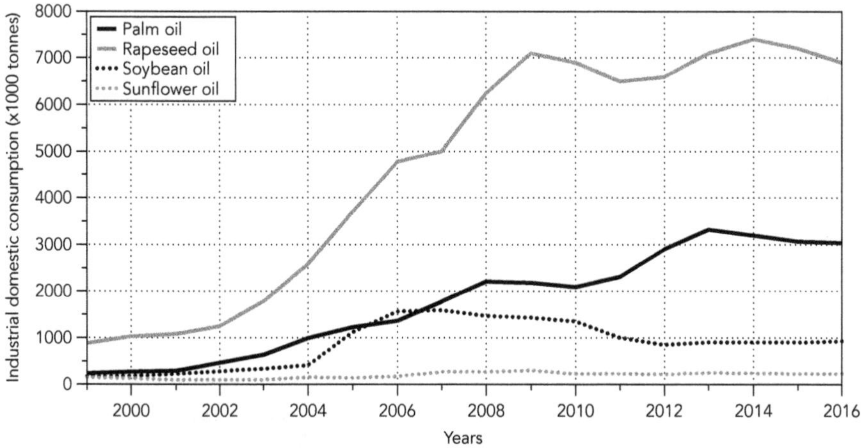

Figure 9 Trends in industrial domestic consumption of oils in Europe.

very high gasoline and diesel excise duties compared to countries such as the United States). The aim of the defiscalisation policy was to promote locally produced rapeseed oil, especially on fallow ground. In 2003, Europe made a major commitment to support biofuels with dedicated EU legislation. In 2004, total defiscalisation of biodiesel in Germany accelerated several years' growth of rapeseed oil and rapeseed biodiesel (Fig. 9). German rapeseed production quickly became dominant in Europe, followed by France (which carried out partial defiscalisation). Rapeseed plantations are quite sensitive to weather and disease and, in Northern Europe, the crop is prone to temporary tensions between supply and demand, with subsequent volatility in rapeseed oil prices. The slow or delayed response of rapeseed production to price variations creates opportunities for cheaper vegetable oil imports from international markets, typically soybean and palm oil. After the initial increase in industrial domestic consumption of rapeseed oil, a surge in soybean oil imports from the United States was observed. The European Union did set up specific anti-dumping policies to address this issue, transforming the surge into a temporary peak which is visible from 2004 to 2011 (Fig. 9). This led to the substitution of large soybean oil imports from the United States by lower volumes of imports from Argentina. At the same time, cheaper palm oil (already imported in large quantities for European food) helped to offset local variations in vegetable oil supplies within Europe. Rapeseed oil production levelled off from 2009, because of diminishing biodiesel defiscalisation. In Germany and France, defiscalisation rates began to decrease in 2007 but it appears that the critical level was crossed in 2009 (Table 3).

Table 3 Amount of biodiesel defiscalisation per hL, in France

Year	2007	2008	2009	2010	2011	2012	2013	2014	2015
Euros/hL	25	22	15	11	8	8	8	4.5	3

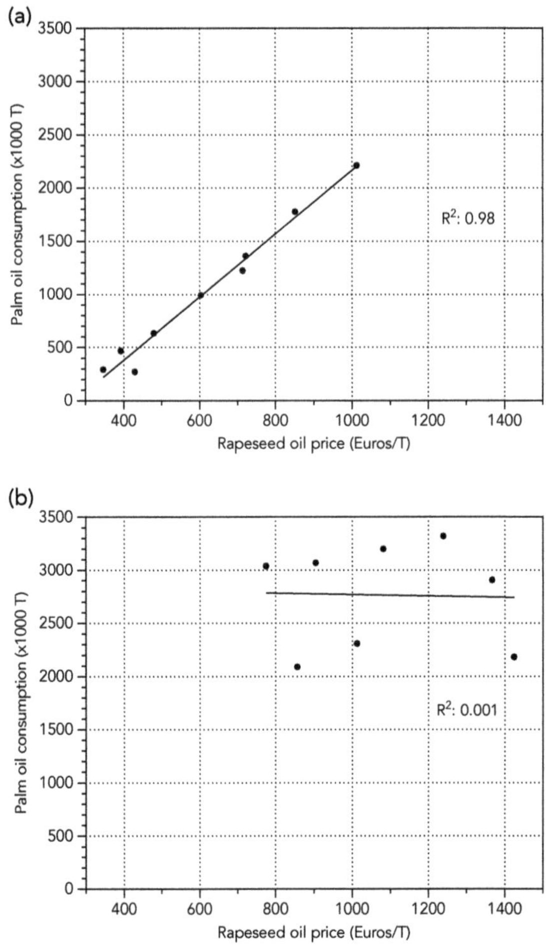

Figure 10 Variation of palm oil industrial domestic consumption in Europe according to rapeseed oil price. (a) from 2000 to 2008. (b) from 2009 to 2016.

During this period, there was a European debate as to whether palm oil was an economic substitute for rapeseed oil with mixed evidence according to the years of reference. What was critical was the price of rapeseed oil and the level of defiscalisation for biodiesel. A simple econometric test called 'cross price elasticity of demand' and shown as E_c (Equation 1) shows that if two goods are substitutes for each other, independent or complementary to each other:

$$E_c = \frac{\Delta Q_A / Q_A}{\Delta P_B / P_B} \quad (1)$$

with Q_A being quantity of commodity A and P_B being price of commodity B.

The cross price elasticity of palm oil demand in Europe was 3.1 from 2000 to 2008 on average and 0.07 from 2009 to 2016. This meant that, from 2000 to 2008, the quantity of palm oil changed by 3.1% when the price of rapeseed oil changed by 1%. Each time the price of rapeseed oil increased for any reason, such as a temporary reduction in production anywhere in Europe, palm oil imports and consumption were stimulated. This shows that industrial palm oil consumption in Europe was completely driven by rapeseed oil prices between 2000 and 2008. It was then a perfect economic substitute to rapeseed oil. During the second period, the quantity of palm oil changed by 0.07% when the price of rapeseed oil changed by 1%. When the defiscalisation of biodiesel decreased, the market saturated, rapeseed oil production levelled off and the consumption of palm oil became completely independent from the price of rapeseed oil (Fig. 10). At the same time, there was still a large international supply of cheaper soybean and palm oil where prices were driven by non-European markets. In Europe, palm oil has come to act as a substitute when biodiesel is defiscalised, and, if not, as a commodity independent of the price of competing rapeseed oil.

While the main European producers of rapeseed oil are Germany and France, the main European consumers of palm oil for biodiesel are Italy and Spain, with palm oil representing 95% and 90%, respectively, of their vegetable oil use for biodiesel (Sihvonen, 2016b). From a bioclimatic perspective, rapeseed is as exotic to these countries as palm oil is to Germany and France. While subsidising rapeseed oil production makes sense in Germany and France, it does not in Italy and Spain and market forces prevail. In Italy and Spain, it is more rational to import palm oil available relatively cheaply at any harbour, rather than transport it by road from Northern Europe and compete with German and French consumption of rapeseed oil.

6 The geopolitics of agribusiness: competition between productions systems

Soybean, rapeseed and sunflower are originally temperate-grown oilseeds. These commodities are dominated by agribusiness mega-corporations all originating from western countries. They form the 'ABCD Empire'. There are four of these large groups: Archers Daniel Midlands, Bunge, Cargill and Louis Dreyfus, and they dominate the trade in agricultural commodities. Despite its temperate origin, soybean is also grown in subtropical and tropical areas in Brazil thanks to new varieties. Even in Brazil, the majority of soybean plantations are directly or indirectly under the control of the 'ABCD Empire'. Conversely, palm oil production and trade are more diverse, shared between a number of smaller corporations originating in Southeast Asia (Table 4). Figure 11 illustrates the divide between north and south, or between temperate and tropical systems of production. The fact that soybean varieties have been improved to grow in subtropical and tropical conditions blurs the picture slightly. However, with 60% of Brazilian soybean being dominated by Archers Daniel Midlands and Bunge (Anon., 2015), it remains undeniable that oil crops are a source of competition between western and Southeast Asian agribusiness corporations.

Another factor in this competition is the difference in productivity per hectare for these crops. Figure 11 shows that the area required by palm oil is disproportionately small compared to its production. A clearer picture of the 'space footprint' of palm oil versus other major oils is given in Fig. 12. Today, productivity per hectare differentiates

Table 4 Agribusiness corporations controlling 70% of the palm oil world market in 2010 (Bureau Van Dijk, 2015)

Company	Country of headquarters	Market capitalisation (million USD)	Land bank (ha)	% of world market	% of world land bank
Wilmar	Singapore	14 565	243 138	25	6
Sime Darby	Malaysia	12 175	986 824	21	26
IOI Corp	Malaysia	7147	217 918	12	6
KL Kepong	Malaysia	6203	270 040	11	7
Indofood	Indonesia	5410	224 083	9	6
Golden Agri	Singapore	3360	451 063	6	12
Astra Agro	Indonesia	2427	297 862	4	8

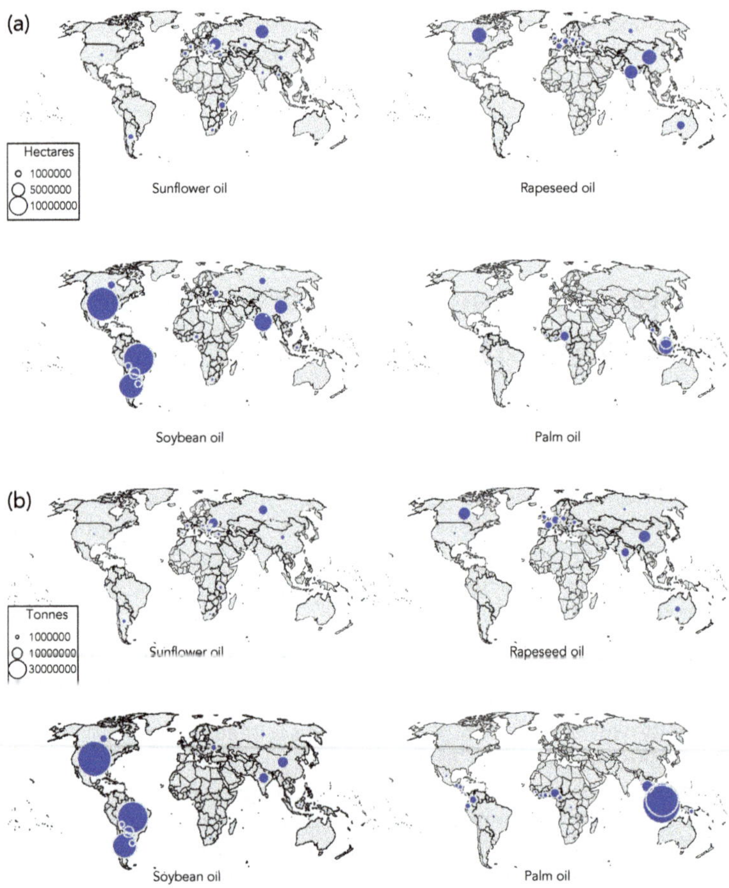

Figure 11 Global location of major vegetable oil production in 2013. (a) Hectarage. (b) Tonnage.

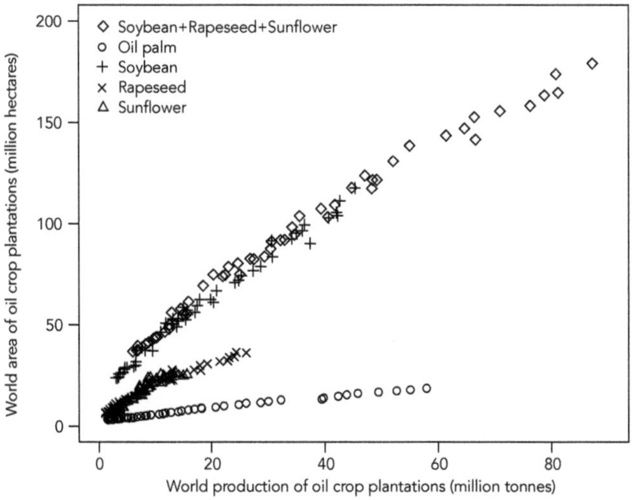

Figure 12 Progression of global plantation area versus production of the major oil crops.

the vegetable oils in terms of cost per tonne. It makes palm oil the most competitive in any temperate biodiesel business. The production of Brazilian soybean oil hardly affects this situation since Brazilian biofuel production revolves around ethanol. In the future, and as agricultural land becomes scarcer, the difference in area required to produce one tonne of oil could become the main competitive factor. In recent decades, the size of plantations for all vegetable oils has increased tremendously to meet growing global demand (Fig. 13). Conversely, the hectarage of palm oil has grown quite modestly and, especially in comparison to its main competitors in biodiesel production (Fig. 13).

7 Conclusions

A global perspective of vegetable oils means looking at their market share on a global scale (Fig. 14). Supported by US policies, the global market share of soybean oil peaked at the end of the 1970s and has decreased ever since. That of rapeseed oil increased until the 1990s and started to decrease at the turn of the millennium. European biodiesel policies subsequently encouraged a revival, but rapeseed oil seems to have stagnated since 2009. After a peak at the end of the 1960s, the global sunflower market share has been constantly on the decline. A striking figure is the steady and seemingly irresistible growth in the market share of palm oil since 1970 (Fig. 14). Looking at the competition between southern and northern production systems, palm oil is clearly threatening the other oil production systems.

However, as noted earlier, fossil fuel prices and domestic interests play a major role in the use, or not, of vegetable oils for biofuel. Comparative cost competitiveness of fossil fuel remains critical in the development of palm oil-based biodiesel, but energy independence is also a major factor to consider. For example, Indonesia, once a net producer of fossil fuel oil, is now a net importer. A new Energy Bill passed in 2015 committed to a mandate

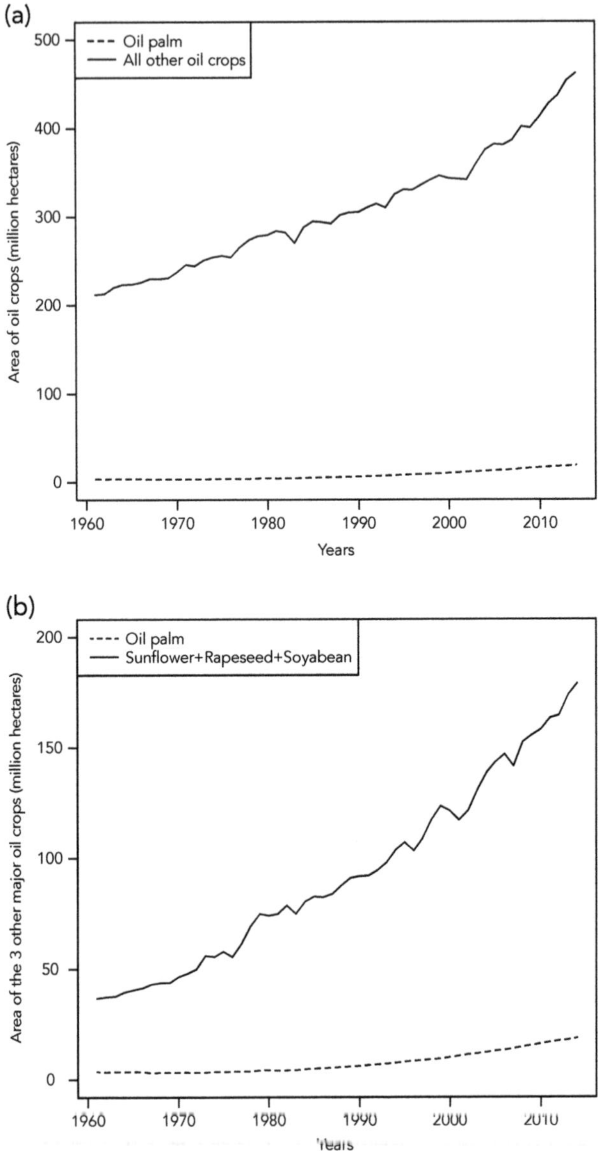

Figure 13 Oil palm planted area in perspective with other oil crops from 1961 to 2015. (a) All oil crops. (b) Sunflower, rapeseed and soybean.

of 30% of biodiesel blend in all diesel consumption by 2025. This means that between 7 and 8 million tonnes of palm oil-based biodiesel should be blended into Indonesian fuel per year by this date. Such quantities would completely transform the balance of global biodiesel economics. It remains to be seen if the Indonesian bill will be implemented soon.

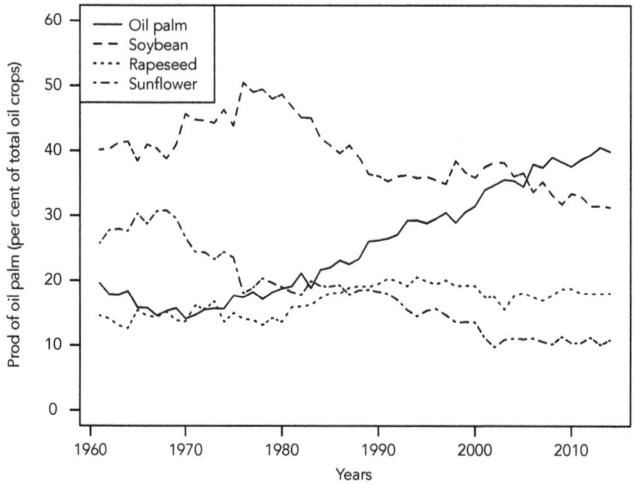

Figure 14 Progression of global palm oil production percentage versus other vegetable oils percentages from 1961 to 2015.

8 References

Anon (2008), *The State of Food and Agriculture – Biofuels: Prospects, Risks and Opportunities*, Rome, Italy: FAO.

Anon (2014), *Biomass Meeting the Biotechnology Challenge*, Paris, France: Total.

Anon (2015), The ABCD's empire face new competitors, pp. 1–4. Available at: https://campus.hesge.ch/commodity_trading/archives/13238 (Accessed 22 January 2017).

Bailis, R. and Baka, J. (2011), Constructing sustainable biofuels: Governance of the emerging biofuel economy. *Annals of the Association of American Geographers*, 101(4), 827–38.

Bureau Van Dijk. (2015), Osiris database. *osiris.bvdinfo.com*, (DVD). Available at: https://osiris.bvdinfo.com/version-20151014/Login.serv?product=osirisneo&SetLanguage=en.

Capareda, S. C. and Press, C. (2014), *Introduction to Biomass Energy Conversions*, Boca Raton: CRC Press/Taylor & Francis Group.

Diesel, R. (1912), *Die Entstehung des Dieselmotors*, Berlin, Germany: Springer.

Dings, J. (2016), Cars and trucks burn almost half of palm oil used in Europe. *transportenvironment.orq*, p. 4. Available at: https://www.transportenvironment.org/sites/te/files/publications/2016_05_TE_EU_vegetable_oil_biodiesel_market_FINAL_0_0.pdf (Accessed 22 January 2017).

Ernie Ginting, V. P. (2015), The Indonesia alternatives aviation fuels roadmap and readiness. In: *International Green Aviation Conference*. Denpasar, Indonesia, p. 15.

Fargione, J., Hill, J., Tilman, D., et al. (2008), Land clearing and the biofuel carbon debt. *Science*, 319(5867), 1235–8.

Flach, B. et al. (2016), *EU Biofuels Annual 2016*, S. Phillips (Ed.), The Hague, Netherlands.

Gomez, L. D., Steele King, C. G. and McQueen Mason, S. J. (2008), Sustainable liquid biofuels from biomass: the writing's on the walls. *New Phytologist*, 178(3), 473–85.

Haberl, H., Heinz Erb, K., Krausmann, F., et al. (2007), Quantifying and mapping the human appropriation of net primary production in earth's terrestrial ecosystems. *Proceedings of the National Academy of Sciences*, 104(31), 12942–7.

International Energy Agency. (2013), Renewable energy outlook. In: *World Energy Outlook*. Paris, France: worldenergyoutlook.org.
Krausmann, F. and Kowalski, M. F. (2008), The global sociometabolic transition. *Journal of Industrial Ecology*. 12(5–6), 637–56.
Lee, K. T. and Ofori-Boateng, C. (2013), *Sustainability of Biofuel Production from Oil Palm Biomass*, Springer Science & Business Media.
Meyer, S. D. (2009), Biofuels role in global food and energy security. In: *2009 Western Hemisphere Economic Officers' Conference*. Saint-Louis, USA, p. 15.
Naik, S. N, Goud, V. V., Rout, P. K., et al. (2010), Production of first and second generation biofuels: A comprehensive review. *Renewable and Sustainable Energy Reviews*, 14(2), 578–97.
Neslen, A. (2016), Leaked figures show spike in palm oil use for biodiesel in Europe | Environment | The Guardian. *The Guardian*, pp. 1–3.
OECD. (2015), *OECD-FAO Agricultural Outlook, 2015*, Paris, France: OECD. Available at: http://www.oecd-ilibrary.org/agriculture-and-food/oecd-fao-agricultural-outlook-2015_agr_outlook-2015-en.
Ong, H. C. (2012), *Techno-economic Analysis of Biodiesel Production from Palm, Jatropha curcas and Calophyllum Inophyllum for Road Transport in Malaysia*.
Rival, A. and Levang, P. (2014), *Palms of Controversies: Oil Palm and Development Challenges*, Bogor, Indonesia: Center for International Forestry Research (CIFOR).
Roda, J.-M., Goralski, M., Benoist, A., et al. (2015), *Sustainability of biojet-fuel in Malaysia*, J.-M. Roda (Ed.), Paris, France: CIRAD.
Sihvonen, J. (2016a), Drivers are the top consumers of palm oil in Europe. *transportenvironment.org*, pp. 1–4.
Sihvonen, J. (2016b), Europe keeps burning more palm oil in its diesel cars and trucks. *transport-environment.org*. Available at: https://www.transportenvironment.org/sites/te/files/2016_11_Briefing_Palm_oil_use_continues_to_grow.pdf (Accessed 22 January 2017).
Wilkinson, J., Afiff, S., Carriquiry, M., et al. (2013), *Biofuels and Food Security*, Rome, Italy: HLPE/FAO.

Index

Lightning Source UK Ltd.
Milton Keynes UK
UKHW02n1155051018
330004UK00002B/2/P

9 781786 761040